The

International Scientific Series.

VOL. XLIX.

ORIGIN

OF

CULTIVATED PLANTS

BY

ALPHONSE DE CANDOLLE

FOREIGN ASSOCIATE OF THE ACADEMY OF SCIENCES OF THE INSTITUTE OF FRANCE;
FOREIGN MEMBER OF THE ROYAL SOCIETIES OF LONDON, EDINBURGH,
AND DUBLIN; OF THE ACADEMIES OF ST. PETERSBURG,
STOCKHOLM, BERLIN, MUNICH, BRUSSELS, COPENHAGEN, AMSTERDAM,
ROME, TURIN, MADRID, BOSTON, ETC.

HAFNER PUBLISHING COMPANY
New York and London
1967

Reprint of the 2nd Edition
1886

First Printing 1959
Second Printing 1964
Third Printing 1967

B59·02703

Library of Congress Catalog Card Number 59-7042

Printed in the U.S.A.

NOBLE OFFSET PRINTERS, INC.
NEW YORK 3, N. Y.

AUTHOR'S PREFACE.

THE knowledge of the origin of cultivated plants is interesting to agriculturists, to botanists, and even to historians and philosophers concerned with the dawnings of civilization.

I discussed this question of origin in a chapter in my work on geographical botany; but the book has become scarce, and, moreover, since 1855 important facts have been discovered by travellers, botanists, and archæologists. Instead of publishing a second edition, I have drawn up an entirely new and more extended work, which treats of the origin of almost double the number of species belonging to the tropics and the temperate zones. It includes almost all plants which are cultivated, either on a large scale for economic purposes, or in orchards and kitchen gardens.

I have always aimed at discovering the condition and the habitat of each species before it was cultivated. It was needful for this end to distinguish from among innumerable varieties that which should be regarded as the most ancient, and to find out from what quarter of

the globe it came. The problem is more difficult than it
appears at first sight.   In the last century and up to
the middle of the present, authors made little account
of it, and the most able have contributed to the propagation of erroneous ideas. I believe that three out
of four of Linnæus' indications of the original home of
cultivated plants are incomplete or incorrect. His statements have since been repeated, and in spite of what
modern writers have proved touching several species,
they are still repeated in periodicals and popular works.
It is time that mistakes, which date in some cases from
the Greeks and Romans, should be corrected. The actual
condition of science allows of such correction, provided
we rely upon evidence of varied character, of which
some portion is quite recent, and even unpublished; and
this evidence should be sifted as we sift evidence in historical research.   It is one of the rare cases in which
a science founded on observation should make use of
testimonial proof. It will be seen that this method
leads to satisfactory results, since I have been able to
determine the origin of almost all the species, sometimes
with absolute certainty, and sometimes with a high
degree of probability.

I have also endeavoured to establish the number of
centuries or thousands of years during which each
species has been in cultivation, and how its culture
spread in different directions at successive epochs.

A few plants cultivated for more than two thousand
years, and even some others, are not now known in a

spontaneous, that is, wild condition, or at any rate this condition is not proved. Questions of this nature are subtle. They, like the distinction of species, require much research in books and in herbaria. I have even been obliged to appeal to the courtesy of travellers or botanists in all parts of the world to obtain recent information. I shall mention these in each case with the expression of my grateful thanks.

In spite of these records, and of all my researches, there still remain several species which are unknown wild. In the cases where these come from regions not completely explored by botanists, or where they belong to genera as yet insufficiently studied, there is hope that the wild plant may be one day discovered. But this hope is fallacious in the case of well-known species and countries. We are here led to form one of two hypotheses; either these plants have since history began so changed in form in their wild as well as in their cultivated condition that they are no longer recognized as belonging to the same species, or they are extinct species. The lentil and the chick-pea probably no longer exist in nature; and other species, as wheat, maize, the broad bean, and carthamine, which are very rarely found wild, appear to be in course of extinction. The number of cultivated plants with which I am here concerned, being two hundred and forty-nine, the three, four, or five species, extinct or nearly extinct, is a large proportion, representing a thousand species, out of the whole number of phanerogams. This destruction of forms must have taken place during the short period of a few hundred centuries, on continents where they might have spread, and under circumstances which are commonly considered unvarying. This shows how the history of cultivated plants is allied to the most important problems of the general history of organized beings.

GENEVA, 1882.

CONTENTS.

PART I.
GENERAL REMARKS.

CHAPTER   PAGE

I. IN WHAT MANNER AND AT WHAT EPOCHS CULTIVATION BEGAN IN DIFFERENT COUNTRIES 1

II. METHODS FOR DISCOVERING OR PROVING THE ORIGIN OF SPECIES 8

PART II.
ON THE STUDY OF SPECIES, CONSIDERED AS TO THEIR ORIGIN, THEIR EARLY CULTIVATION, AND THE PRINCIPAL FACTS OF THEIR DIFFUSION.

I. PLANTS CULTIVATED FOR THEIR SUBTERRANEAN PARTS, SUCH AS ROOTS, TUBERCLES, OR BULBS 29

II. PLANTS CULTIVATED FOR THEIR STEMS OR LEAVES 83

III. PLANTS CULTIVATED FOR THEIR FLOWERS, OR FOR THE ORGANS WHICH ENVELOP THEM 161

IV. PLANTS CULTIVATED FOR THEIR FRUITS 168

V. PLANTS CULTIVATED FOR THEIR SEEDS 313

PART III.
SUMMARY AND CONCLUSION.

I. GENERAL TABLE OF SPECIES, WITH THEIR ORIGIN AND THE EPOCH OF THEIR EARLIEST CULTIVATION 436

II. GENERAL OBSERVATIONS AND CONCLUSIONS 447

INDEX 463

ORIGIN

OF

CULTIVATED PLANTS

—◦◦—

PART I.

General Remarks.

CHAPTER I.

IN WHAT MANNER AND AT WHAT EPOCHS CULTIVATION BEGAN IN DIFFERENT COUNTRIES.

THE traditions of ancient peoples, embellished by poets, have commonly attributed the first steps in agriculture and the introduction of useful plants, to some divinity, or at least to some great emperor or Inca. Reflection shows that this is hardly probable, and observation of the attempts at agriculture among the savage tribes of our own day proves that the facts are quite otherwise.

In the progress of civilization the beginnings are usually feeble, obscure, and limited. There are reasons why this should be the case with the first attempts at agriculture or horticulture. Between the custom of gathering wild fruits, grain, and roots, and that of the regular cultivation of the plants which produce them, there are several steps. A family may scatter seeds around its dwelling, and provide itself the next year with the same product in the forest. Certain fruit trees may exist near a dwelling without our knowing whether they were planted, or whether the hut was built beside them in order to profit by them. War and the chase often interrupt attempts at cultivation. Rivalry and mistrust cause the imitation of one tribe by another to make but slow progress. If some great personage command the cultivation of a plant, and institute some ceremonial to show its utility, it is probably because obscure and unknown men have previously spoken of it, and that successful experiments have been already made. A longer or shorter succession of local and short-lived experiments must have occurred before such a display, which is calculated to impress an already numerous public. It is easy to understand that there must have been determining causes to excite these attempts, to renew them, to make them successful.

The first cause is that such or such a plant, offering some of those advantages which all men seek, must be within reach. The lowest savages know the plants of their country; but the example of the Australians and Patagonians shows that if they do not consider them productive and easy to rear, they do not entertain the idea of cultivating them. Other conditions are sufficiently evident: a not too rigorous climate; in hot countries, the moderate duration of drought; some degree of security and settlement; lastly, a pressing necessity, due to insufficient resources in fishing, hunting, or in the production of indigenous and nutritious plants, such as the chestnut, the date-palm, the banana, or the breadfruit tree. When men can live without work it is what they like best. Besides, the element of hazard in hunting and fishing attracts primitive, and sometimes civilized man, more than the rude and regular labour of cultivation.

I return to the species which savages are disposed to cultivate. They sometimes find them in their own country, but often receive them from neighbouring peoples, more favoured than themselves by natural conditions, or already possessed of some sort of civilization. When a people is not established on an island, or in some place difficult of access, they soon adopt certain plants, discovered elsewhere, of which the advantage is evident, and are thereby diverted from the cultivation of the poorer species of their own country. History shows us that wheat, maize, the sweet potato, several species of the genus Panicum, tobacco, and other plants, especially annuals, were widely diffused before the historical period. These useful species opposed and arrested the timid attempts made here and there on less productive or less agreeable plants. And we see in our own day, in various countries, barley replaced by wheat, maize preferred to buckwheat and many kinds of millet, while some vegetables and other cultivated plants fall into disrepute because other species, sometimes brought from a distance, are more profitable. The difference in value, however great, which is found among plants already improved by culture, is less than that which exists between cultivated plants and others completely wild. Selection, that great factor which Darwin has had the merit of introducing so happily into science, plays an important part when once agriculture is established; but in every epoch, and especially in its earliest stage, the choice of species is more important than the selection of varieties.

The various causes which favour or obstruct the beginnings of agriculture, explain why certain regions have been for thousands of years peopled by husbandmen, while others are still inhabited by nomadic tribes. It is clear that, owing to their well-known qualities and to the favourable conditions of climate, it was at an early period found easy to cultivate rice and several leguminous plants in Southern Asia, barley and wheat in Mesopotamia and in Egypt, several species of Panicum in Africa, maize, the potato, the sweet potato, and manioc in America. Centres were thus formed whence the most useful species were diffused. In the north of Asia, of Europe, and of America, the climate is unfavourable, and the indigenous plants are unproductive ; but as hunting and fishing offered their resources, agriculture must have been introduced there late, and it was possible to dispense with the good species of the south without great suffering. It was different in Australia, Patagonia, and even in the south of Africa. The plants of the temperate region in our hemisphere could not reach these countries by

reason of the distance, and those of the intertropical
zone were excluded by great drought or by the absence of
a high temperature. At the same time, the indigenous
species are very poor. It is not merely the want of
intelligence or of security which has prevented the inhabitants from cultivating them. The nature of the
indigenous flora has so much to do with it, that the
Europeans, established in these countries for a hundred
years, have only cultivated a single species, the *Tetragonia*, an insignificant green vegetable. I am aware
that Sir Joseph Hooker[1] has enumerated more than a
hundred Australian species which may be used in some
way ; but as a matter of fact they were not cultivated
by the natives, and, in spite of the improved methods of
the English colonists, no one does cultivate them. This
clearly demonstrates the principle of which I spoke just
now, that the choice of species is more important than
the selection of varieties, and that there must be valuable
qualities in a wild plant in order to lead to its cultivation.

In spite of the obscurity of the beginnings of cultivation in each region, it is certain that they occurred at
very different periods. One of the most ancient examples
of cultivated plants is in a drawing representing figs,
found in Egypt in the pyramid of Gizeh. The epoch of
the construction of this monument is uncertain. Authors
have assigned a date varying between fifteen hundred and
four thousand two hundred years before the Christian era.
Supposing it to be two thousand years, its actual age
would be four thousand years. Now, the construction
of the pyramids could only have been the work of a
numerous, organized people, possessing a certain degree of
civilization, and consequently an established agriculture,
dating from some centuries back at least. In China, two
thousand seven hundred years before Christ, the Emperor
Chenming instituted the ceremony at which every year
five species of useful plants are sown—rice, sweet potato,
wheat, and two kinds of millet.[2] These plants must

---

[1] Hooker, *Flora Tasmaniæ*, i. p. cx.
[2] Bretschneider, *On the Study and Value of Chinese Botanical Works*,.
p. 7.

have been cultivated for some time in certain localities before they attracted the emperor's attention to such a degree. Agriculture appears, then, to be as ancient in China as in Egypt. The constant relations between Egypt and Mesopotamia lead us to suppose that an almost contemporaneous cultivation existed in the valleys of the Euphrates and the Nile. And it may have been equally early in India and in the Malay Archipelago. The history of the Dravidian and Malay peoples does not reach far back, and is sufficiently obscure, but there is no reason to believe that cultivation has not been known among them for a very long time, particularly along the banks of the rivers.

The ancient Egyptians and the Phœnicians propagated many plants in the region of the Mediterranean, and the Aryan nations, whose migrations towards Europe began about 2500, or at latest 2000 years B.C., carried with them several species already cultivated in Western Asia. We shall see, in studying the history of several species, that some plants were probably cultivated in Europe and in the north of Africa prior to the Aryan migration. This is shown by names in languages more ancient than the Aryan tongues; for instance, Finn, Basque, Berber, and the speech of the Guanchos of the Canary Isles. However, the remains, called kitchen-middens, of ancient Danish dwellings, have hitherto furnished no proof of cultivation or any indication of the possession of metal.[1] The Scandinavians of that period lived principally by fishing and hunting, and perhaps eked out their subsistence by indigenous plants, such as the cabbage, the nature of which does not admit any remnant of traces in the dung-heaps and rubbish, and which, moreover, did not require cultivation. The absence of metals does not in these northern countries argue a greater antiquity than the age of Pericles, or even the palmy days of the Roman republic. Later, when bronze

---

[1] De Naidaillac, *Les Premiers Hommes et les Temps Préhistoriques*, i. pp. 266, 268. The absence of traces of agriculture among these remains is, moreover, corroborated by Heer and Cartailhac, both well versed in the discoveries of archæology.

was known in Sweden—a region far removed from the then civilized countries—agriculture had at length been introduced. Among the remains of that epoch was found a carving of a cart drawn by two oxen and driven by a man.[1]

The ancient inhabitants of Eastern Switzerland, at a time when they possessed instruments of polished stone and no metals, cultivated several plants, of which some were of Asiatic origin. Heer [2] has shown, in his admirable work on the lake-dwellings, that the inhabitants had intercourse with the countries south of the Alps. They may also have received plants cultivated by the Iberians, who occupied Gaul before the Kelts. At the period when the lake-dwellers of Switzerland and Savoy possessed bronze, their agriculture was more varied. It seems that the lake-dwellers of Italy, when in possession of this metal, cultivated fewer species than those of Savoy,[3] and this may be due either to a greater antiquity or to local circumstances. The remains of the lake-dwellers of Laybach and of the Mondsee in Austria prove likewise a completely primitive agriculture ; no cereals have been found at Laybach, and but a single grain of wheat at the Mondsee.[4] The backward condition of agriculture in this eastern part of Europe is contrary to the hypothesis, based on a few words used by ancient historians, that the Aryans sojourned first in the region of the Danube, and that Thrace was civilized before Greece. In spite of this example, agriculture appears in general to have been more ancient in the temperate parts of Europe than we should be inclined to believe from the Greeks, who were disposed, like certain modern

---

[1] M. Montelius, from Cartailhac, *Revue,* 1875, p. 237.

[2] Heer, *Die Pflanzen der Pfahlbauten,* in 4to, Zurich, 1865. See the article on " Flax."

[3] Perrin, *Étude Préhistorique de la Savoie,* in 4to, 1870 ; Castelfranco, *Notizie intorno alla Stazione lacustre di Lagozza ;* and Sordelli, *Sulle piante della torbiera della Lagozza,* in the *Actes de la Soc. Ital. des Scien. Nat.,* 1880.

[4] Much, *Mittheil d. Anthropol. Ges. in Wien,* vol. vi. ; Sacken, *Sitzber. Akad. Wien.,* vol. vi.   Letter of Heer on these works and analysis of them in Naidaillac, i. p. 247.

writers, to attribute the origin of all progress to their own nation.

In America, agriculture is perhaps not quite so ancient as in Asia and Egypt, if we are to judge from the civilization of Mexico and Peru, which does not date even from the first centuries of the Christian era. However, the widespread cultivation of certain plants, such as maize, tobacco, and the sweet potato, argues a considerable antiquity, perhaps two thousand years or thereabouts. History is at fault in this matter, and we can only hope to be enlightened by the discoveries of archæology and geology.

CHAPTER II.

METHODS FOR DISCOVERING OR PROVING THE ORIGIN OF SPECIES.

1. *General reflections.* As most cultivated plants have been under culture from an early period, and the manner of their introduction into cultivation is often little known, different means are necessary in order to ascertain their origin. For each species we need a research similar to those made by historians and archæologists—a varied research, in which sometimes one process is employed, sometimes another; and these are afterwards combined and estimated according to their relative value. The naturalist is here no longer in his ordinary domain of observation and description; he must support himself by historical proof, which is never demanded in the laboratory; and botanical facts are required, not with respect to the physiology of plants—a favourite study of the present day—but with regard to the distinction of species and their geographical distribution.

I shall, therefore, have to make use of methods of which some are foreign to naturalists, others to persons versed in historical learning. I shall say a few words of each, to explain how they should be employed and what is their value.

2. *Botany.* One of the most direct means of discovering the geographical origin of a cultivated species, is to seek in what country it grows spontaneously, and without the help of man. The question appears at the first glance to be a simple one. It seems, indeed, that

by consulting floras, works upon species in general, or herbaria, we ought to be able to solve it easily in each particular case. Unfortunately it is, on the contrary, a question which demands a special knowledge of botany, especially of geographical botany, and an estimate of botanists and of collectors, founded on a long experience. Learned men, occupied with history or with the interpretation of ancient authors, are liable to grave mistakes when they content themselves with the first testimony they may happen to light upon in a botanical work. On the other hand, travellers who collect plants for a herbarium are not always sufficiently observant of the places and circumstances in which they find them. They often neglect to note down what they have remarked on the subject. We know, however, that a plant may have sprung from others cultivated in the neighbourhood; that birds, winds, etc., may have borne the seeds to great distances; that they are sometimes brought in the ballast of vessels or mixed with their cargoes. Such cases present themselves with respect to common species, much more so with respect to cultivated plants which abound near human dwellings. A collector or traveller had need be a keen observer to judge if a plant has sprung from a wild stock belonging to the flora of the country, or if it is of foreign origin. When the plant is growing near dwellings, on walls, among rubbish-heaps, by the wayside, etc., we should be cautious in forming an opinion.

It may also happen that a plant strays from cultivation, even to a distance from suspicious localities, and has nevertheless but a short duration, because it cannot in the long run support the conditions of the climate or the struggle with the indigenous species. This is what is called in botany an *adventive* species. It appears and disappears, a proof that it is not a native of the country. Every flora offers numerous examples of this kind. When these are more abundant than usual, the public is struck by the circumstance. Thus, the troops hastily summoned from Algeria into France in 1870, disseminated by fodder and otherwise a number of

African and southern species which excited wonder, but of which no trace remained after two or three winters. Some collectors and authors of floras are very careful in noting these facts. Thanks to personal relations with some of them, and to frequent references to their herbaria and botanical works, I flatter myself I am acquainted• with them. I shall, therefore, willingly cite their testimony in doubtful cases. For certain countries and certain species I have addressed myself directly to these eminent naturalists. I have appealed to their memory, to their notes, to their herbaria, and from the answers they have been so kind as to return, I have been enabled to add unpublished documents to those found in works already made public. My sincere thanks are due for information of this nature received from Mr. C. B. Clarke on the plants of India, from M. Boissier on those of the East, from M. Sagot on the species of French Guiana, from M. Cosson on those of Algeria, from MM. Decaisne and Bretschneider on the plants of China, from M. Pancic on the cereals of Servia, from Messrs. Bentham and Baker on the specimens of the herbarium at Kew, lastly from M. Edouard André on the plants of America. This zealous traveller was kind enough to lend me some most interesting specimens of species cultivated in South America, which he found presenting every appearance of indigenous plants.

A more difficult question, and one which cannot be solved at once, is whether a plant growing wild, with all the appearance of the indigenous species, has existed in the country from a very early period, or has been introduced at a more or less ancient date.

For there are naturalized species, that is, those that are introduced among the plants of the ancient flora, and which, although of foreign origin, persist there in such a manner that observation alone cannot distinguish them, so that historical records or botanical considerations, whether simple or geographical, are needed for their detection. In a very general sense, taking into consideration the lengthened periods with which science is concerned, nearly all species, especially in the regions lying outside the tropics, have been once naturalized; that is to say, they have, from geographical and physical circumstances, passed from one region to another. When, in 1855, I put forward the idea that conditions anterior to our epoch determined the greater number of the facts of the actual distribution of plants—this was the sense of several of the articles, and of the conclusion of my two volumes of geographical botany [1]—it was received with considerable surprise. It is true that general considerations of palæontology had just led Dr. Unger,[2] a German savant, to adopt similar ideas, and before him Edward Forbes had, with regard to some species of the southern counties of the British Isles, suggested the hypothesis of an ancient connection with Spain.[3] But the proof that it is impossible to explain the habitations of the whole number of present species by means of the conditions existing for some thousands of years, made a greater impression, because it belonged more especially to the department of botanists, and did not relate to only a few plants of a single country. The hypothesis suggested by Forbes became an assured fact and capable of general application, and is now a truism of science. All that is written on geographical or zoological botany rests upon this basis, which is no longer contested.

This principle, in its application to each country and each species, presents a number of difficulties; for when a cause is once recognized, it is not always easy to discover how it has affected each particular case. Luckily, so far as cultivated plants are concerned, the questions which occur do not make it necessary to go back to very ancient times, nor to dates which cannot be defined by a given number of years or centuries. No doubt the modern specific forms date from a period earlier than the great extension of glaciers in the northern hemi-

---

[1] Alph. de Candolle, *Géographie Botanique Raisonnée*, chap. x. p. 1055; chap. xi., xix., xxvii.

[2] Unger, *Versuch einer Geschichte der Pflanzenwelt*, 1852.

[3] Forbes, *On the Connection between the Distribution of the Existing Fauna and Flora of the British Isles, with the Geological Changes which have affected their Area*, in 8vo, *Memoirs of the Geological Survey*, vol. i. 1846.

sphere—a phenomenon of several thousand years' duration, if we are to judge from the size of the deposits transported by the ice ; but cultivation began after this epoch, and even in many instances within historic time. We have little to do with previous events. Cultivated species may have changed their abode before cultivation, or in the course of a longer time they may have changed their form ; this belongs to the general study of all organized life, and we are concerned only with the examination of each species since its cultivation or in the time immediately before it. This is a great simplification.

The question of age, thus limited, may be approached by means of historical or other records, of which I shall presently speak, and by the principles of geographical botany.

I shall briefly enumerate these, in order to show in what manner they can aid in the discovery of the geographical origin of a given plant.

As a rule, the abode of each species is constant, or nearly constant. It is, however, sometimes disconnected ; that is to say, that the individuals of which it is composed are found in widely separated regions. These cases, which are extremely interesting in the study of the vegetable kingdom and of the surface of the globe, are far from forming the majority. Therefore, when a culti-vated species is found wild, frequently in Europe, more rarely in the United States, it is probable that, in spite of its indigenous appearance in America, it has become naturalized after being accidentally transported thither.

The genera of the vegetable kingdom, although usually composed of several species, are often confined to a single region. It follows, that the more species included in a genus all belonging to the same quarter of the globe, the more probable it is that one of the species, apparently indigenous in another part of the world, has been transported thither and has become naturalized there, by escaping from cultivation. This is especially the case with tropical genera, because they are more often restricted either to the old or to the new world.

Geographical botany teaches us what countries have genera and even species in common, in spite of a certain distance, and what, on the contrary, are very different, in spite of similarity of climate or inconsiderable distance. It also teaches us what species, genera, and families are scattered over a wide area, and the more limited extent of others. These data are of great assistance in determining the probable origin of a given species. Naturalized plants spread rapidly. I have quoted examples elsewhere [1] of instances within the last two centuries, and similar facts have been noted from year to year. The rapidity of the recent invasion of *Anacharis Alsinastrum* into the rivers of Europe is well known, and that of many European plants in New Zealand, Australia, California, etc., mentioned in several floras or modern travels.

The great abundance of a species is no proof of its antiquity. *Agave Americana*, so common on the shores of the Mediterranean, although introduced from America, and our cardoon, which now covers a great part of the Pampas of La Plata, are remarkable instances in point. As a rule, an invading species makes rapid way, while extinction is, on the contrary, the result of the strife of several centuries against unfavourable circumstances.[2]

The designation which should be adopted for allied species, or, to speak scientifically, allied forms, is a problem often presented in natural history, and more often in the category of cultivated species than in others. These plants are changed by cultivation. Man adopts new and convenient forms, and propagates them by artificial means, such as budding, grafting, the choice of seeds, etc. It is clear that, in order to discover the origin of one of these species, we must eliminate as far as possible the forms which appear to be artificial, and concentrate our attention on the others. A simple reflection may guide this choice, namely, that a cultivated species varies chiefly in those parts for which it is cultivated. The others remain unmodified, or present trifling alterations,

---

[1] A. de Candolle, *Géographie Botanique Raisonnée*, chap. vii. and x.
[2] *Ibid.*, chap. viii. p. 804.

of which the cultivator takes no note, because they are useless to him. We may expect, therefore, to find the fruit of a wild fruit tree small and of a doubtfully agreeable flavour, the grain of a cereal in its wild state small, the tubercles of a wild potato small, the leaves of indigenous tobacco narrow, etc., without, however, going so far as to imagine that the species developed rapidly under cultivation, for man would not have begun to cultivate it if it had not from the beginning presented some useful or agreeable qualities.

When once a cultivated plant has been reduced to such a condition as permits of its being reasonably compared with analogous spontaneous forms, we have still to decide what group of nearly similar plants it is proper to designate as constituting a species. Botanists alone are competent to pronounce an opinion on this question, since they are accustomed to appreciate differences and resemblances, and know the confusion of certain works in the matter of nomenclature. This is not the place to discuss what may reasonably be termed a species. I have stated in some of my articles the principles which seem to me the best. As their application would often require a study which has not been made, I have thought it well occasionally to treat quasi-specific forms as a group which appears to me to correspond to a species, and I have sought the geographical origin of these forms as though they were really specific.

To sum up: botany furnishes valuable means of guessing or proving the origin of cultivated plants and for avoiding mistakes. We must, however, by no means forget that practical observation must be supplemented by research in the study. After gaining information from the collector who sees the plants in a given spot or district, and who draws up a flora or a catalogue of species, it is indispensable to study the known or probable geographical distribution in books and in herbaria, and to reflect upon the principles of geographical botany and on the questions of classification, which cannot be done by travelling or collecting. Other researches, of which I shall speak presently, must be combined with those of botany if we would arrive at satisfactory conclusions.

3. *Archæology and Palæontology.* The most direct proof which can be conceived of the ancient existence of a species in a given country is to see its recognizable fragments in old buildings or deposits, of a more or less certain date.

The fruits, seeds, and different portions of plants taken from ancient Egyptian tombs, and the drawings which surround them in the pyramids, have given rise to most important researches, which I shall often have to mention. Nevertheless, there is a possible source of error; the fraudulent introduction of modern plants into the sarcophagi of the mummies. This was easily discovered in the case of some grains of maize, for instance, a plant of American origin, which were introduced by the Arabs; but species cultivated in Egypt within the last two or three thousand years may have been added, which would thus appear to have belonged to an earlier period. The tumuli or mounds of North America, and the monuments of the ancient Mexicans and Peruvians, have furnished records about the plants cultivated in that part of the world. Here we are concerned with an epoch subsequent to the pyramids of Egypt.

The deposits of the Swiss lake-dwellings have been the subject of important treatises, among which that of Heer, quoted just now, holds the first place. Similar works have been published on the vegetable remains found in other lakes or peat mosses of Switzerland, Savoy, Germany, and Italy. I shall quote them with reference to several species. Dr. Gross has been kind enough to send me seeds and fruits taken from the lake-dwellings of Neuchâtel; and my colleague, Professor Heer, has favoured me with several facts collected at Zurich since the publication of his work. I have already said that the rubbish-heaps of the Scandinavian countries, called kitchen-middens, have furnished no trace of cultivated vegetables.

The tufa of the south of France contains leaves and other remains of plants, which have been discovered by

MM. Martins, Planchon, de Saporta, and other savants. Their date is not, perhaps, always earlier than that of the first lacustrine deposits, and it is possible that it agrees with that of ancient Egyptian monuments, and of ancient Chinese books. Lastly, the mineralogic strata, with which geologists are specially concerned, tell us much about the succession of vegetable forms in different countries; but here we are dealing with epochs far anterior to agriculture, and it would be a strange and certainly a most valuable chance if a modern cultivated species were discovered in the European tertiary epoch. No such discovery has hitherto been made with any certainty, though uncultivated species have been recognized in strata prior to the glacial epoch of the northern hemisphere. For the rest, if we do not succeed in finding them, the consequences will not be clear, since it may be said, either that such a plant came at a later date from a different region, or that it had formerly another form which renders its recognition impossible in a fossil state.

4. *History.* Historical records are important in order to determine the date of certain cultures in each country. They also give indications as to the geographical origin of plants when they have been propagated by the migrations of ancient peoples, by travellers, or by military expeditions.

The assertions of authors must not, however, be accepted without examination.

The greater number of ancient historians have confused the fact of the cultivation of a species in a country with that of its previous existence there in a wild state. It has been commonly asserted, even in our own day, that a species cultivated in America or China is a native of America or China. A no less common error is the belief that a species comes originally from a given country because it has come to us from thence, and not direct from the place in which it is really indigenous. Thus the Greeks and Romans called the peach the Persian apple, because they had seen it cultivated in Persia, where it probably did not grow wild. It was a native of China, as I have elsewhere shown. They called the pomegranate, which had spread gradually from garden to garden from Persia to Mauritania, the apple of Carthage (*Malum Punicum*). Very ancient authors, such as Herodotus and Berosius, are yet more liable to error, in spite of their desire to be accurate.

We shall see, when we speak of maize, that historical documents which are complete forgeries may deceive us about the origin of a species. It is curious, for it seems to be no one's interest to lie about such agricultural facts. Fortunately, facts of botany and archæology enable us to detect errors of this nature.

The principal difficulty, which commonly occurs in the case of ancient historians, is to find the exact translation of the names of plants, which in their books always bear the common names. I shall speak presently of the value of these names and how the science of language may be brought to bear on the questions with which we are occupied, but I must first indicate those historical notions which are most useful in the study of cultivated plants.

Agriculture came originally, at least so far as the principal species are concerned, from three great regions, in which certain plants grew, regions which had no communication with each other. These are—China, the southwest of Asia (with Egypt), and intertropical America. I do not mean to say that in Europe, in Africa, and elsewhere savage tribes may not have cultivated a few species locally, at an early epoch, as an addition to the resources of hunting and fishing; but the great civilizations based upon agriculture began in the three regions I have indicated. It is worthy of note that in the old world agricultural communities established themselves along the banks of the rivers, whereas in America they dwelt on the high lands of Mexico and Peru. This may perhaps have been due to the original situation of the plants suitable for cultivation, for the banks of the Mississippi, of the Amazon, of the Orinoco, are not more unhealthy than those of the rivers of the old world.

A few words about each of the three regions.

China had already possessed for some thousands of years a flourishing agriculture and even horticulture, when she entered for the first time into relations with Western Asia, by the mission of Chang-Kien, during the reign of the Emperor Wu-ti, in the second century before the Christian era. The records, known as Pent-sao, written in our Middle Ages, state that he brought back the bean, the cucumber, the lucern, the saffron, the sesame, the walnut, the pea, spinach, the water-melon, and other western plants,[1] then unknown to the Chinese. Chang-Kien, it will be observed, was no ordinary ambassador. He considerably enlarged the geographical knowledge, and improved the economic condition of his countrymen. It is true that he was constrained to dwell ten years in the West, and that he belonged to an already civilized people, one of whose emperors had, 2700 B.C., consecrated with imposing ceremonies the cultivation of certain plants. The Mongolians were too barbarous, and came from too cold a country, to have been able to introduce many useful species into China; but when we consider the origin of the peach and the apricot, we shall see that these plants were brought into China from Western Asia, probably by isolated travellers, merchants or others, who passed north of the Himalayas. A few species spread in the same way into China from the West before the embassy of Chang-Kien.

Regular communication between China and India only began in the time of Chang-Kien, and by the circuitous way of Bactriana;[2] but gradual transmissions from place to place may have been effected through the Malay Peninsula and Cochin-China. The writers of Northern China may have been ignorant of them, and especially since the southern provinces were only united to the empire in the second century before Christ.[3]

Regular communications between China and Japan only took place about the year 57 of our era, when an ambassador was sent; and the Chinese had no real knowledge of their eastern neighbours until the third

---

[1] Bretschneider, *On the Study and Value*, etc., p. 15.
[2] *Ibid.*   [3] *Ibid.*, p. 23.

century, when the Chinese character was introduced into Japan.[1]

The vast region which stretches from the Ganges to Armenia and the Nile was not in ancient times so isolated as China. Its inhabitants exchanged cultivated plants with great facility, and even transported them to a distance. It is enough to remember that ancient migrations and conquests continually intermixed the Turanian, Aryan, and Semitic peoples between the Caspian Sea, Mesopotamia, and the Nile. Great states were formed nearly at the same time on the banks of the Euphrates and in Egypt, but they succeeded to tribes which had already cultivated certain plants. Agriculture is older in that region than Babylon and the first Egyptian dynasties, which date from more than four thousand years ago. The Assyrian and Egyptian empires afterwards fought for supremacy, and in their struggles they transported whole nations, which could not fail to spread cultivated species. On the other hand, the Aryan tribes who dwelt originally to the north of Mesopotamia, in a land less favourable to agriculture, spread westward and southward, driving out or subjugating the Turanian and Dravidian nations. Their speech, and those which are derived from it in Europe and Hindustan, show that they knew and transported several useful species.[2] After these ancient events, of which the dates are for the most part uncertain, the voyages of the Phœnicians, the wars between the Greeks and Persians, Alexander's expedition into India, and finally the Roman rule, completed the spread of cultivation in the interior of Western Asia, and even introduced it into Europe and the north of Africa, wherever the climate permitted.

Later, at the time of the crusades, very few useful plants yet remained to be brought from the East. A

---

[1] *Atsuma-gusa. Recueil pour servir à la connaissance de l'extrême Orient*, Turretini, vol. vi., pp. 200, 293.

[2] There are in the French language two excellent works, which give the sum of modern knowledge with regard to the East and Egypt. The one is the *Manuel de l'Histoire Ancienne de l'Orient*, by François Lenormand, 3 vols. in 12mo, Paris, 1869; the other, *L'Histoire Ancienne des Peuples de l'Orient*, by Maspero, 1 vol. in 8vo, Paris, 1878.

few varieties of fruit trees which the Romans did not
possess, and some ornamental plants, were, however, then
brought to Europe.

The discovery of America in 1492 was the last great
event which caused the diffusion of cultivated plants
into all countries. The American species, such as the
potato, maize, the prickly pear, tobacco, etc., were first
imported into Europe and Asia. Then a number of
species from the old world were introduced into America.
The voyage of Magellan (1520–1521) was the first direct
communication between South America and Asia. In the
same century the slave trade multiplied communications
between Africa and America. Lastly, the discovery of
the Pacific Islands in the eighteenth century, and the
growing facility of the means of communication, combined
with a general idea of improvement, produced that more
general dispersion of useful plants of which we are
witnesses at the present day.

5. *Philology.* The common names of cultivated plants
are usually well known, and may afford indications touching the history of a species, but there are examples
in which they are absurd, based upon errors, or vague
and doubtful, and this involves a certain caution in
their use.

I could quote a number of such names in all languages;
it is enough to mention, in French, *blé de Turquie*, maize,
a plant which is not a wheat, and which comes from
America; in English, Jerusalem artichoke (*Helianthus
tuberosus*), which does not come from Jerusalem, but
from North America, and is no artichoke.

A number of names given to foreign plants by
Europeans when they are settled in the colonies, express false or insignificant analogies. For instance, the
New Zealand flax resembles the true flax as little as
possible; it is merely that a textile substance is obtained
from its leaves. The *mahogany apple* (cashew) of the
French West India Isles is not an apple, nor even the
fruit of a pomaceous tree, and has nothing to do with
mahogany.

Sometimes the common names have changed, in

passing from one language to another, in such a manner as to give a false or absurd meaning. Thus the tree of Judea of the French (*Cercis Siliquastrum*) has become the Judas tree in English. The fruit called by the Mexicans *ahuaca*, is become the *avocat* (lawyer) of the French colonists.

Not unfrequently names of plants have been taken by the same people at successive epochs or in different provinces, sometimes as generic, sometimes as specific names. The French word *blé*, for instance, may mean several species of the genus Triticum, and even of very different nutritious plants (maize and wheat), or a given species of wheat.

Several common names have been transferred from one plant to another through error or ignorance. Thus the confusion made by early travellers between the sweet potato (*Convolvulus Batatas*) and the potato (*Solanum tuberosum*) has caused the latter to be called potato in English and *patatas* in Spanish.

If modern, civilized peoples, who have great facilities for comparing species, learning their origin and verifying their names in books, have made such mistakes, it is probable that ancient nations have made many and more grave errors. Scholars display vast learning in explaining the philological origin of a name, or its modifications in derived languages, but they cannot discover popular errors or absurdities. It is left for botanists to discover and point them out. We may note, in passing, that the double or compound names are the most doubtful. They may consist of two mistakes ; one in the root or principal name, the other in the addition or accessory name, destined almost always to indicate the geographical origin, some visible quality, or some comparison with other species. The shorter a name is, the better it merits consideration in questions of origin or antiquity; for it is by the succession of years, of the migrations of peoples, and of the transport of plants, that the addition of often erroneous epithets takes place. Similarly, in symbolic writing, like that of the Chinese and the Egyptians, unique and simple signs indicate long-known species, not imported from foreign countries, while complicated signs are doubtful or indicate a foreign origin. We must not forget, however, that the signs have often been rebuses, based on chance resemblances in the words, or on superstitious and fanciful ideas.

The identity of a common name for a given species in several languages may have two very different explanations. It may be because a plant has been spread by a people which has been dispersed and scattered. It may also result from the transmission of a plant from one people to another with the name it bore in its original home. The first case is that of the hemp, of which the name is similar, at least as to the root, in all the tongues derived from the primitive Aryan stock. The second is seen in the American name of tobacco, the Chinese of tea, which have spread into a number of countries, without any philological or ethnographic filiation. This case has occurred oftener in modern than in ancient times, because the rapidity of communications allows of the simultaneous introduction of a plant and of its name, even where the distance is great.

The diversity of names for the same species may also spring from various causes. As a rule, it indicates an early existence in different countries, but it may also arise from the mixture of races, or from names of varieties which take the place of the original name. Thus in England we find, according to the county, a Keltic, Saxon, Danish, or Latin name ; and flax bears in Germany the names of *flachs* and *lein*, words which are evidently of different origin.

When we desire to make use of the common names to gather from them certain probabilities regarding the origin of species, it is necessary to consult dictionaries and the dissertations of philologists ; but we must take into account the chances of error in these learned men, who, since they are neither cultivators nor botanists, may have made mistakes in the application of a name to a species.

The most considerable collection of common names is

that of Nemnich, published in 1793.[1] I have another in manuscript which is yet more complete, drawn up in our library by an old pupil of mine, Moritzi, by means of floras and of several books of travel written by botanists. There are, besides, dictionaries of the names of the species in given countries or in some special language. This kind of glossary does not often contain explanations of etymology; but in spite of what Hehn [2] may say, a naturalist possessed of an ordinary general education can recognize the connection or the fundamental differences between certain names in different languages, and need not confound modern with ancient languages. It is not necessary to be initiated into the mysteries of suffixes or affixes, of dentals and labials. No doubt the researches of a philologist into etymologies are more profound and valuable, but this is rarely necessary when our researches have to do with cultivated plants. Other sciences are more useful, especially that of botany; and philologists are more often deficient in these than naturalists are deficient in philology, for the very evident reason that more place is given to languages than to natural history in general education. It appears to me, moreover, that philologists, notably those who are occupied with Sanskrit, are always too eager to find the etymology of every name. They do not allow sufficiently for human stupidity, which has in all time given rise to absurd words, without any real basis, and derived only from error or superstition.

The filiation of modern European tongues is known to every one. That of ancient languages has, for more than half a century, been the object of important labours. Of these I cannot here give even a brief notice. It is sufficient to recall that all modern European languages are derived from the speech of the Western Aryans, who came from Asia, with the exception of Basque (derived from the Iberian language), Finnish, Turkish, and Hun-

[1] Nemnich, *Allgemeines polyglotten-Lexicon der Naturgeschichte*, 2 vols in 4to.

[2] Hehn, *Kulturpflanzen und Hausthiere in ihren Uebergang aus Asien*, in 8vo, 3rd edit. 1877.

garian, into which, moreover, words of Aryan origin have been introduced. On the other hand, several modern languages of India, Ceylon, and Java, are derived from the Sanskrit of the Eastern Aryans, who left Central Asia after the Western Aryans. It is supposed, with sufficient probability, that the first Western Aryans came into Europe 2500 B.C., and the Eastern Aryans into India a thousand years later.

Basque (or Iberian), the speech of the Guanchos of the Canary Isles, of which a few plant names are known, and Berber, are probably connected with the ancient tongues of the north of Africa.

Botanists are in many cases forced to doubt the common names attributed to plants by travellers, historians, and philologists. This is a consequence of their own doubts respecting the distinction of species and of the well-known difficulty of ascertaining the common name of a plant. The uncertainty becomes yet greater in the case of species which are more easily confounded or less generally known, or in the case of the languages of little-civilized nations. There are, so to speak, degrees of languages in this respect, and the names should be accepted more or less readily according to these degrees.

In the first rank, for certainty, are placed those languages which possess botanical works. For instance, it is possible to recognize a species by means of a Greek description by Dioscorides or Theophrastus, and by the less complete Latin texts of Cato, Columella, or Pliny. Chinese books also give descriptions. Dr. Bretschneider, of the Russian legation at Pekin, has written some excellent papers upon these books, from which I shall often quote.[1]

The second degree is that of languages possessing a literature composed only of theological and poetical works, or of chronicles of kings and battles. Such works

---

[1] Bretschneider, *On the Study and Value of Chinese Botanical Works, with Notes on the History of Plants and Geographical Botany from Chinese Sources*, in 8vo, 51 pp., with illustrations, Foochoo, without date, but the preface bears the date Dec. 1870. *Notes on Some Botanical Questions*, in 8vo, 14 pp., 1880.

make mention here and there of plants, with epithets or
reflections on their mode of flowering, their ripening,
their use, etc., which allow their names to be divined,
and to be referred to modern botanical nomenclature.
With the added help of a knowledge of the flora of the
country, and of the common names in the languages
derived from the dead language, it is possible to discover
approximately the sense of some words.   This is the case
with Sanskrit,[1] Hebrew,[2] and Armenian.[3]

Lastly, a third category of dead languages offers no
certainty, but merely presumptions or hypothetical and
rare indications.   It comprehends those tongues in which
there is no written work, such as Keltic, with its dialects,
the ancient Sclavonic, Pelasgic, Iberian, the speech of
the primitive Aryans, Turanians, etc.   It is possible to
guess certain names or their approximate form in these
dead languages by two methods, both of which should
be employed with caution.

The first and best is to consult the languages derived,
or which we believe to be derived, directly from the
ancient tongues, as Basque for the Iberian language,
Albanian for the Pelasgic, Breton, Erse, and Gaelic for
Keltic.   The danger lies in the possibility of mistake in
the filiation of the languages, and especially in a mistaken
belief in the antiquity of a plant-name which may have

[1] Wilson's dictionary contains names of plants, but botanists have
more confidence in the names indicated by Roxburgh in his *Flora
Indica* (edit. of 1832, 3 vols. in 8vo), and in Piddington's *English Index
to the Plants of India*, Calcutta, 1832.   Scholars find a greater number
of words in the texts, but they do not give sufficient proof of the sense
of these words.   As a rule, we have not in Sanskrit what we have in
Hebrew, Greek, and Chinese—a quotation of phrases concerning each
word translated into a modern language.

[2] The best work on the plant-names in the Old Testament is that of
Rosenmüller, *Handbuch der biblischen Alterkunde*, in 8vo, vol. iv., Leipzig,
1830.   A good short work, in French, is *La Botanique de la Bible*, by
Fred. Hamilton, in 8vo, Nice, 1871.

[3] Reynier, a Swiss botanist, who had been in Egypt, has given the
sense of many plant-names in the Talmud.   See his volumes entitled
Economie Publique et Rurale des Arabes et des Juifs, in 8vo, 1820 ;
and *Economie Publique et Rurale des Egyptiens et des Carthaginois*,
in 8vo, Lausanne, 1823.   The more recent works of Duschak and Löw
are not based upon a knowledge of Eastern plants, and are unintelligible
to botanists because of names in Syriac and Hebrew characters.

been introduced by another people. Thus the Basque
language contains many words which seem to have been
taken from the Latin at the time of the Roman rule.
Berber is full of Arab words, and Persian of words of
every origin, which probably did not exist in Zend.
The other method consists in reconstructing a dead
language which had no literature, by means of those
which are derived from it; for instance, the speech of
the Western Aryans, by means of the words common to
several European languages which have sprung from it.
Fick's dictionary will hardly serve for the words of
ancient Aryan languages, for he gives but few plant-
names, and his arrangement renders it unintelligible to
those who have no knowledge of Sanskrit. Adolphe
Pictet's work [1] is far more important to naturalists, and
a second edition, augmented and improved, has been
published since the author's death. Plant-names and
agricultural terms are explained and discussed in this
work, in a manner all the more satisfactory that an
accurate knowledge of botany is combined with philology.
If the author attributes perhaps too much importance
to doubtful etymologies, he makes up for it by other
knowledge, and by his excellent method and lucidity.

The plant-names of the Euskarian or Basque language
have been considered from the point of view of their
probable etymology by the Comte de Charencey, in *Les
Actes de la Société Philologique* (vol. i. No. 1, 1869). I
shall have occasion to quote this work, of which the
difficulties were great, in the absence of all literature
and of all derived languages.

6. *The necessity for combining the different methods.*
The various methods of which I have spoken are of
unequal value. It is clear that when we have archæological records about a given species, like those of the
Egyptian monuments, or of the Swiss lake-dwellings,
these are facts of remarkable accuracy. Then come
the data furnished by botany, especially those on the
spontaneous existence of a species in a given country.

---

[1] Adolphe Pictet, *Les Origines des Peuples Indo-Européens*, 3 vols. in
8vo, Paris, 1878.

These, if examined with care, may be very important. The assertions contained in the works of historians or even of naturalists respecting an epoch at which science was only beginning, have not the same value. Lastly, the common names are only an accessory means, especially in modern languages, and a means which, as we have seen, is not entirely trustworthy. So much may be said in a general way, but in each particular case one method or the other may be more or less important.

Each can only lead to probabilities, since we are dealing with facts of ancient date which are beyond the reach of direct and actual observation. Fortunately, if the same probability is attained in three or four different ways, we approach very near to certainty. The same rule holds good for researches into the history of plants as for researches into the history of nations. A good author consults historians who have spoken of events, the archives in which unpublished documents are found, the inscriptions on ancient monuments, the newspapers, private letters, finally memoirs and even tradition. He gathers probabilities from every source, and then compares these probabilities, weighs and discusses them before deciding. It is a labour of the mind which requires intelligence and judgment. This labour differs widely from observation employed in natural history, and from pure reason which is proper to the exact sciences. Nevertheless, when, by several methods, we reach the same probability, I repeat that the latter is very nearly a certainty. We may even say that it is as much a certainty as historical science can pretend to attain.

I have the proof of this when I compare my present work with that which I composed by the same methods in 1855. For the species which I then studied, I have now more authorities and better authenticated facts, but my conclusions on the origin of each species have scarcely altered. As they were already based on a combination of methods, probabilities have usually become certainties, and I have not been led to conclusions absolutely contrary to those previously formed.

Archæological, philological, and botanical data become

more and more numerous. By their means the history
of cultivated plants is perfected, while the assertions of
ancient authors lose instead of gaining in importance.
From the discoveries of antiquaries and philologists,
moderns are better acquainted than the Greeks with
Chaldea and ancient Egypt. They can prove mistakes
in Herodotus. Botanists on their side correct Theophrastus, Dioscorides, and Pliny from their knowledge of
the flora of Greece and Italy, while the study of classical
authors to which learned men have applied themselves
for three centuries has already furnished all that it has to
give. I cannot help smiling when, at the present day,
savants repeat well-known Greek and Latin phrases, and
draw from them what they call conclusions. It is trying
to extract juice from a lemon which has already been
repeatedly squeezed. We must say it frankly, the works
which repeat and commentate on the ancient authors
of Greece and Rome without giving the first place to
botanical and archæological facts, are no longer on a
level with the science of the day. Nevertheless, I could
name several German works which have attained to the
honour of a third edition. It would have been better to
reprint the earlier publications of Fraas and Lenz, of
Targioni and Heldreich, which have always given more
weight to the modern data of botany, than to the vague
descriptions of classic authors; that is to say, to facts
than to words and phrases,

PART II.

On the Study of Species, considered as to their Origin, their early Cultivation, and the Principal Facts of their Diffusion.[1]

CHAPTER I.

PLANTS CULTIVATED FOR THEIR SUBTERRANEAN PARTS, SUCH AS ROOTS, TUBERCLES, OR BULBS.[2]

Radish.—*Raphanus sativus*, Linnæus.

The radish is cultivated for what is called the root, which is, properly speaking, the lower part of the stem with the tap root.[3] Every one knows how the size, shape, and colour of those organs which become fleshy vary according to the soil or the variety.

There is no doubt that the species is indigenous in the temperate regions of the old world; but, as it has been cultivated in gardens from the earliest historic times, from China and Japan to Europe, and as it sows

---

[1] A certain number of species whose origin is well known, such as the carrot, sorrel, etc., are mentioned only in the summary at the beginning of the last part, with an indication of the principal facts concerning them.

[2] Some species are cultivated sometimes for their roots and sometimes for their leaves or seeds. In other chapters will be found species cultivated sometimes for their leaves (as fodder) or for their seeds, etc. I have classed them according to their commonest use. The alphabetical index refers to the place assigned to each species.

[3] See the young state of the plant when the part of the stem below the cotyledons is not yet swelled. Turpin gives a drawing of it in the *Annales des Sciences Naturelles*, series 1, vol. xxi. pl. 5.

itself frequently round cultivated plots, it is difficult to
fix upon its starting-point.

Formerly *Raphanus sativus* was confounded with
kindred species of the Mediterranean region, to which
certain Greek names were attributed; but Gay, the
botanist, who has done a good deal towards eliminating these analogous forms,[1] considered *R. sativus* as a
native of the East, perhaps of China. Linnæus also supposed this plant to be of Chinese origin, or at least that
variety which is cultivated in China for the sake of extracting oil from the seeds.[2] Several floras of the south
of Europe mention the species as subspontaneous or
escaped from cultivation, never as spontaneous. Ledebour had seen a specimen found near Mount Ararat, had
sown the seeds of it and verified the species.[3] However,
Boissier,[4] in 1867, in his *Eastern Flora*, says that it is
only subspontaneous in the cultivated parts of Anatolia,
near Mersivan (according to Wied), in Palestine (on his
own authority), in Armenia (according to Ledebour), and
probably elsewhere, which agrees with the assertions
found in European floras.[5] Buhse names a locality, the
Ssahend mountains, to the south of the Caucasus, which
appears to be far enough from cultivation. The recent
Flora of British India,[6] and the earlier *Flora of Cochin-
China* by Loureiro, mention the radish only as a cultivated species. Maximowicz saw it in a garden in the
north-east of China.[7] Thunberg speaks of it as a plant
of general cultivation in Japan, and growing also by
the side of the roads,[8] but the latter fact is not repeated
by modern authors, who are probably better informed.[9]

Herodotus (*Hist.*, l. 2, c. 125) speaks of a radish which
he calls *surmaia*, used by the builders of the pyramid of

[1] In A. de Candolle, *Géogr. Bot. Raisonnée*, p. 826.
[2] Linnæus, *Spec. Plant*, p. 935.
[3] Ledebour, *Fl. Ross.*, i. p. 225.
[4] Boissier, *Fl. Orient*, i. p. 400.
[5] Buhse, *Aufzählung Transcaucasien*, p. 30.
[6] Hooker, *Flora of British India*, i. p. 166.
[7] Maximowicz, *Primitiæ Floræ Amurensis*, p. 47.
[8] Thunberg, *Fl. Jap.*, p. 263.
[9] Franchet and Savatier, *Enum. Plant. Jap.*, i. p. 39.

Cheops, according to an inscription upon the monument. Unger [1] copied from Lepsius' work two drawings from the temple of Karnak, of which the first, at any rate, appears to represent the radish.

From all this we gather, first, that the species spreads easily from cultivation in the west of Asia and the south of Europe, while it does not appear with certainty in the flora of Eastern Asia; and secondly, that in the regions south of the Caucasus it is found without any sign of culture, so that we are led to suppose that the plant is wild there. From these two reasons it appears to have come originally from Western Asia between Palestine, Anatolia, and the Caucasus, perhaps also from Greece; its cultivation spreading east and west from a very early period.

The common names support these hypotheses. In Europe they offer little interest when they refer to the quality of the root (*radis*), or to some comparison with the turnip (*ravanello* in Italian, *rabica* in Spanish, etc.), but the ancient Greeks coined the special name *raphanos* (easily reared). The Italian word *ramoraccio* is derived from the Greek *armoracia*, which was used for *R. sativus* or some allied species. Modern interpreters have erroneously referred this name to *Cochlearia Armoracia* or horse-radish, which I shall come to presently. Semitic [2] languages have quite different names (*fugla* in Hebrew, *fuil, fidgel, figl*, etc., in Arab.). In India, according to Roxburgh,[3] the common name of a variety with an enormous root, as large sometimes as a man's leg, is *moola* or *moolee*, in Sanskrit *mooluka*. Lastly, for Cochin-China, China, and Japan, authors give various names which differ very much one from the other. From this diversity a cultivation which ranged from Greece to Japan must be very ancient, but nothing can thence be concluded as to its original home as a spontaneous plant.

A totally different opinion exists on the latter point,

---

[1] Unger, *Pflanzen des Alten Ægyptens*, p. 51, figs. 24 and 29.
[2] In my manuscript dictionary of common names, drawn from the floras of thirty years ago.
[3] Roxburgh, *Fl. Ind.*, iii. p. 126.

which we must also examine. Several botanists[1] suspect
that *Raphanus sativus* is simply a particular condition,
with enlarged root and non-articulated fruit, of *Raphanus raphanistrum*, a very common plant in the temperate cultivated districts of Europe and Asia, and
which is also found in a wild state in sand and light
soil near the sea—for instance, at St. Sebastian, in Dalmatia, and at Trebizond.[2] Its usual haunts are in deserted
fields; and many common names which signify wild
radish, show the affinity of the two plants. I should not
insist upon this point if their supposed identity were a
mere presumption, but it rests upon experiments and
observations which it is important to know.

In *R. raphanistrum* the siliqua is articulated, that
is to say, contracted at intervals, and the seeds placed
each in a division. In *R. sativus* the siliqua is continuous, and forms a single cavity. Some botanists had
made this difference the basis of two distinct genera,
Raphanistrum and *Raphanus*. But three accurate observers, Webb, Gay, and Spach, have noticed among
plants of *Raphanus sativus*, raised from the same seed,
both unilocular and articulated pods, some of them
bilocular, others plurilocular. Webb[3] arrived at the
same results when he afterwards repeated these experiments, and he observed yet another fact of some importance : the radish which sows itself by chance, and is
not cultivated, produced the siliquæ of *Raphanistrum*.[4]
Another difference between the two plants is in the
root, fleshy in *R. sativus*, slender in *R. raphanistrum;* but this changes with cultivation, as appears
from the experiments of Carrière, the head gardener of
the nurseries of the Natural History Museum in Paris.[5]
It occurred to him to sow the seeds of the slender-

---

[1] Webb, *Phytogr. Canar.*, p. 83 ; *Iter. Hisp.*, p. 71 ; Bentham, *Fl.
Hong Kong*, p. 17 ; Hooker, *Fl. Brit. Ind.*, i. p. 166.
[2] Willkomm and Lange, *Prod. Fl. Hisp.*, iii. p. 748 ; Viviani, *Flor.
Dalmat.*, iii. p. 104 ; Boissier, *Fl. Orient.*, i. p. 401.
[3] Webb, *Phytographia Canariensis*, i. p. 83.
[4] Webb, *Iter. Hispaniense*, 1838, p. 72.
[5] Carrière, *Origine des Plantes Domestiques démontrée par la Culture
du Radis Sauvage*, in 8vo, 24 pp., 1869.

rooted *Raphanistrum* in both stiff and light soil, and in
the fourth generation he obtained fleshy radishes, of
varied colour and form like those of our gardens. He
even gives the figures, which are really curious and conclusive. The pungent taste of the radish was not
wanting. To obtain these changes, Carrière sowed in
September, so as to make the plant almost biennial
instead of annual. The thickening of the root was the
natural result, since many biennial plants have fleshy
roots.

The inverse experiment remains to be tried—to sow
cultivated radishes in a poor soil. Probably the roots
would become poòrer and poorer, while the siliquæ would
become more and more articulated.

From all the experiments I have mentioned, *Raphanus sativus* might well be a variety of *R. raphanistrum*, an unstable variety determined by the
existence of several generations in a fertile soil. We
cannot suppose that ancient uncivilized peoples made
essays like those of Carrière, but they may have noticed
plants of *Raphanistrum* grown in richly manured soil,
with more or less fleshy roots ; and this soon suggested
the idea of cultivating them.

I have, however, one objection to make, founded on
geographical botany. *Raphanus raphanistrum* is a
European plant which does not exist in Asia.[1] It cannot, therefore, be this species that has furnished the inhabitants of India, China, and Japan with the radishes
which they have cultivated for centuries. On the other
hand, how could *R. raphanistrum*, which is supposed
to have been modified in Europe, have been transmitted
in ancient times across the whole of Asia ? The transport
of cultivated plants has commonly proceeded from Asia
into Europe. Chang-Kien certainly brought vegetables
from Bactriana into China in the second century B.C.,
but the radish is not named among the number.

Horse-radish—*Cochlearia Armoracia*, Linnæus.

This Crucifer, whose rather hard root has the taste of

---

[1] Ledebour, *Fl. Ross.*; Boissie,r *Fl. Orient.* Works on the flora of the
valley of the Amur.

mustard, was sometimes called in French *cran*, or *cranson de Bretagne*. This was an error caused by the old botanical name *Armoracia*, which was taken for a corruption of *Armorica* (Brittany). *Armoracia* occurs in Pliny, and was applied to a crucifer of the Pontine province, which was perhaps *Raphanus sativus*. After I had formerly [1] pointed out this confusion, I expressed myself as follows on the mistaken origin of the species :— *Cochlearia Armoracia* is not wild in Brittany, a fact now established by the researches of botanists in the west of France. The Abbé Delalande mentions it in his little work, entitled *Hœdic et Houat*,[2] in which he gives so interesting an account of the customs and productions of these two little islands of Brittany. He quotes the opinion of M. le Gall, who, in an unpublished flora of Morbihan, declares the plant foreign to Brittany. This proof, however, is less strong than others, since the south coast of the peninsula of Brittany is not yet sufficiently known to botanists, and the ancient Armorica extended over a portion of Normandy where the wild horse-radish is now found.[3] This leads me to speak of the original home of the species. English botanists mention it as wild in Great Britain, but are doubtful about its origin. Watson [4] considers it as introduced by cultivation. The difficulty of extirpating it, he says, from places where it is cultivated, is well known to gardeners. It is therefore not surprising that this plant should take possession of waste ground, and persist there so as to appear indigenous. Babington [5] mentions only one spot where the species appears to be really wild, namely, Swansea. We will try to solve the problem by further arguments.

Cochlearia Armoracia is a plant belonging to the temperate, and especially to the eastern regions of Europe. It is diffused from Finland to Astrakhan, and to the

[1] A. de Candolle, *Géographie Botanique Raisonnée*, p. 654.
[2] Delalande, *Hœdic et Houat*, 8vo pamphlet, Nantes, 1850, p. 109.
[3] Hardouin, Renou, and Leclerc, *Catalogue du Calvados*, p. 85 ; De Brebisson, *Fl. de Normandie*, p. 25.
[4] Watson, *Cybele*, i. p. 159.
[5] Babington, *Manual of Brit. Bot.*, 2nd edit., p. 28.

desert of Cuman.[1] Grisebach mentions also several localities in Turkey in Europe, near Enos, for instance, where it abounds on the sea-shore.[2] The further we advance towards the west of Europe, the less the authors of floras appear sure that the plant is indigenous, and the localities assigned to it are more scattered and doubtful. The species is rarer in Norway than in Sweden,[3] in the British Isles than in Holland, where a foreign origin is not attributed to it.[4] The specific names confirm the impression of its origin in the east rather than in the west of Europe; thus the name *chren*[5] in Russia recurs in all the Sclavonic languages, *krenai* in Lithuanian, *chren* in Illyrian,[6] etc. It has introduced itself into a few German dialects, round Vienna,[7] for instance, where it persists, in spite of the spread of the German tongue. We owe to it also the French names *cran* or *cranson*. The word used in Germany, *Meerretig,* and in Holland, *meer-radys,* whence the Italian Swiss dialect has taken the name *méridi,* or *mérédi,* means sea-radish, and is not primitive like the word *chren.* It comes probably from the fact that the plant grows well near the sea, a circumstance common to many of the *Cruciferæ,* and which should be the case with this species, for it is wild in the east of Russia where there is a good deal of salt soil. The Swedish name *peppar-rot*[8] suggests the idea that the species came into Sweden later than the introduction of pepper by commerce into the north of Europe. However, the name may have taken the place of an older one, which has remained unknown to us. The English name of horse-radish is not of such an original nature as to lead to a belief in the existence of the species in the country before the Saxon conquest. It means a very strong

---

[1] Ledebour, *Fl. Ross.,* i. p. 159.
[2] Grisebach, *Spicilegium Fl. Rumel.,* i. p. 265.
[3] Fries, *Summa,* p. 30.
[4] Miquel, *Disquisitio pl. regn. Batav.*
[5] Moritzi, *Dict. Inéd. des Noms Vulgaires.*
[6] Moritzi, *ibid. ;* Viviani, *Fl. Dalmat.,* iii. p. 322
[7] Neilreich, *Fl. Wien,* p. 502.
[8] Linnæus, *Fl. Suecica,* No. 540.

radish. The Welsh name *rhuddygl maurth* [1] is only the translation of the English word, whence we may infer that the Kelts of Great Britain had no special name, and were not acquainted with the species. In the west of France, the name *raifort*, which is the commonest, merely means strong root. Formerly it bore in France the names of German, or Capuchin mustard, which shows a foreign and recent origin. On the contrary, the word *chren* is in all the Sclavonic languages, a word which has penetrated into some German and French dialects under the forms of *kreen, cran,* and *cranson,* and which is certainly of a primitive nature, and shows the antiquity of the species in temperate Eastern Europe. It is therefore most probable that cultivation has propagated and naturalized the plant westward from the east for about a thousand years.

Turnips—*Brassica species et varietates radice incrassata.*

The innumerable varieties and subvarieties of the turnip known as swedes, Kohl-rabi, etc., may be all attributed to one of the four species of Linnæus—*Brassica napus, Br. oleracea, Br. rapa, Br. campestris*—of which the two last should, according to modern authors, be fused into one. Other varieties of the species are cultivated for the leaves (cabbages), for the inflorescence (cauliflowers), or for the oil which is extracted from the seed (colza, rape, etc.). When the root or the lower part of the stem [2] is fleshy, the seed is not abundant, nor worth the trouble of extracting the oil; when those organs are slender, the production of the seed, on the contrary, becomes more important, and decides the economic use of the plant. In other words, the store of nutritious matter is placed sometimes in the lower, sometimes in the upper part of the plant, although the organization of the flower and fruit is similar, or nearly so.

---

[1] H. Davies, *Welsh Botanology*, p. 63.

[2] In turnips and swedes the swelled part is, as in the radish, the lower part of the stem, below the cotyledons, with a more or less persistent part of the root. (See Turpin, *Ann. Sc. Natur.*, ser. 1, vol. xxi.) In the Kohl-rabi (*Brassica oleracea caulo-rapa*) it is the stem.

Touching the question of origin, we need not occupy ourselves with the botanical limits of the species, and with the classification of the races, varieties, and sub-varieties,[1] since all the *Brassicæ* are of European and Siberian origin, and are still to be seen in these regions wild, or half wild, in some form or other.

Plants so commonly cultivated and whose germination is so easy often spread round cultivated places ; hence some uncertainty regarding the really wild nature of the plants found in the open country. Nevertheless, Linnæus mentions that *Brassica napus* grows in the sand on the sea-coast in Sweden (Gothland), Holland, and England, which is confirmed, as far as Sweden is concerned, by Fries,[2] who, with his usual attention to questions of this nature, mentions *Br. Campestris*, L. (type of the *Rapa* with slender roots), as really wild in the whole Scandinavian peninsula, in Finland and Denmark. Ledebour[3] indicates it in the whole of Russia, Siberia, and the Caspian Sea.

The floras of temperate and southern Asia mention rapes and turnips as cultivated plants, never as escaped from cultivation.[4] This is already an indication of foreign origin. The evidence of philology is no less significant.

There is no Sanskrit name for these plants, but only modern Hindu and Bengalee names, and those only for *Brassica rapa* and *B. oleracea*.[5] Kæmpfer[6] gives Japanese names for the turnip—*busei*, or more commonly *aona*—but there is nothing to show that these names are ancient. Bretschneider, who has made a careful study of Chinese authors, mentions no *Brassica*. Apparently they do not occur in any of the ancient works on botany and agriculture, although several varieties are now cultivated in China.

It is just the reverse in Europe. The old languages

---

[1] This classification has been the subject of a paper by Augustin Pyramus de Candolle, *Transactions of the Horticultural Society*, vol. v.

[2] Fries, *Summa Veget. Scand.*, i. p. 29.

[3] Ledebour, *Fl. Ross.*, i. p. 216.

[4] Boissier, *Flora Orientalis ;* Sir J. Hooker, *Flora of British India ;* Thunberg, *Flora Japonica ;* Franchet and Savatier, *Enumeratio Plantarum Japonicarum.*

[5] Piddington, *Index.*   [6] Kæmpfer, *Amœn.*, p. 822.

have a number of names which seem to be original. *Brassica rapa* is called *meipen* or *erfinen*[1] in Wales; *repa* and *rippa* in several Slav tongues,[2] which answers to the Latin *rapa*, and is allied to the *neipa* of the Anglo-Saxons. The *Brassica napus* is in Welsh *bresych yr yd ;* in Erse *braisscagh buigh*, according to Threlkeld,[3] who sees in *braisscagh* the root of the Latin *Brassica*. A Polish name, *karpiele*, a Lithuanian, *jellazoji*,[4] are also given, without speaking of a host of other names, transferred sometimes in popular speech from one species to another. I shall speak of the names of *Brassica oleracea* when I come to vegetables.

The Hebrews had no names for cabbages, rapes, and turnips,[5] but there are Arab names : *selgam* for the *Br. napus*, and *subjum* or *subjumi* for *Br. rapa ;* words which recur in Persian and even in Bengali, transferred perhaps from one species to another. The cultivation of these plants has therefore been diffused in the south-west of Asia since Hebrew antiquity.

Finally, every method, whether botanical, historical, or philological, leads us to the following conclusions :—

Firstly, the *Brassicæ* with fleshy roots were originally natives of temperate Europe.

Secondly, their cultivation was diffused in Europe before, and in Asia after, the Aryan invasion.

Thirdly, the primitive slender-rooted form of *Brassica napus*, called *Br. campestris*, had probably from the beginning a more extended range, from the Scandinavian peninsula towards Siberia and the Caucasus. Its cultivation was perhaps introduced into China and Japan, through Siberia, at an epoch which appears not to be much earlier than Greco-Roman civilization.

Fourthly, the cultivation of the various forms or species of *Brassica* was diffused throughout the south-west of Asia at an epoch later than that of the ancient Hebrews.

[1] Davies, *Welsh Botanology*, p. 65.
[2] Moritzi, *Dict. MS.*, compiled from published floras.
[3] Threlkeld, *Synopsis Stirpium Hibernicarum*, 1 vol. in 8vo, 1727.
[4] Moritzi, *Dict. MS.*
[5] Rosenmüller, *Biblische Naturgeschichte*, vol. i., gives none.

Skirret—*Sium Sisarum*, Linnæus.

This vivacious Umbellifer, furnished with several diverging roots in the form of a carrot, is believed to come from Eastern Asia. Linnæus indicates China, doubtfully; and Loureiro,[1] China and Cochin-China, where he says it is cultivated. Others have mentioned Japan and the Corea, but in these countries there are species which it is easy to confound with the one in question, particularly *Sium Ninsi* and *Panax Ginseng.* Maximowicz,[2] who has seen these plants in China and in Japan, and who has studied the herbariums of St. Petersburgh, recognizes only the Altaic region of Siberia and the North of Persia as the home of the wild *Sium Sisarum.* I am very doubtful whether it is to be found in the Himalayas or in China, since modern works on the region of the river Amoor and on British India make no mention of it.

It is doubtful whether the ancient Greeks and Romans knew this plant. The names *Sisaron* of Dioscorides, *Siser* of Columella and of Pliny,[3] are attributed to it. Certainly the modern Italian name *sisaro* or *sisero* seems to confirm this idea; but how could these authors have failed to notice that several roots descend from the base of the stem, whereas all the other umbels cultivated in Europe have but a single tap-root ? It is just possible that the *siser* of Columella, a cultivated plant, may have been the parsnip ; but what Pliny says of the *siser* does not apply to it. According to him it was a medicinal plant, *inter medica dicendum.*[4] He says that Tiberius caused a quantity to be brought every year from Germany, which proves, he adds, that it thrives in cold countries.

If the Greeks had received the plant direct from Persia, Theophrastus would probably have known it. It came perhaps from Siberia into Russia, and thence into Germany, in which case the anecdote about Tiberius might well apply to the skirret. I cannot find any

[1] Linnæus, *Species*, p. 361; Loureiro, *Fl. Cochinchinensis*, p. 225.

[2] Maximowicz, *Diagnoses Plantarum Japonicæ et Manshuriæ*, in *Mélanges Biologiques du Bulletin de l'Acad., St. Petersburg*, decad 13, p. 18.

[3] Dioscorides, *Mat. Med.*, l. 2, c. 139; Columella, l. 11, c. 3, 18, 35; Lenz, *Bot. der Alten*, p. 560.

[4] Pliny, *Hist. Plant.*, l. 19, c. 5.

Russian name, certainly, but the Germans have original names, *Krizel* or *Grizel*, *Görlein* or *Gierlein*, which indicate an ancient cultivation, more than the ordinary name *Zuckerwurzel*, or sugar-root.[1] The Danish name has the same meaning—*sokerot*, whence the English *skirret*. The name *sisaron* is not known in modern Greece; nor was it known there even in the Middle Ages, and the plant is not now cultivated in that country.[2] There are reasons for doubt as to the true sense of the words *sisaron* and *siser*. Some botanists of the sixteenth century thought that *sisaron* was perhaps the *parsnip* proper, and Sprengel[3] supports this idea.

The French names *chervis* and *girole*[4] would perhaps teach us something if we knew their origin. Littré derives *chervis* from the Spanish *chirivia*, but the latter is more likely derived from the French. Bauhin[5] mentions the low Latin names *servillum*, *chervillum*, or *servillam*, words which are not in Ducange's dictionary. This may well be the origin of *chervis*, but whence came *servillum* or *chervillum?*

Arracacha or **Arracacia**—*Arracacha esculenta*, de Candolle.

An umbel generally cultivated in Venezuela, New Granada, and Ecuador as a nutritious plant. In the temperate regions of those countries it bears comparison with the potato, and even yields, we are assured, a lighter and more agreeable *fecula*. The lower part of the stem is swelled into a bulb, on which, when the plant thrives well, tubercles, or lateral bulbs, form themselves, and persist for several months, which are more prized than the central bulb, and serve for future planting.[6]

The species is probably indigenous in the region where

[1] Nemnich, *Polygl. Lexicon*, ii. p. 1313.
[2] Lenz, *Bot. der Alten*, p. 560; Heldreich, *Nutzpflanzen Griechenlands;* Langkavel, *Bot. der Späteren Griechen.*
[3] Sprengel, *Dioscoridis*, etc., ii. p. 462.
[4] Olivier de Serres, *Théâtre de l'Agriculture*, p. 471.
[5] Bauhin, *Hist. Pl.*, iii. p. 154.
[6] The best information about the cultivation of this plant was given by Bancroft to Sir W. Hooker, and may be found in the *Botanical Magazine*, pl. 3092. A. P. de Candolle published, in *La 5ᵉ Notice sur les Plantes Rares des Jardin Bot. de Genève*, an illustration showing the principal bulb.

it is cultivated, but I do not find in any author a positive assertion of the fact. The existing descriptions are drawn from cultivated stocks. Grisebach indeed says that he has seen (presumably in the herbarium at Kew) specimens gathered in New Granada, in Peru, and in Trinidad,[1] but he does not say whether they were wild. The other species of the same genus, to the number of a dozen, grow in the same districts of America, which renders the above-mentioned origin more probable.

The introduction of the arracacha into Europe has been attempted several times without success. The damp climate of England accounts for the failure of Sir William Hooker's attempts ; but ours, made at two different times, under very different conditions, have met with no better success. The lateral bulbs did not form, and the central bulb died in the house where it was placed for the winter. The bulbs presented to different botanical gardens in France and Italy and elsewhere shared the same fate. It is clear that if the plant is in America really equal to the potato in productiveness and taste, this will never be the case in Europe. Its cultivation does not in America spread as far as Chili and Mexico, like that of the potato and sweet potato, which confirms the difficulty of propagation observed elsewhere.

Madder—*Rubia tinctorum*, Linnæus.

The madder is certainly wild in Italy, Greece, the Crimea, Asia Minor, Syria, Persia, Armenia, and near Lenkoran.[2] As we advance westward in the south of Europe, the wild, indigenous nature of the plant becomes more and more doubtful. There is uncertainty even in France. In the north and east the plant appears to be "naturalized in hedges and on walls,"[3] or "subspontaneous," escaped from former cultivation.[4] In Provence and Languedoc it is more spontaneous or wild, but here also it may have spread from a somewhat extensive

[1] Grisebach, *Flora of British West-India Islands.*
[2] Bertoloni, *Flora Italica*, ii. p. 146 ; Decaisne, *Recherches sur la Garance*, p. 68 ; Boissier, *Flora Orientalis*, iii. p. 17 ; Ledebour, *Flora Rossica*, ii. p. 405.
[3] Cosson and Germain, *Flore des Environs de Paris*, ii. p. 365.
[4] Kirschleger, *Flore d'Alsace*, i. p. 359.

cultivation. In the Iberian peninsula it is mentioned as
"subspontaneous."[1] It is the same in the north of Africa.[2]
Evidently the natural, ancient, and undoubted habitation
is western temperate Asia and the south-east of Europe.
It does not appear that the plant has been found beyond
the Caspian Sea in the land formerly occupied by the
Indo-Europeans, but this region is still little known.
The species only exists in India as a cultivated plant,
and has no Sanskrit name.[3]

Neither is there any known Hebrew name, while the
Greeks, Romans, Slavs, Germans, and Kelts had various
names, which a philologist could perhaps trace to one
or two roots, but which nevertheless indicate by their
numerous modifications an ancient date. Probably the
wild roots were gathered in the fields before the idea of
cultivating the species was suggested. Pliny, however,
says [4] that it was cultivated in Italy in his time, and it
is possible that the custom was of older date in Greece
and Asia Minor.

The cultivation of madder is often mentioned in
French records of the Middle Ages.[5] It was afterwards
neglected or abandoned, until Althen reintroduced it
into the neighbourhood of Avignon in the middle of the
eighteenth century. It flourished formerly in Alsace,
Germany, Holland, and especially in Greece, Asia Minor,
and Syria, whence the exportation was considerable ; but
the discovery of dyes extracted from inorganic substances
has suppressed this cultivation, to the great detriment of
the provinces which drew large profits from it.

Jerusalem Artichoke—*Helianthus tuberosus*, Linnæus.

It was in the year 1616 that European botanists first
mentioned this Composite, with a large root better
adapted for the food of animals than of man. Columna [6]
had seen it in the garden of Cardinal Farnese, and called
it *Aster peruanus tuberosus*. Other authors of the same

---

[1] Willkomm and Lange, *Prodromus Floræ Hispanicæ*, ii. p. 307.
[2] Ball, *Spicilegium Floræ Maroccanæ*, p. 483; Munby, *Catal. Plant.*
Alger., edit. 2, p. 17.
[3] Piddington, *Index*.   [4] Plinius, lib. 19, cap. 3.
[5] De Gasparin, *Traité d'Agriculture*, iv. p. 253.
[6] Columna, *Ecphrasis*, ii. p. 11.

century gave it epithets showing that it was believed to come from Brazil, or from Canada, or from the Indies, that is to say, America. Linnæus[1] adopted, on Parkinson's authority, the opinion of a Canadian origin, of which, however, he had no proof. I pointed out formerly[2] that there are no species of the genus Helianthus in Brazil, and that they are, on the contrary, numerous in North America.

Schlechtendal,[3] after having proved that the Jerusalem artichoke can resist the severe winters of the centre of Europe, observes that this fact is in favour of the idea of a Canadian origin, and contrary to the belief of its coming from some southern region. Decaisne[4] has eliminated from the synonymy of *H. tuberosus* several quotations which had occasioned the belief in a South American or Mexican origin. Like the American botanists, he recalls what ancient travellers had narrated of certain customs of the aborigines of the Northern States and of Canada. Thus Champlain, in 1603, had seen, "in their hands, roots which they cultivate, and which taste like an artichoke." Lescarbot[5] speaks of these roots with the artichoke flavour, which multiply freely, and which he had brought back to France, where they began to be sold under the name of *topinambaux*. The savages, he says, call them *chiquebi*. Decaisne also quotes two French horticulturists of the seventeenth century, Colin and Sagard, who evidently speak of the Jerusalem artichoke, and say it came from Canada. It is to be noted that the name Canada had at that time a vague meaning, and comprehended some parts of the modern United States. Gookin, an American writer on the customs of the aborigines, says that they put pieces of the Jerusalem artichoke into their soups.[6]

Linnæus, *Hortus Cliffortianus*, p. 420.
[2] A. de Candolle, *Géogr. Bot. Raisonnée*, p. 824.
[3] Schlechtendal, *Bot. Zeit.* 1858, p. 113.
[4] Decaisne, *Recherches sur l'Origine de quelques-unes de nos Plantes Alimentaires*, in *Flore des Serres et Jardins*, vol. 23, 1881, p. 112.
[5] Lescarbot, *Histoire de la Nouvelle France*, edit. 3, 1618, t. vi. p. 931.
[6] Pickering, *Chron. Arrang.*, pp. 749, 972.

Botanical analogies and the testimony of contemporaries agree, as we have seen, in considering this plant to be a native of the north-east of America. Dr. Asa Gray, seeing that it is not found wild, had formerly supposed it to be a variety of *H. doronicoides* of Lamarck, but he has since abandoned this idea (*American Journal of Science*, 1883, p. 224). An author gives it as wild in the State of Indiana.[1] The French name *topinambour* comes apparently from some real or supposed Indian name. The English name Jerusalem artichoke is a corruption of the Italian *girasole*, sunflower, combined with an allusion to the artichoke flavour of the root.

Salsify—*Tragopogon porrifolium*, Linnæus.

The salsify was more cultivated a century or two ago than it is now. It is a biennial composite, found wild in Greece, Dalmatia, Italy, and even in Algeria.[2] It frequently escapes from gardens in the west of Europe, and becomes half-naturalized.[3]

Commentators[4] give the name *Tragopogon* (goat's beard) of Theophrastus sometimes to the modern species, sometimes to *Tragopogon crocifolium*, which also grows in Greece. It is difficult to know if the ancients cultivated the salsify or gathered it wild in the country. In the sixteenth century Olivier de Serres says it was a new culture in his country, the south of France. Our word *Salsifis* comes from the Italian *Sassefrica*, that which rubs stones, a senseless term.

Scorzonera—*Scorzonera hispanica*, Linnæus.

This plant is sometimes called the Spanish salsify, from its resemblance to *Tragopogon porrifolium;* but its root has a brown skin, whence its botanical name, and the popular name *écorce noire* in some French provinces.

It is wild in Europe, from Spain, where it abounds, the

[1] *Catalogue of Indiana Plants*, 1881, p. 15.
[2] Boissier, *Fl. Orient.*, iii. p. 745; Viviani, *Fl. Dalmat.*, ii. p. 108; Bertoloni, *Fl. Ital.*, viii. p. 348; Gussone, *Synopsis Fl. Siculæ*, ii. p. 384; Munby, *Catal. Alger.*, edit. 2, p. 22.
[3] A. de Candolle, *Géogr. Bot. Raisonnée*, p. 671.
[4] Fraas, *Synopsis Fl. Class.*, p. 196; Lenz, *Bot. der Alten*, p. 485.

south of France, and Germany, to the region of Caucasus, and perhaps even as far as Siberia, but it is wanting in Sicily and Greece.[1] In several parts of Germany the species is probably naturalized from cultivation.

It seems that this plant has only been cultivated within the last hundred or hundred and fifty years. The botanists of the sixteenth century speak of it as a wild species introduced occasionally into botanical gardens. Olivier de Serres does not mention it.

It was formerly supposed to be an antidote against the bite of adders, and was sometimes called the viper's plant. As to the etymology of the name *Scorzonera*, it is so evident, that it is difficult to understand how early writers, even Tournefort,[2] have declared the origin of the word to be *escorso*, viper in Spanish or Catalan. Viper is in Spanish more commonly *vibora*.

There exists in Sicily a *Scorzonera deliciosa*, Gussone, whose very sugary root is used in the confection of bonbons and sherbets, at Palermo.[3] How is it that its cultivation has not been tried ? It is true that I tasted at Naples *Scorzonera* ices, and found them detestable, but they were perhaps made of the common species (*Scorzonera hispanica*).

Potato—*Solanum tuberosum*, Linnæus.

In 1855 I stated and discussed what was then known about the origin of the potato, and about its introduction into Europe.[4] I will now add the result of the researches of the last quarter of a century. It will be seen that the data formerly acquired have become more certain, and that several somewhat doubtful accessory questions have remained uncertain, though the probabilities in favour of what formerly seemed the truth have grown stronger.

It is proved beyond a doubt that at the time of the discovery of America the cultivation of the potato was

---

[1] Willkomm and Lange, *Prodromus Floræ Hispanicæ*, ii. p. 223 . De Candolle, *Flore Française*, iv. p. 59 ; Koch, *Synopsis Fl. Germ.*, edit; 2, p. 488 ; Ledebour, *Fl. Ross.*, ii. p. 794; Boissier, *Fl. Orientalis*, iii. p. 767 ; Bertoloni, *Fl. Ital.*, viii. p. 365.

[2] Tournefort, *Eléments de Botanique*, p. 379.

[3] Gussone, *Synopsis Floræ Siculæ*.

[4] A. de Candolle, *Géogr. Bot. Raisonnée*, pp. 810, 816.

practised, with every appearance of ancient usage, in the temperate regions extending from Chili to New Granada, at altitudes varying with the latitude. This appears from the testimony of all the early travellers, among whom I shall name Acosta for Peru,[1] and Pedro Cieca, quoted by de l'Ecluse,[2] for Quito.

In the eastern temperate region of South America, on the heights of Guiana and Brazil, for instance, the potato was not known to the aborigines, or if they were acquainted with a similar plant, it was *Solanum Commersonii*, which has also a tuberous root, and is found wild in Montevideo and in the south of Brazil. The true potato is certainly now cultivated in the latter country, but it is of such recent introduction that it has received the name of the English Batata.[3]   According to Humboldt it was unknown in Mexico,[4] a fact confirmed by the silence of subsequent authors, but to a certain degree contradicted by another historical fact.   It is said that Sir Walter Raleigh, or rather Thomas Herriott, his companion in several voyages, brought back to Ireland, in 1585 or 1586, some tubers of the Virginian potato.[5] Its name in its own country was *openawk*.   From Herriott's description of the plant, quoted by Sir Joseph Banks,[6] there is no doubt that it was the potato, and not the batata, which at that period was sometimes confounded with it.   Besides, Gerard[7] tells us that he received from Virginia the potato which he cultivated in his garden, and of which he gives an illustration which agrees in all points with *Solanum tuberosum*. He was so proud of it that he is represented, in his portrait at the beginning of the work, holding in his hand a flowering branch of this plant.

[1] Acosta, p. 163, *verso*.
[2] De l'Ecluse (or Clusius), *Rariarum Plantarum Historiæ*, 1601, lib. 4, p. lxxix., with illustration.
[3] De Martius, *Flora Brasil.*, vol. x. p. 12.
[4] Von Humboldt, *Nouvelle Espagne*, edit. 2, vol. ii. p. 451 ; *Essai sur la Géographie des Plantes*, p. 29.
[5] At that epoch Virginia was not distinguished from Carolina.
[6] Banks, *Trans. Hort. Soc.*, 1805, vol. i. p. 8.
[7] Gerard, *Herbal*, 1597, p. 781, with illustration.

The species could scarcely have been introduced into Virginia or Carolina in Raleigh's time (1585), unless the ancient Mexicans had possessed it, and its cultivation had been diffused among the aborigines to the north of Mexico. Dr. Roulin, who has carefully studied the works on North America, has assured me that he has found no signs of the potato in the United States before the arrival of the Europeans. Dr. Asa Gray also told me so, adding that Mr. Harris, one of the men most intimately acquainted with the language and customs of North American tribes, was of the same opinion. I have read nothing to the contrary in recent publications, and we must not forget that a plant so easy of cultivation would have spread itself even among nomadic tribes, had they possessed it. It seems to me most likely that some inhabitants of Virginia—perhaps English colonists— received tubers from Spanish or other travellers, traders or adventurers, during the ninety years which had elapsed since the discovery of America. Evidently, dating from the conquest of Peru and Chili, in 1535 to 1585, many vessels could have carried tubers of the potato as provisions, and Sir Walter Raleigh, making war on the Spaniards as a privateer, may have pillaged some vessel which contained them. This is the less improbable, since the Spaniards had introduced the plant into Europe before 1585.

Sir Joseph Banks[1] and Dunal[2] were right to insist upon the fact that the potato was first introduced by the Spaniard, since for a long time the credit was generally given to Sir Walter Raleigh, who was the second introducer, and even to other Englishmen, who had introduced, not the potato but the *batata* (sweet potato), which is more or less confounded with it.[3] A celebrated botanist, de l'Ecluse,[4] had nevertheless defined the facts in a

---

[1] Banks, *Trans. Hort. Soc.*, 1805, vol. i. p. 8.

[2] Dunal, *Hist. Nat. des Solanum*, in 4to.

[3] The plant imported by Sir John Hawkins and Sir Francis Drake was clearly the sweet potato, Sir J. Banks says; whence it results that the questions discussed by Humboldt touching the localities visited by these travellers do not apply to the potato.

[4] De l'Ecluse, *Rariarum Plantarum Historia*, 1601, lib. 4, p. lxxviii.

remarkable manner. It is he who published the first good description and illustration of the potato, under the significant name of *Papas Peruanorum*. From what he says, the species has little changed under the culture of nearly three centuries, for it yielded in the beginning as many as fifty tubers of unequal size, from one to two inches long, irregularly ovoid, reddish, ripening in November (at Vienna). The flower was more or less pink externally, and reddish within, with five longitudinal stripes of green, as is often seen now. No doubt numerous varieties have been obtained, but the original form has not been lost. De l'Ecluse compares the scent of the flower with that of the lime, the only difference from our modern plant. He sowed seeds which produced a white-flowered variety, such as we sometimes see now.

The plants described by de l'Ecluse were sent to him in 1588, by Philippe de Sivry, Seigneur of Waldheim and Governor of Mons, who had received them from some one in attendance on the papal legate in Belgium. De l'Ecluse adds that the species had been introduced into Italy from Spain or America (*certum est vel ex Hispania, vel ex America habuisse*), and he wonders that, although the plant had become so common in Italy that it was eaten like a turnip and given to the pigs, the learned men of the University of Padua only became acquainted with it by means of the tuber which he sent them from Germany. Targioni [1] has not been able to discover any proof that the potato was as widely cultivated in Italy at the end of the sixteenth century as de l'Ecluse asserts, but he quotes Father Magazzini of Vallombrosa, whose posthumous work, published in 1623, mentions the species as one previously brought, without naming the date, from Spain or Portugal by barefooted friars. It was, therefore, towards the end of the sixteenth or at the beginning of the seventeenth century that the cultivation of the potato became known in Tuscany. Independently of what de l'Ecluse and the agriculturist of Vallombrosa

---

[1] Targioni-Tozzetti, *Lezzioni*, ii. p. 10; *Cenni Storici sull' Introduzione di Varie Piante nell' Agricoltura di Toscana*, 1 vol. in 8vo, Florence, 1853 p. 37.

say of its introduction from the Iberian peninsula, it is not at all likely that the Italians had any dealings with Raleigh's companions.

No one can doubt that the potato is of American origin; but in order to know from what part of that vast continent it was brought, it is necessary to know if the plant is found wild there, and in what localities.

To answer this question clearly, we must first remove two causes of error : the confusion of allied species of the genus *Solanum* with the potato ; and the other, the mistakes made by travellers as to the wild character of the plant.

The allied species are *Solanum Commersonii* of Dunal, of which I have already spoken; *S. maglia* of Molina, a Chili species; *S. immite* of Dunal, a native of Peru ; and *S. verrucosum* [1] of Schlechtendal, which grows in Mexico. These three kinds of *Solanum* have smaller tubers than *S. tuberosum,* and differ also in other characteristics indicated in special works on botany. Theoretically, it may be believed that all these, and other forms growing in America, are derived from a single earlier species, but in our geological epoch they present themselves with differences which seem to me to justify specific distinctions, and no experiments have proved that by crossing one with another a product would be obtained of which the seed (not the tubers) would propagate the race. Leaving these more or less doubtful questions of species, let us try to ascertain whether the common form of *Solanum tuberosum* has been found wild, and merely remark that the abundance of tuberous solanums growing in the temperate regions of America, from Chili or Buenos Ayres as far as Mexico, confirms the fact of an American origin. If we knew nothing more, this would be a strong presumption in favour of this country being the original home of the potato.

The second cause of error is very clearly explained

---

[1] *Solanum verrucosum,* whose introduction into the neighbourhood of Gex, near Geneva, I mentioned in 1855, has since been abandoned because its tubers are too small, and because it does not, as it was hoped, withstand the *potato-fungus.*

by the botanist Weddell,[1] who has carefully explored
Bolivia and the neighbouring countries. "When we
reflect," he says, "that on the arid Cordillera the Indians
often establish their little plots of cultivation on points
which would appear almost inaccessible to the great
majority of our European farmers, we understand that
when a traveller chances to visit one of these cultivated
plots, long since abandoned, and finds there a plant of
Solanum tuberosum which has accidentally persisted, he
gathers it in the belief that it is really wild; but of this
there is no proof."

We come now to facts. These abound concerning the
wild character of the plant in Chili.

In 1822, Alexander Caldcleugh,[2] English consul,
sent to the London Horticultural Society some tubers of
the potato which he had found in the ravines round
Valparaiso. He says that these tubers are small, sometimes red, sometimes yellowish, and rather bitter in taste.[3]
"I believe," he adds, "that this plant exists over a great
extent of the littoral, for it is found in the south of
Chili, where the aborigines call it *maglia*." This is
probably a confusion with *S. maglia* of botanists; but
the tubers of Valparaiso, planted in London, produced
the true potato, as we see from a glance at Sabine's
coloured figure in the *Transactions of the Horticultural
Society*. The cultivation of this plant was continued
for some time, and Lindley certified anew, in 1847, its
identity with the common potato.[4] Here is the account
of the Valparaiso plant, given by a traveller to Sir
William Hooker.[5] "I noticed the potato on the shore
as far as fifteen leagues to the north of this town, and to
the south, but I do not know how far it extends. It

---

[1] *Chloris Andina*, in 4to, p. 103.

[2] Sabine, *Trans. Hort. Soc.*, vol. v. p. 249.

[3] No importance should be attached to this flavour, nor to the watery
quality of some of the tubers, since in hot countries, even in the south
of Europe, the potato is often poor. The tubers, which are subterranean ramifications of the stem, are turned green by exposure to the
light, and are rendered bitter.

[4] *Journal Hort. Soc.*, vol. iii. p. 66.

[5] Hooker, *Botanical Miscellanies*, 1831, vol. ii. p. 203.

grows on cliffs and hills near the sea, and I do not remember to have seen it more than two or three leagues from the coast. Although it is found in mountainous places, far from cultivation, it does not exist in the immediate neighbourhood of the fields and gardens where it is planted, excepting when a stream crosses these enclosures and carries the tubers into uncultivated places." The potato described by these two travellers had white flowers, as is seen in some cultivated European varieties, and like the plant formerly reared by de l'Ecluse. We may assume that this is the natural colour of the species, or at least one of the most common in its wild state.

Darwin, in his voyage in the *Beagle*, found the potato growing wild in great abundance on the sand of the sea-shore, in the archipelago of Southern Chili, and growing with a remarkable vigour, which may be attributed to the damp climate. The tallest plants attained to the height of four feet. The tubers were small as a rule, though one of them was two inches in diameter. They were watery, insipid, but with no bad taste when cooked. "The plant is undoubtedly wild," says the author,[1] "and its specific identity has been confirmed first by Henslow, and afterwards by Sir Joseph Hooker in his *Flora Antarctica*.[2]

A specimen in the herbarium collected by Claude Gay, considered by Dunal to be *Sclanum tuberosum*, bears this inscription : " From the centre of the Cordilleras of Talcagouay, and of Cauquenes, in places visited only by botanists and geologists." The same author, Gay, in his *Flora Chilena*,[3] insists upon the abundance of the wild potato in Chili, even among the Araucanians in the mountains of Malvarco, where, he says, the soldiers of Pincheira used to go and seek it for food. This evidence sufficiently proves its wild state in Chili, so that I may omit other less convincing testimony—for instance, that of Molina and Meyen, whose specimens from Chili have not been examined.

The climate of the coast of Chili is continued upon

---

[1] *Journal of the Voyage,* etc., edit. 1852, p. 285.
[2] Vol. i. part 2, p. 329. [3] Vol. v. p. 74.

the heights as we follow the chain of the Andes, and the cultivation of the potato is of ancient date in the temperate regions of Peru, but the wild character of the species there is not so entirely proved as in the case of Chili.[1] Pavon declared he found it on the coast at Chancay, and near Lima. The heat of these districts seems very great for a species which requires a temperate or even a rather cold climate. Moreover, the specimen in Boissier's herbarium, gathered by Pavon, belongs, according to Dunal,[2] to another species, to which he has given the name of *S. immite*. I have seen the authentic specimen, and have no doubt that it belongs to a species distinct from the *S. tuberosum*. Sir W. Hooker[3] speaks of McLean's specimen, gathered in the hills round Lima, without any information as to whether it was found wild. The specimens (more or less wild) which Matthews sent from Peru to Sir W. Hooker belong, according to Sir Joseph,[4] to varieties which differ a little from the true potato. Mr. Hemsley,[5] who has seen them recently in the herbarium at Kew, believes them to be "distinct forms, not more distinct, however, than certain varieties of the species."

Weddell,[6] whose caution in this matter we already know, expresses himself as follows:—"I have never found *Solanum tuberosum* in Peru under such circumstances as left no doubt that it was indigenous; and I even declare that I do not attach more belief to the wild nature of other plants found scattered on the Andes outside Chili, hitherto considered as indigenous."

On the other hand, M. Ed. André[7] collected with great care, in two elevated and wild districts of Columbia, and in another near Lima, specimens which he believed he might attribute to *S. tuberosum*. M. André has been kind enough to lend them to me. I have compared them attentively with the types of Dunal's species in

---

[1] Ruiz and Pavon, *Flora Peruviana*, ii. p. 38.
[2] Dunal, *Prodromus*, xiii., sect. i. p. 22.
[3] Hooker, *Bot. Miscell.*, ii.   [4] Hooker, *Fl. Antarctica.*
[5] *Journal Hort. Soc.*, new series, vol. v.
[6] Weddell, *Chloris Andina*, p. 103.
[7] André, in *Illustration Horticole*, 1877, p. 114.

my herbarium and in that of M. Boissier. None of these Solanaceæ belong, in my opinion, to *S. tuberosum,* although that of La Union, near the river Cauca, comes nearer than the rest. None—and this is yet more certain —answers to *S. immite* of Dunal. They are nearer to *S. columbianum* of the same author than to *S. tuberosum* or *S. immite.* The specimen from Mount Quindio presents a singular characteristic—it has pointed ovoid berries.[1]

In Mexico the tuberous Solanums attributed to *S. tuberosum,* or, according to Hemsley,[2] to allied forms, do not appear to be identical with the cultivated plant. They belong to *S. Fendleri,* which Dr. Asa Gray considered at first as a separate species, and afterwards[3] as a variety of *S. tuberosum* or of *S. verrucosum.*

We may sum up as follows :—

1. The potato is wild in Chili, in a form which is still seen in our cultivated plants.

2. It is very doubtful whether its natural home extends to Peru and New Granada.

3. Its cultivation was diffused before the discovery of America from Chili to New Granada.

4. It was introduced, probably in the latter half of the sixteenth century, into that part of the United States now known as Virginia and North Carolina.

5. It was imported into Europe between 1580 and 1585, first by the Spaniards, and afterwards by the English, at the time of Raleigh's voyages to Virginia.[4]

Batata, or Sweet Potato—*Convolvulus batatas,* Linnæus; *Batatas edulis,* Choisy.

The roots of this plant, swelled into tubers, resemble potatoes, whence it arose that sixteenth-century navigators applied the same name to these two very different species. The sweet potato belongs to the Convolvulus family, the potato to the Solanum family ; the fleshy

---

[1] The form of the berries in *S. columbianum* and *S. immite* is not yet known.

[2] Hemsley, *Journal Hort. Soc.,* new series, vol. v.

[3] Asa Gray, *Synoptical Flora of North America,* ii. p. 227.

[4] See, for the successive introduction into the different parts of Europe, Clos, *Quelques Documents sur l'Histoire de la Pomme de Terre,* in 8vo, 1874, in *Journal d'Agric. Pratiq. du Midi de la France.*

parts of the former are roots, those of the latter subterranean branches.[1] The sweet potato is sugary as well
as farinaceous. It is cultivated in all countries within
or near the tropics, and perhaps more in the new than
in the old world.[2]

Its origin is, according to a great number of authors,
doubtful. Humboldt,[3] Meyen,[4] and Boissier[5] hold to its
American, Boyer,[6] Choisy,[7] etc., to its Asiatic origin. The
same diversity is observed in earlier works. The question
is the more difficult since the Convolvulaceæ is one of the
most widely diffused families, either from a very early
epoch or in consequence of modern transportation.

There are powerful arguments in favour of an
American origin. The fifteen known species of the
genus *Batatas* are all found in America; eleven in that
continent alone, four both in America and the old
world, with possibility or probability of transportation.
The cultivation of the common sweet potato is widely
diffused in America. It dates from a very early epoch.
Marcgraff[8] mentions it in Brazil under the name of
jetica. Humboldt says that the name *camote* comes
from a Mexican word. The word *Batatas* (whence comes
by a mistaken transfer the word potato) is given as
American. Sloane and Hughes[9] speak of the sweet
potato as of a plant much cultivated, and having several
varieties in the West Indies. They do not appear to
suspect that it had a foreign origin. Clusius, who was
one of the first to mention the sweet potato, says he had
eaten some in the south of Spain, where it was supposed
to have come from the new world.[10] He quotes the

---

[1] Turpin gives figures which clearly show these facts. *Mém. du
Muséum,* vol. xix. plates 1, 2, 5.

[2] Dr. Sagot gives interesting details on the method of cultivation,
the product, etc., in the *Journal Soc. d'Hortic. de France,* second series,
vol. v. pp. 450–458.

[3] Humboldt, *Nouvelle Espagne,* edit. 2, vol. ii. p. 470.

[4] Meyen, *Grundrisse Pflanz. Geogr.,* p. 373.

[5] Boissier, *Voyage Botanique en Espagne.*

[6] Boyer, *Hort. Maurit.,* p. 225.   [7] Choisy, in *Prodromus,* p. 338.

[8] Marcgraff, *Bres.,* p. 16, with illustration.

[9] Sloane, *Hist. Jam.,* i. p. 150; Hughes, *Barb.,* p. 223.

[10] Clusius, *Hist.,* ii. p. 77.

names *Batatas, camotes, amotes, ajes*,[1] which were foreign
to the languages of the old world. The date of his
book is 1601. Humboldt[2] says that, according to
Gomara, Christopher Columbus, when he appeared for
the first time before Queen Isabella, offered her various
productions from the new world, sweet potatoes among
others. Thus, he adds, the cultivation of this plant was
already common in Spain from the beginning of the sixteenth century. Oviedo,[3] writing in 1526, had seen the
sweet potato freely cultivated by the natives of St.
Domingo, and had introduced it himself at Avila, in Spain.
Rumphius[4] says positively that, according to the general
opinion, sweet potatoes were brought by the Spanish
Americans to Manilla and the Moluccas, whence the
Portuguese diffused it throughout the Malay Archipelago.
He quotes the popular names, which are not Malay, and
which indicate an introduction by the Castillians.
Lastly, it is certain that the sweet potato was unknown
to the Greeks, Romans, and Arabs; that it was not
cultivated in Egypt even eighty years ago,[5] a fact which
it would be hard to explain if we supposed its origin to
be in the old world.

On the other hand, there are arguments in favour of an
Asiatic origin. The Chinese *Encyclopœdia of Agriculture* speaks of the sweet potato, and mentions different
varieties;[6] but Bretschneider[7] has proved that the
species is described for the first time in a book of the
second or third century of our era. According to
Thunberg,[8] the sweet potato was brought to Japan by
the Portuguese. Lastly, the plant cultivated at Tahiti,
in the neighbouring islands, and in New Zealand, under
the names *umara, gumarra,* and *gumalla,* described by
Forster[9] under the name of *Convolvulus chrysorhizus,* is,

---

[1] *Ajes* was a name for the yam (Humboldt, *Nouvelle Espagne*).
[2] Humboldt, *ibid.*
[3] Oviedo, Ramusio's translation, vol. iii. pt. 3.
[4] Rumphius, *Amboin.,* v. p. 368.
[5] Forskal, p. 54; Delile, *Ill.*
[6] D'Hervey Saint-Denys, *Rech. sur l'Agric. des Chin.,* 1850, p. 109.
[7] *Study and Value of Chinese Botanical Works,* p. 13.
[8] Thunberg, *Flora Japon.,* p. 84.   [9] Forster, *Plantœ Escul.,* p. 56.

according to Sir Joseph Hooker, the sweet potato.[1] Seemann[2] remarks that these names resemble the Quichuen name of the sweet potato in America, which is, he says, *cumar*. The cultivation of the sweet potato became general in Hindustan in the eighteenth century.[3] Several popular names are attributed to it, and even, according to Piddington,[4] a Sanskrit name, *ruktalu*, which has no analogy with any name known to me, and is not in Wilson's Sanskrit Dictionary. According to a note given me by Adolphe Pictet, *ruktalu* seems a Bengalee name composed from the Sanskrit *alu* (*Rukta* plus *álu*, the name of *Arum campanulatum*). This name in modern dialects designates the yam and the potato. However, Wallich[5] gives several names omitted by Piddington. Roxburgh[6] mentions no Sanskrit name. Rheede[7] says the plant was cultivated in Malabar, and mentions common Indian names.

The arguments in favour of an American origin seem to me much stronger. If the sweet potato had been known in Hindustan at the epoch of the Sanskrit language it would have become diffused in the old world, since its propagation is easy and its utility evident. It seems, on the contrary, that this cultivation remained long unknown in the Sunda Isles, Egypt, etc. Perhaps an attentive examination might lead us to share the opinion of Meyer,[8] who distinguished the Asiatic plant from the American species. However, this author has not been generally followed, and I suspect that if there is a different Asiatic species it is not, as Meyer believed, the sweet potato described by Rumphius, which the latter says was brought from America, but the Indian plant of Roxburgh.

Sweet potatoes are grown in Africa; but either the cultivation is rare, or the species are different. Robert Brown[9] says that the traveller Lockhardt had not seen

---

[1] Hooker, *Handbook of New Zealand Flora*, p. 194.
[2] Seemann, *Journal of Bot.*, 1866, p. 328.
[3] Roxburgh, edit. Wall., ii. p. 69.   [4] Piddington, *Index*.
[5] Wallich, *Flora Ind.*   [6] Roxburgh, edit. 1832, vol. i. p. 483.
[7] Rheede, *Mal.*, vii. p. 95.   [8] Meyer, *Primitiœ Fl. Esseq.*, p. 103.
[9] R. Brown, *Bot. Congo*, p. 55.

the sweet potato of whose cultivation the Portuguese missionaries make mention. Thonning[1] does not name it. Vogel brought back a species cultivated on the western coast, which is certainly, according to the authors of the *Flora Nigritiana, Batatas paniculata* of Choisy. It was, therefore, a plant cultivated for ornament or for medicinal purposes, for its root is purgative.[2] It might be supposed that in certain countries in the old or new world *Ipomœa tuberosa,* L., had been confounded with the sweet potato; but Sloane[3] tells us that its enormous roots are not eatable.[4]

Ipomœa mammosa, Choisy (*Convolvulus mammosus,* Loureiro; *Batata mammosa,* Rumphius), is a Convolvulaceous plant with an edible root, which may well be confounded with the sweet potato, but whose botanical character is nevertheless distinct. This species grows wild near Amboyna (Rumphius), where it is also cultivated. It is prized in Cochin-China.

As for the sweet potato (*Batatas edulis*), no botanist, as far as I know, has asserted that he found it wild himself, either in India or America.[5] Clusius[6] affirms upon hearsay that it grows wild in the new world and in the neighbouring islands.

In spite of the probability of an American origin, there remains, as we have seen, much that is unknown or uncertain touching the original home and the transport of this species, which is a valuable one in hot countries. Whether it was a native of the new or of the old world, it is difficult to explain its transportation from America to China at the beginning of our era, and

---

[1] Schumacher and Thonning, *Besk. Guin.*

[2] Wallich, in Roxburgh, *Fl. Ind.,* ii. p. 63.

[3] Sloane, *Jam.,* i. p. 152.

[4] Several Convolvulaceæ have large roots, or more properly root-stocks, but in this case it is the base of the stem with a part of the root which is swelled, and this root-stock is always purgative, as in the Jalap and Turbith, while in the sweet potato it is the lateral roots, a different organ, which swell.

[5] No. 701 of Schomburgh, coll. 1, is wild in Guiana. According to Choisy, it is a variety of the *Batatas edulis;* according to Bentham (Hook, *Jour. Bot.,* v. p. 352), of the *Batatas paniculata.* My specimen, which is rather imperfect, seems to me to be different from both.

[6] Clusius, *Hist.,* ii. p. 77.

to the South Sea Islands at an early epoch, or from Asia and from Australia to America at a time sufficiently remote for its cultivation to have been early diffused from the Southern States to Brazil and Chili. We must assume a prehistoric communication between Asia and America, or adopt another hypothesis, which is not inapplicable to the present case. The order *Convolvulaceæ* is one of those rare families of dicotyledons in which certain species have a widely extended area, extending even to distant continents.[1] A species which can at the present day endure the different climates of Virginia and Japan may well have existed further north before the epoch of the great extension of glaciers in our hemisphere, and prehistoric men may have transported it southward when the climatic conditions altered. According to this hypothesis, cultivation alone preserved the species, unless it is at last discovered in some spot in its ancient habitation—in Mexico or Columbia, for instance.[2]

Beetroot—*Beta vulgaris* and *B. maritima*, Linnæus; *Beta vulgaris*, Moquin.

This plant is cultivated sometimes for its fleshy root (red beet), sometimes for its leaves, which are used as a vegetable (white beet), but botanists are generally agreed in not dividing the species. It is known from other examples that plants slender rooted by nature easily become fleshy rooted from the effects of soil or cultivation.

The slender-rooted variety grows wild in sandy soil, and especially near the sea in the Canary Isles, and all along the coasts of the Mediterranean Sea, and as far as the Caspian Sea, Persia, and Babylon,[3] perhaps even as

[1] A. de Candolle, *Géogr. Bot. Raisonné*, pp. 1041–1043, and pp. 516–518.

[2] Dr. Bretschneider, after having read the above, wrote to me from Pekin that the cultivated sweet potato is of origin foreign to China, according to Chinese authors. The handbook of agriculture of Nung-chang-tsuan-shu, whose author died in 1633, asserts this fact. He speaks of a sweet potato wild in China, called *chu*, the cultivated species being *kan-chu*. The *Min-shu*, published in the sixteenth century, says that the introduction took place between 1573 and 1620. The American origin thus receives a further proof.

[2] Moquin-Tandon, in *Prodromus*, vol. xiii. pt. 2, p. 55; Boissier, *Flora Orientalis*, iv. p. 898; Ledebour, *Fl. Rossica*, iii. p. 692.

far as the west of India, whence a specimen was brought by Jaquemont, although it is not certain that it was growing wild. Roxburgh's Indian flora, and Aitchison's more recent flora of the Punjab and of the Sindh, only mention the plant as a cultivated species. It has no Sanskrit name,[1] whence it may be inferred that the Aryans had not brought it from western temperate Asia, where it exists. The nations of Aryan race who had previously migrated into Europe probably did not cultivate it, for I find no name common to the Indo-European languages. The ancient Greeks, who used the leaves and roots, called the species *teutlion;*[2] the Romans, *beta.* Heldreich[3] gives also the ancient Greek name *sevkle,* or *sfekelie,* which resembles the Arab name *selg, silq,*[4] among the Nabatheans. The Arab name has passed into the Portuguese *selga.* No Hebrew name is known. Everything shows that its cultivation does not date from more than three or four centuries before the Christian era.

The red and white roots were known to the ancients, but the number of varieties has greatly increased in modern times, especially since the beetroot has been cultivated on a large scale for the food of cattle and for the production of sugar. It is one of the plants most easily improved by selection, as the experiments of Vilmorin have proved.[5]

Manioc—*Manihot utilissima,* Pohl; *Jatropha manihot,* Linnæus.

The manioc is a shrub belonging to the Euphorbia family, of which several roots swell in their first year; they take the form of an irregular ellipse, and contain a fecula (tapioca) with a more or less poisonous juice.

It is commonly cultivated in the equatorial or tropical regions, especially in America from Brazil to the West Indies. In Africa the cultivation is less general, and seems to be more recent. In certain Asiatic colonies it is

---

[1] Roxburgh, *Flora Indica,* ii. p. 59 ; Piddington, *Index.*
[2] Theophrastus and Dioscorides, quoted by Lenz, *Botanik der Griechen und Römer,* p. 446 ; Fraas, *Synopsis Fl. Class.,* p. 233.
[3] Heldreich, *Die Nutzpflanzen Griechenlands,* p. 22.
[4] Alawâm, *Agriculture nabathéenne,* from E. Meyer, *Geschichte der Botanik,* iii. p. 75.
[5] *Notice sur l'Amélioration des Plantes par le Semis,* p. 15.

decidedly of modern introduction. It is propagated by budding.

Botanists are divided in opinion whether the innumerable varieties of manioc should be regarded as forming one, two, or several different species. Pohl [1] admitted several besides his *Manihot utilissima*, and Dr. Müller,[2] in his monograph on the Euphorbiaceæ, places the variety *aipi* in an allied species, *M. palmata*, a plant cultivated with the others in Brazil, and of which the root is not poisonous. This last character is not so distinct as might be believed from certain books and even from the assertions of the natives. Dr. Sagot,[3] who has compared a dozen varieties of manioc cultivated at Cayenne, says expressly, "There are maniocs more poisonous than others, but I doubt whether any are entirely free from noxious principles."

It is possible to account for these singular differences of properties in very similar plants by the example of the potato. The *Manihot* and *Solanum tuberosum* both belong to suspected families (*Euphorbiaceæ* and *Solanaceæ*). Several of their species are poisonous in some of their organs; but the fecula, wherever it is found, is never harmful, and the same holds good of the cellular tissue, freed from all deposit; that is to say, reduced to cellulose. In the preparation of cassava, or manioc flour, great care is taken to scrape the outer skin of the root, then to pound or crush the fleshy part so as to express the more or less poisonous juice, and finally the paste is submitted to a baking which expels the volatile parts.[4] Tapioca is the pure fecula without the mixture of the tissues which still exist in the cassava. In the potato the outer pellicle contracts noxious qualities when it is allowed to become green by exposure to the light, and it is well known that unripe or diseased tubers, containing too small a proportion of fecula with

---

[1] Pohl, *Plantarum Brasiliæ Icones et Descriptiones*, in fol., vol. i.
[2] J. Müller, in *Prodromus*, xv., sect. 2, pp. 1062–1064.
[3] Sagot, *Bull. de la Soc. Bot. de France*, Dec. 8, 1871.
[4] I give the essentials of the preparation; the details vary according to the country. See on this head: Aublet, *Guyane*, ii. p. 67; Decourtilz, *Flora des Antilles*, iii. p. 113; Sagot, etc.

much sap, are not good to eat, and would cause positive harm to persons who consumed any quantity of them. All potatoes, and probably all maniocs, contain something harmful, which is observed even in the products of distillation, and which varies with several causes ; but only matter foreign to the fecula should be mistrusted.

The doubts about the number of species into which the cultivated manihots should be divided are no source of difficulty regarding the question of geographic origin. On the contrary, we shall see that they are an important means of proving an American origin.

The Abbé Raynal had formerly spread the erroneous opinion that the manioc was imported into America from Africa. Robert Brown [1] denied this in 1818, but without giving reasons in support of his opinion ; and Humboldt,[2] Moreau de Jonnes,[3] and Saint Hilaire [4] insisted upon its American origin. It can hardly be doubted for the following reasons :—

1. Maniocs were cultivated by the natives of Brazil, Guiana, and the warm region of Mexico before the arrival of the Europeans, as all early travellers testify. In the West Indies this cultivation was, according to Acosta,[5] common enough in the sixteenth century to inspire the belief that it was also there of a certain antiquity.

2. It is less widely diffused in Africa, especially in regions at a distance from the west coast. It is known that manioc was introduced into the Isle of Bourbon by the Governour Labourdonnais.[6] In Asiatic countries, where a plant so easy to cultivate would probably have spread had it been long known on the African continent, it is mentioned here and there as an object of curiosity of foreign origin.[7]

[1] R. Brown, *Botany of the Congo*, p. 50.
[2] Humboldt, *Nouvelle Espagne*, edit. 2, vol. ii. p. 398.
[3] *Hist. de l'Acad. des Sciences*, 1824.
[4] Guillemin, *Archives de Botanique*, i. p. 239.
[5] Acosta, *Hist. Nat. des Indes*, French trans., 1598, p. 163.
[6] Thomas, *Statistique de Bourbon*, ii. p. 18.
[7] The catalogue of the botanical gardens of Buitenzorg, 1866, p. 222, says expressly that the *Manihot utilissima* comes from Bourbon and America.

3. The natives of America had several ancient names
for the varieties of manioc, especially in Brazil,[1] which
does not appear to have been the case in Africa, even on
the coast of Guinea.[2]

4. The varieties cultivated in Brazil, in Guiana, and
in the West Indies are very numerous, whence we may
presume a very ancient cultivation. This is not the case
in Africa.

5. The forty-two known species of the genus *Manihot*,
without counting *M. utilissima*, are all wild in America;
most of them in Brazil, some in Guiana, Peru, and
Mexico; not one in the old world.[3] It is very unlikely that
a single species, and that the cultivated one, was a native
both of the old and of the new world, and all the more so
since in the family *Euphorbiaceæ* the area of the woody
species is usually restricted, and since phanerogamous
plants are very rarely common to Africa and America.

The American origin of the manioc being thus
established, it may be asked how the species has been
introduced into Guinea and Congo. It was probably
the result of the frequent communications established in
the sixteenth century by Portuguese merchants and
slave-traders.

The *Manihot utilissima* and the allied species or
variety called *aipi*, which is also cultivated, have not
been found in an undoubtedly wild state. Humboldt
and Bonpland, indeed, found upon the banks of the
Magdalena a plant of *Manihot utilissima* which they
called *almost* wild,[4] but Dr. Sagot assures me that it has
not been found in Guiana, and that botanists who have
explored the hot region in Brazil have not been more
fortunate. We gather as much from the expressions
of Pohl, who has carefully studied these plants, and who
was acquainted with the collections of Martius, and had

---

[1] *Aypi, mandioca, manihot, manioch, yuca,* etc., in Pohl, *Icones and
Desc.,* i. pp. 30, 33. Martius, *Beiträge z. Ethnographie,* etc., *Braziliens,*
ii. p. 122, gives a number of names.

[2] Thonning (in Schumacher, *Besk. Guin.*), who is accustomed to
quote the common names, gives none for the manioc.

[3] J. Müller, in *Prodromus,* xv., sect. 1, p. 1057.

[4] Kunth, in Humboldt and B., *Nova Genera,* ii. p. 108.

no doubt of their American origin. If he had observed a wild variety identical with those which are cultivated, he would not have suggested the hypothesis that the manioc is obtained from his *Manihot pusilla*[1] of the province of Goyaz, a plant of small size, and considered as a true species or as a variety of *Manihot palmata*.[2] Martius declared in 1867, that is after having received a quantity of information of a later date than his journey, that the plant was not known in a wild state.[3] An early traveller, usually accurate, Piso,[4] speaks of a wild *mandihoca*, of which the Tapuyeris, the natives of the coast to the north of Rio Janeiro, ate the roots. "It is," he says, "very like the cultivated plant;" but the illustration he gives of it appears unsatisfactory to authors who have studied the maniocs. Pohl attributes it to his *M. aipi*, and Dr. Müller passes it over in silence. For my part, I am disposed to believe what Piso says, and his figure does not seem to me entirely unsatisfactory. It is better than that by Vellozo, of a wild manioc which is doubtfully attributed to *M. aipi*.[5] If we do not accept the origin in eastern tropical Brazil, we must have recourse to two hypotheses: either the cultivated maniocs are obtained from one of the wild species modified by cultivation, or they are varieties which exist only by the agency of man after the disappearance of their fellows from modern wild vegetation.

Garlic—*Allium sativum*, Linnæus.

Linnæus, in his *Species Plantarum*, indicates Sicily as the home of the common garlic; but in his *Hortus Cliffortianus*, where he is usually more accurate, he does not give its origin. The fact is that, according to all the most recent and complete floras of Sicily, Italy, Greece, France, Spain, and Algeria, garlic is not considered to be indigenous, although specimens have been gathered here and there which had more or less the appearance of

[1] Pohl, *Icones et Descr.*, i. p. 36, pl. 26.   [2] Müller, in *Prodromus*.
[3] De Martius, *Beiträge zur Ethnographie*, etc., i. pp. 19, 136.
[4] Piso, *Historia Naturalis Braziliæ*, in folio, 1658, p. 55, *cum icone*.
[5] *Jatropia Sylvestris Vell. Fl. Flum.*, 16, t. 83.   See Müller, in D. C. *Prodromus*, xv. p. 1063.

being so. A plant so constantly cultivated and so easily propagated may spread from gardens and persist for a considerable time without being wild by nature. I do not know on what authority Kunth[1] mentions that the species is found in Egypt. According to authors who are more accurate[2] in their accounts of the plants of that country, it is only found there under cultivation. Boissier, whose herbarium is so rich in Eastern plants, possesses no wild specimens of it. The only country where garlic has been found in a wild state, with the certainty of its really. being so, is the desert of the Kirghis of Sungari ; bulbs were brought thence and cultivated at Dorpat,[3] and specimens were afterwards seen by Regel.[4] The latter author also says that he saw a specimen which Wallich had gathered as wild in British India ; but Baker,[5] who had access to the rich herbarium at Kew, does not speak of it in his review of the "*Alliums* of India, China, and Japan."

Let us see whether historical and philological records confirm the fact of an origin in the south-west of Siberia alone.

Garlic has been long cultivated in China under the name of *suan*. It is written in Chinese by a single sign, which usually indicates a long known and even a wild species.[6] The floras of Japan [7] do not mention it, whence I gather that the species was not wild in Eastern Siberia and Dahuria, but that the Mongols brought it into China.

According to Herodotus, the ancient Egyptians made great use of it. Archæologists have not found the proof of this in the monuments, but this may be because the plant was considered unclean by the priests.[8]

---

[1] Kunth, *Enum.*, iv. p. 381.
[2] Schweinfurth and Ascherson, *Aufzählung*, p. 294.
[3] Ledebour, *Flora Altaica*, ii. p. 4 ; *Flora Rossica*, iv. p. 16².
[4] Regel, *Allior. Monogr.*, p. 44.
[5] Baker, in *Journal of Bot.*, 1874, p. 295.
[6] Bretschneider, *Study and Value*, etc., pp. 15, 4, and 7.
[7] Thunberg, *Fl. Jap.* ; Franchet and Savatier, *Enumeratio*, 1876, vol. ii.
[8] Unger, *Pflanzen des Alten Ægyptens*, p. 42.

There is a Sanskrit name, *mahoushouda*,[1] become *loshoun* in Bengali, and to which appears to be related the Hebrew name *schoum* or *schumin*,[2] which has produced the Arab *thoum* or *toum*. The Basque name *baratchouria* is thought by de Charencey [3] to be allied with Aryan names. In support of his hypothesis I may add that the Berber name, *tiskert*, is quite different, and that consequently the Iberians seem to have received the plant and its name rather from the Aryans than from their probable ancestors of Northern Africa. The Lettons call it *kiplohks*, the Esthonians *krunslauk*, whence probably the German *Knoblauch*. The ancient Greek name appears to have been *scorodon*, in modern Greek *scordon*. The names given by the Slavs of Illyria are *bili* and *cesan*. The Bretons say *quinen*,[4] the Welsh *craf, cenhinnen*, or *garlleg*, whence the English *garlic*. The Latin *allium* has passed into the languages of Latin origin.[5] This great diversity of names intimates a long acquaintance with the plant, and even an ancient cultivation in Western Asia and in Europe. On the other hand, if the species has existed only in the land of the Kirghis, where it is now found, the Aryans might have cultivated it and carried it into India and Europe; but this does not explain the existence of so many Keltic, Slav, Greek, and Latin names which differ from the Sanskrit. To explain this diversity, we must suppose that its original abode extended farther to the west than that known at the present day, an extension anterior to the migrations of the Aryans.

If the genus Allium were once made, as a whole, the object of such a serious study as that of Gay on some

---

[1] Piddington, *Index.*

[2] Hiller, *Hierophyton;* Rosenmüller, *Bibl. Alterthum*, vol. iv.

[3] De Charencey, *Actes de la Soc. Phil.*, 1st March, 1869.

[4] Davies, *Welsh Botanology.*

[5] All these common names are found in my dictionary compiled by Moritzi from floras. I could have quoted a larger number, and mentioned the probable etymologies, as given by philologists—Hehn, for instance, in his *Kulturpflanzen aus Asien*, p. 171 and following; but this is not necessary to show its origin and early cultivation in several different countries.

of its species,[1] perhaps it might be found that certain
wild European forms, included by authors under *A.
arenarium*, L., *A. arenarium*, Sm., or *A. scorodoprasum*,
L., are only varieties of *A. sativum*. In that case everything would agree to show that the earliest peoples of
Europe and Western Asia cultivated such form of the
species just as they found it from Tartary to Spain,
giving it names more or less different.

ONION—*Allium Cepa*, Linnæus.

I will state first what was known in 1855 ;[2] I will
then add the recent botanical observations which confirm
the inferences from philological data.

The onion is one of the earliest of cultivated species.
Its original country is, according to Kunth, unknown.[3]
Let us see if it is possible to discover it. The modern
Greeks call *Allium Cepa*, which they cultivate in
abundance, *krommunda*.[4] This is a good reason for believing that the *krommuon* of Theophrastus [5] is the same
species, as sixteenth-century writers already supposed.[6]
Pliny[7] translated the word by *cœpa*. The ancient Greeks
and Romans knew several varieties, which they distinguished by the names of countries : *Cyprium, Cretense,
Samothraciae*, etc. One variety cultivated in Egypt [8] was
held to be so excellent that it received divine honours,
to the great amusement of the Romans.[9]   Modern
Egyptians designate *A. Cepa* by the name of *basal* [10] or
bussul,[11] whence it is probable that the *bezalim* of the
Hebrews is the same species, as commentators have said.[12]
There are several distinct names—*palandu, latarka, sa-
kandaka*,[13] and a number of modern Indian names. The
species is commonly cultivated in India, Cochin-China,

---

[1] *Annales des Sc. Nat.*, 3rd series, vol. viii.
[2] A. de Candolle, *Géogr. Bot. Raisonnée*, ii. p. 828.
[3] Kunth, *Enumer.*, iv. p. 394.
[4] Fraas, *Syn. Fl. Class.*, p. 291.
[5] Theophraśtus, *Hist.*, l. 7, c. 4.
[6] J. Bauhin, *Hist.*, ii. p. 548.   [7] Pliny, *Hist.*, l. 19, c. 6.   [8] *Ibid.*
[9] Juvenalis, *Sat.* 15.   [10] Forskal, p. 65.
[11] Ainslie's *Mat. Med. Ind.*, i. p. 269.
[12] Hiller, *Hieroph.*, ii. p. 36; Rosenmüller, *Handbk. Bibl. Alterk.*, iv.
p. 96.
[13] Piddington, *Index ;* Ainslie's *Mat. Med. Ind.*

China,[1] and even in Japan.[2] It was largely consumed by the ancient Egyptians. The drawings on their monuments often represent this species.[3] Thus its cultivation in Southern Asia and the eastern region of the Mediterranean dates from a very early epoch. Moreover, the Chinese, Sanskrit, Hebrew, Greek, and Latin names have no apparent connection. From this last fact we may deduce the hypothesis that its cultivation was begun after the separation of the Indo-European nations, the species being found ready to hand in different countries at once. This, however, is not the present state of things, for we hardly find even vague indications of the wild state of *A. Cepa*. I have not discovered it in European or Caucasian floras; but Hasselquist[4] says, "It grows in the plains near the sea in the environs of Jericho." Dr. Wallich mentioned in his list of Indian plants, No. 5072, specimens which he saw in districts of Bengal, without mentioning whether they were cultivated. This indication, however insufficient, together with the antiquity of the Sanskrit and Hebrew names, and the communication which is known to have existed between the peoples of India and of Egypt, lead me to suppose that this plant occupied a vast area in Western Asia, extending perhaps from Palestine to India. Allied species, sometimes mistaken for *A. Cepa*, exist in Siberia.[5]

The specimens collected by Anglo-Indian botanists, of which Wallich gave the first idea, are now better known. Stokes discovered *Allium Cepa* wild in Beluchistan. He says, "wild on the Chehil Tun." Griffith brought it from Afghanistan and Thomson from Lahore, to say nothing of other collectors, who are not explicit as to the wild or cultivated nature of their specimens.[6] Boissier possesses a wild specimen found in the mountainous regions of the Khorassan. The umbels are smaller than in the

---

[1] Roxburgh, *Fl. Ind.*, ii. ; Loureiro, *Fl. Cochin.*, p. 249.
[2] Thunberg, *Fl. Jap.*, p. 132.
[3] Unger, *Pflanzen d. Alt. Ægypt.*, p. 42, figs. 22, 23, 24
[4] Hasselquist, *Voy. and Trav.*, p. 279.
[5] Ledebour, *Fl. Rossica*, iv. p. 169.
[6] Aitchison, *A Catalogue of the Plants of the Punjab and the Sindh*, in 8vo, 1869, p. 19; Baker, in *Journal of Bot.*, 1874, p. 295.

cultivated plant, but there is no other difference. Dr. Regel, jun., found it to the south of Kuldscha, in Western Siberia.[1] Thus my former conjectures are completely justified ; and it is not unlikely that its habitation extends even as far as Palestine, as Hasselquist said.

The onion is designated in China by a single sign (pronounced *tsung*), which may suggest a long existence there as an indigenous plant.[2] I very much doubt, however, that the area extends so far to the east.

Humboldt[3] says that the Americans have always been acquainted with onions, in Mexican *xonacatl*. "Cortes," he says, "speaking of the comestibles sold at the market of the ancient Tenochtillan, mentions onions, leeks, and garlic." I cannot believe, however, that these names applied to the species cultivated in Europe. Sloane, in the seventeenth century, had only seen one *Allium* cultivated in Jamaica (*A. Cepa*), and that was in a garden with other European vegetables.[4] The word *xonacatl* is not in Hernandez, and Acosta[5] says distinctly that the onions and garlics of Peru are of European origin. The species of the genus Allium are rare in America.

Spring, or **Welsh Onion**—*Allium fistulosum*, Linnæus.

This species was for a long time mentioned in floras and works on horticulture as of unknown origin; but Russian botanists have found it wild in Siberia towards the Altaï mountains, on the Lake Baïkal in the land of the Kirghis.[6] The ancients did not know the plant.[7] It must have come into Europe through Russia in the Middle Ages, or a little later. Dodoens,[8] an author of the sixteenth century, has given a figure of it, hardly recognizable, under the name of *Cepa oblonga*.

Shallot—*Allium ascalonicum*, Linnæus.

It was believed, according to Pliny,[9] that this plant

---

[1] *Ill. Hortic.*, 1877, p. 167.
[2] Bretschneider, *Study and Value*, etc., pp. 47 and 7.
[3] *Nouvelle Espagne*, 2nd edit., ii. p. 476.
[4] Sloane, *Jam.*, i. p. 75.
[5] Acosta, *Hist. Nat. des Indes*, French trans., p. 163.
[6] Ledebour, *Flora Rossica*, iv. p. 169.
[7] Lenz, *Botanik. der Alten Griechen und Römer*, p. 295.
[8] Dodoens, *Pemptades*, p. 687.   [9] Pliny, *Hist.*, l. 19, c. 6.

took its name from Ascalon, in Judæa; but Dr. Fournier [1] thinks that the Latin author mistook the meaning of the word *Askalônion* of Theophrastus. However this may be, the word has been retained in modern languages under the form of *échalote* in French, *chalote* in Spanish, *scalogno* in Italian, *Aschaluch* or *Eschlauch* in German.

In 1855 I had spoken of the species as follows: [2]—

"According to Roxburgh, [3] *Allium ascalonicum* is much cultivated in India. The Sanskrit name *pulandu* is attributed to it, a word nearly identical with *palandu*, attributed to *A. Cepa.* [4] Evidently the distinction between the two species is not clear in Indian or Anglo-Indian works.

"Loureiro says he saw *Allium ascalonicum* cultivated in Cochin-China, [5] but he does not mention China, and Thunberg does not indicate this species in Japan. Its cultivation, therefore, is not universal in the east of Asia. This fact, and the doubt about the Sanskrit name, lead me to think that it is not ancient in Southern Asia. Neither, in spite of the name of the species, am I convinced that it existed in Western Asia. Rauwolf, Forskal, and Delile do not mention it in Siberia, in Arabia, or in Egypt. Linnæus [6] mentions Hasselquist as having found the species in Palestine. Unfortunately, he gives no details about the locality, nor about its wild condition. In the *Travels* of Hasselquist [7] I find a *Cepa montana* mentioned as growing on Mount Tabor and on a neighbouring mountain, but there is nothing to prove that it was this species. In his article on the onions and garlics of the Hebrews he mentions only *Allium Cepa*, then *A. porrum* and *A. sativum*. Sibthorp did not find it in Greece, [8] and Fraas [9] does not mention it as now cultivated

---

[1] He will treat of this in a publication entitled *Cibaria*, which will shortly appear.

[2] *Géog. Bot. Raisonnée*, p. 829.

[3] Roxburgh, *Fl. Ind.*, edit. 1832, vol. ii. p. 142.

[4] Piddington, *Index.*

[5] Loureiro, *Fl. Cochin.*, p. 251.

[6] Linnæus, *Species*, p. 429.

[7] Hasselquist, *Voy. and Trav.*, 1766, pp. 281, 282.

[8] Sibthorp, *Prodr.*   [9] Fraas, *Syn. Fl. Class.*, p. 291.

in that country.   According to Koch,[1] it is naturalized
among the vines near Fiume.   However, Viviani[2] only
speaks of it as a cultivated plant in Dalmatia.

"From all these facts I am led to believe that
Allium ascalonicum is not a species.   It is enough to
render its primitive existence doubtful, to remark : (1)
that Theophrastus and ancient writers in general have
spoken of it as a form of the *Allium Cepa,* having the
same importance as the varieties cultivated in Greece,
Thrace, and elsewhere ; (2) that its existence in a wild
state cannot be proved ; (3) that it is little cultivated,
or not all, in the countries where it is supposed to have
had its origin, as in Syria, Egypt, and Greece ; (4) that
it is commonly without flowers, whence the name of *Cepa
sterilis* given by Bauhin, and the number of its bulbs is
an allied fact ; (5) when it does flower, the organs of the
flower are similar to those of *A. Cepa,* or at least no
difference has been hitherto discovered, and according to
Koch[3] the only difference in the whole plant is that the
stalk and leaves are less swelled, although fistulous."

Such was formerly my opinion.[4]   The facts published
since 1855 do not destroy my doubts, but, on the contrary,
justify them.   Regel, in 1875, in his monograph of the
genus Allium, declares he has only seen the shallot as a
cultivated species.   Aucher Eloy has distributed a plant
from Asia Minor under the name of *A. ascalonicum,* but
judging from my specimen this is certainly not the
species.   Boissier tells me that he has never seen *A.
ascalonicum* in the East, and it is not in his herbarium.
The plant from the Morea which bears this name in the
flora of Bory and Chaubard is quite a different species,
which he has named *A. gomphrenoides.*   Baker,[5] in his
review of the Alliums of India, China, and Japan,
mentions *A. ascalonicum* in districts of Bengal and of
the Punjab, from specimens of Griffith and Aitchison ;
but he adds, "They are probably cultivated plants."

[1] Koch, *Syn. Fl. Germ.,* 2nd edit:, p. 833.
[2] Viviani, *Fl. Dalmat.,* p. 138.   [3] Koch, *Syn. Fl. Germ.*
[4] A. de Candolle, *Géogr. Bot. Raisonnée,* p. 829.
[5] Baker, in *Journ. of Bot.,* 1874, p. 295.

He attributes to *A. ascalonicum Allium sulvia*, Ham., of Nepal, a plant little known, and whose wild character is uncertain. The shallot produces many bulbs, which may be propagated or preserved in the neighbourhood of cultivation, and thus cause mistakes as to its origin.

Finally, in spite of the progress of botanical investigations in the East and in India, this form of Allium has not been found wild with certainty. It appears to me, therefore, more probable than ever that it is a modification of *A. Cepa*, dating from about the beginning of the Christian era—a modification less considerable than many of those observed in other cultivated plants, as, for instance, in the cabbage.

Rocambole—*Allium scorodoprasum*, Linnæus.

If we cast a glance at the descriptions and names of *A. scorodoprasum* in works on botany since the time of Linnæus, we shall see that the only point on which authors are agreed is the common name of *rocambole*. As to the distinctive characters, they sometimes approximate the plant to *Allium sativum*, sometimes regard it as altogether distinct. With such different definitions, it is difficult to know in what country the plant, well known in its cultivated state as the *rocambole*, is found wild. According to Cosson and Germain,[1] it grows in the environs of Paris. According to Grenier and Godron,[2] the same form grows in the east of France. Burnat says he found the species undoubtedly wild in the Alpes-Maritimes, and he gave specimens of it to Boissier. Willkomm and Lange do not consider it to be wild in Spain,[3] though one of the French names of the cultivated plant is *ail* or *eschalote d'Espagne*. Many other European localities seem to me doubtful, since the specific characters are so uncertain. I mention, however, that, according to Ledebour,[4] the plant which he calls *A. scorodoprasum* is very common in Russia from Finland to the Crimea. Boissier received a specimen of it

---

[1] Cosson and Germain, *Flore*, ii. p. 553.
[2] Grenier and Godron, *Flore de France*, iii. p. 197.
[3] Willkomm and Lange, *Prodr. Fl. Hisp.*, i. p. 885.
[4] Ledebour, *Flora Rossica*, iv. p. 163.

from Dobrutscha, sent by the botanist Sintenis. The natural habitat of the species borders, therefore, on that of *Allium sativum*, or else an attentive study of all these forms will show that a single species, comprising several varieties, extends over a great part of Europe and the bordering countries of Asia.

The cultivation of this species of onion does not appear to be of ancient date. It is not mentioned by Greek and Roman authors, nor in the list of plants recommended by Charlemagne to the intendants of his gardens.[1] Neither does Olivier de Serres speak of it. We can only give a small number of original common names among ancient peoples. The most distinctive are in the North. *Skovlög* in Denmark, *keipe* and *rackenboll* in Sweden.[2] *Rockenbolle*, whence comes the French name, is German. It has not the meaning given by Littré. Its etymology is *Bolle*, onion, growing among the rocks, *Rocken*.[3]

Chives—*Allium schœnoprasum*, Linnæus.

This species occupies an extensive area in the northern hemisphere. It is found all over Europe, from Corsica and Greece to the south of Sweden, in Siberia as far as Kamtschatka, and also in North America, but only near the Lakes Huron and Superior and further north [4]—a remarkable circumstance, considering its European habitat. The variety found in the Alps is the nearest to the cultivated form.[5]

The ancient Greeks and Romans must certainly have known the species, since it is wild in Italy and Greece. Targioni believes it to be the *Scorodon schiston* of Theophrastus; but we are dealing with words without descriptions, and authors whose specialty is the interpretation of Greek text, like Fraas and Lenz, are prudent enough to affirm nothing. If the ancient names are doubtful, the fact of the cultivation of the plant at this epoch is yet more so. It is possible that the custom of gathering it in the fields existed.

---

[1] Le Grand d'Aussy, *Histoire de la Vie des Français*, vol. i. p. 122.
[2] Nemnich, *Polyglott. Lexicon*, p. 187.   [3] *Ibid.*
[4] Asa Gray, *Botany of the Northern States*, edit. 5, p. 534.
[5] De Candolle, *Flore Française*, iv. p. 227.

Colocasia—*Arum esculentum,* Linnæus; *Colocasia antiquorum,* Schott.[1]

This species is cultivated in the damp districts of the tropics, for the swelled lower portion of the stem, which forms an edible rhizome similar to the subterraneous part of the iris. The petioles and the young leaves are also utilized as a vegetable. Since the different forms of the species have been properly classed, and since we have possessed more certain information about the floras of the south of Asia, we cannot doubt that this plant is wild in India, as Roxburgh[2] formerly, and Wight[3] and others have more recently asserted; likewise in Ceylon,[4] Sumatra,[5] and several islands of the Malay Archipelago.[6]

Chinese books make no mention of it before a work of the year 100 B.C.[7] The first European navigators saw it cultivated in Japan and as far as the north of New Zealand,[8] in consequence probably of an early introduction, and without the certain co-existence of wild stocks. When portions of the stem or of the tuber are thrown away by the side of streams, they naturalize themselves easily. This was perhaps the case in Japan and the Fiji Islands,[9] judging from the localities indicated. The colocasia is cultivated here and there in the West Indies, and elsewhere in tropical America, but much less than in Asia or Africa, and without the least indication of an American origin.

In the countries where the species is wild there are common names, sometimes very ancient, totally different from each other, which confirms their local origin. Thus the Sanskrit name is *kuchoo,* which persists in modern

---

[1] *Arum Egyptium,* Columma, *Ecphrasis,* ii. p. 1, tab. 1; Rumphius, *Amboin,* vol. v. tab. 109. *Arum colocasia* and *A. esculentum,* Linnæus; *Colocasia antiquorum,* Schott, *Melet.,* i. 18; Engler, in *D. C. Monog. Phaner.,* ii. p. 491.

[2] Roxburgh, *Fl. Ind.,* iii. p. 495.   [3] Wight, *Icones,* t. 786.

[4] Thwaites, *Enum. Plant. Zeylan.,* p. 335.

[5] Miquel, *Sumatra,* p. 258.

[6] Rumphius, *Amboin,* vol. v. p. 318.

[7] Bretschneider, *On the Study and Value,* etc., p. 12.

[8] Forster, *De Plantis Escul.,* p. 58.

[9] Franchet and Savatier, *Enum.,* p. 8; Seemann, *Flora Vitiensis,* p. 284.

Hindu languages—in Bengali, for instance.[1] In Ceylon the wild plant is styled *gahala*, the cultivated plant *kandalla*.[2] The Malay names are *kelady*,[3] *tallus, tallas, tales,* or *taloes*,[4] from which perhaps comes the well-known name of the Otahitans and New Zealanders—*tallo* or *tarro*,[5] *dalo*[6] in the Fiji Islands. The Japanese have a totally distinct name, *imo*,[7] which shows an existence of long duration either indigenous or cultivated.

European botanists first knew the colocasia in Egypt, where it has perhaps not been very long cultivated. The monuments of ancient Egypt furnish no indication of it, but Pliny[8] spoke of it as the *Arum Ægyptium*. Prosper Alpin saw it in the sixteenth century, and speaks of it at length.[9] He says that its name in its country is *culcas*, which Delile[10] writes *qolkas*, and *koulkas*. It is clear that this Arab name of the Egyptian arum has some analogy with the Sanskrit *kuchoo*, which is a confirmation of the hypothesis, sufficiently probable, of an introduction from India or Ceylon. De l'Ecluse[11] had seen the plant cultivated in Portugal, as introduced from Africa, under the name *alcoleaz*, evidently of Arab origin. In some parts of the south of Italy, where the plant has become naturalized, it is, according to Parlatore, called *aro di Egitto*.[12]

The name *colocasia*, given by the Greeks to a plant of which the root was used by the Egyptians, may evidently come from *colcas*, but it has been transferred to a plant differing from the true colcas. Indeed, Dioscorides applies it to the Egyptian bean, or *nelumbo*,[13] which has a large root, or rather rhizome, rather stringy

---

[1] Roxburgh, *Fl. Ind.*
[2] Thwaites, *Enum. Plant. Zeylan.*   [3] Rumphius, *Amboin.*
[4] Miquel, *Sumatra*, p. 258; Hasskarl, *Cat. Horti. Bogor. Alter.*, p. 55.
[5] Forster, *De Plantis Escul.*, p. 58.   [6] Seemann, *Flora Vitiensis.*
[7] Franchet and Savatier, *Enum.*   [8] Pliny, *Hist.*, 1. 19, c. 5.
[9] Alpinus, *Hist. Ægypt. Naturalis*, edit. 2, vol. i. p. 166; ii. p. 192.
[10] Delile, *Fl. Ægypt. Ill.*, p. 28; *De la Colocase des Anciens*, in 8vo, 1846.
[11] Clusius, *Historia*, ii. p. 75.   [12] Parlatore, *Fl. Ital.*, ii. p. 255.
[13] Prosper Alpinus, *Hist. Ægypt. Naturalis;* Columna; Delile, *Ann. du Mus.*, i. p. 375; *De la Colocase des Anciens;* Reynier, *Economie des Egyptiens,* p. 321.

and not good to eat. The two plants are very different, especially in the flower. The one belongs to the *Araceæ*, the other to the *Nymphæaceæ;* the one belongs to the class of *Monocotyledons*, the other to that of the *Dicotyledons*. The nelumbo of Indian origin has ceased to grow in Egypt, while the colocasia of modern botanists has persisted there. If there is any confusion, as seems probable in the Greek authors, it must be explained by the fact that the colcas rarely flowers, at least in Egypt. From the point of view of botanical nomenclature, it matters little that mistakes were formerly made about the plants to which the name colocasia should be applied. Fortunately, modern scientific names are not based upon the doubtful definitions of the ancient Greeks and Romans, and it is sufficient to say now, if the etymology is insisted upon, that colocasia comes from colcas in consequence of an error.

Apé, or **Large-rooted Alocasia**—*Alocasia macrorrhiza,* Schott; *Arum macrorrhizum,* Linnæus.

This araceous plant, which Schott places now in the genus Colocasia, now in the Alocasia, and whose names are far more complicated than might be supposed from those indicated above,[1] is less frequently cultivated than the common colocasia, but in the same manner and nearly in the same countries. Its rhizomes attain the length of a man's arm. They have a distinctly bitter taste, which it is indispensable to remove by cooking.

The aborigines of Otahiti call it *apé,* and those of the Friendly Isles *kappe.*[2] In Ceylon, the common name is *habara,* according to Thwaites.[3] It has other names in the Malay Archipelago, which argues an existence prior to that of the more recent peoples of these regions.

The plant appears to be wild, especially in Otahiti.[4] It is also wild in Ceylon, according to Thwaites, who has studied botany for a long time in that island. It is

---

[1] See Engler, in *D. C. Monographiæ Phanerogarum,* ii. p. 502.
[2] Forster, *De Plantis Esculentis Insularum Oceani Australis,* p. 58.
[3] Thwaites, *Enum. Pl. Zeyl.,* p. 336.
[4] Nadeaud, *Enum. des Plantes Indigènes,* p. 40.

mentioned also in India[1] and in Australia,[2] but its wild condition is not affirmed—a fact always difficult to establish in the case of a species cultivated on the banks of streams, and which is propagated by bulbs. Moreover, it is sometimes confounded with the *Colocasia indica* of Kunth, which grows in the same manner, and is found here and there in cultivated ground ; and this species grows wild, or is naturalized in the ditches and streams of Southern Asia, although its history is not yet well known.

Konjak—*Amorphophallus Konjak*, Koch ; *Amorphophallus Rivieri*, du Rieu, var. *Konjak*, Engler.[3]

The konjak is a tuberous plant of the family Araceæ, extensively cultivated by the Japanese, a culture of which Vidal has given full details in the *Bulletin de la Société d'Acclimatation* of July, 1877. It is considered by Engler as a variety of *Amorphophallus Rivieri*, of Cochin-China, of which horticultural periodicals have given several illustrations in the last few years.[4] It can be cultivated in the south of Europe, like the dahlia, as a curiosity ; but to estimate the value of the bulbs as food, they should be prepared with lime-water, in Japanese fashion, so as to ascertain the amount of fecula which a given area will produce.

Dr. Vidal gives no proof that the Japanese plant is wild in that country. He supposes it to be so from the meaning of the common name, which is, he says, *konniyakou*, or *yamagonniyakou, yama* meaning "mountain." Franchet and Savatier[5] have only seen the plant in gardens. The Cochin-China variety, believed to belong to the same species, grows in gardens, and there is no proof of its being wild in the country.

Yams—*Dioscorea sativa, D. batatas, D. japonica,* and *D. alata.*

The yams, monocotyledonous plants, belonging to

---

[1] Engler, in *D. C. Monog. Phaner.*
[2] Bentham, *Flora Austr.*, viii. p. 155.
[3] Engler, in *D. C. Monogr. Phaner.*, vol. ii. p. 313.
[4] *Gardener's Chronicle*, 1873, p. 610 ; *Flore des Serres et Jardins,* t. 1958, 1959 ; Hooker, *Bot. Mag.*, t. 6195.
[5] Franchet and Savatier, *Enum. Pl. Japoniæ,* ii. p. 7.

the family *Dioscorideœ*, constitute the genus *Dioscorea*, of which botanists have described about two hundred species, scattered over all tropical and sub-tropical countries. They usually have rhizomes, that is, underground stems or branches of stems, more or less fleshy, which become larger when the annual, exposed part of the plant is near its decay.[1] Several species are cultivated in different countries for these farinaceous rhizomes, which are cooked and eaten like potatoes.

The botanical distinction of the species has always presented difficulties, because the male and female flowers are on different individuals, and because the characters of the rhizomes and the lower part of the exposed stems cannot be studied in the herbarium. The last complete work is that of Kunth,[2] published in 1850. It requires revision on account of the number of specimens brought home by travellers in these last few years. Fortunately, with regard to the origin of cultivated species, certain historical and philological considerations will serve as a guide, without the absolute necessity of knowing and estimating the botanical characters of each.

Roxburgh enumerates several *Dioscoreœ*[3] cultivated in India, but he found none of them wild, and neither he nor Piddington [4] mentions Sanskrit names. This last point argues a recent cultivation, or one of originally small extent, in India, arising either from indigenous species as yet undefined, or from foreign species cultivated elsewhere. The Bengali and Hindu generic name is *alu*, preceded by a special name for each species or variety; *kam alu,* for instance, is *Dioscorea alata.* The absence of distinct names in each province also argues a recent cultivation. In Ceylon, Thwaites [5] indicates six wild species, and he adds that *D. sativa,* L., *D. alata,*

---

[1] M. Sagot, *Bull. de la Soc. Bot. de France*, 1871, p. 306, has well described the growth and cultivation of yams, as he has studied them in Cayenne.

[2] Kunth, *Enumeratio*, vol. v.

[3] These are *D. globosa, alata, rubella, fasciculata, purpurea,* of which two or three appear to be merely varieties.

[4] Piddington, *Index.*

[5] Thwaites, *Enum. Plant. Zeyl.*, p. 326.

L., and *D. purpurea*, Roxb., are cultivated in gardens, but are not found wild.

The Chinese yam, *Dioscorea batatas* of Decaisne,[1] extensively cultivated by the Chinese under the name of *Sain-in*, and introduced by M. de Montigny into European gardens, where it remains as a luxury, has not hitherto been found wild in China. Other less-known species are also cultivated by the Chinese, especially the *chou-yu, tou-tchou, chan-yu,* mentioned in their ancient works on agriculture, and which has spherical rhizomes (instead of the pyriform spindles of the *D. batatas*). The names mean, according to Stanislas Julien, mountain arum, whence we may conclude the plant is really a native of the country. Dr. Bretschneider[2] gives three *Dioscoreæ* as cultivated in China (*D. batatas, alata, sativa*), adding, "The *Dioscorea* is indigenous in China, for it is mentioned in the oldest work on medicine, that of the Emperor Schen-nung."

Dioscorea japonica, Thunberg, cultivated in Japan, has also been found in clearings in various localities, but Franchet and Savatier[3] say that it is not positively known to what degree it is wild or has strayed from cultivation. Another species, more often cultivated in Japan, grows here and there in the country according to the same authors. They assign it to *Dioscorea sativa* of Linnæus; but it is known that the famous Swede had confounded several Asiatic and American species under that name, which must either be abandoned or restricted to one of the species of the Indian Archipelago. If we choose the latter course, the true *D. sativa* would be the plant cultivated in Ceylon with which Linnæus was acquainted, and which Thwaites calls the *D. sativa* of Linnæus. Various authors admitted the identity of the Ceylon plant with others cultivated on the Malabar coast, in Sumatra, Java, the Philippine Isles, etc. Blume[4] asserts that *D. sativa*, L., to which

---

[1] Decaisne, *Histoire et Culture de l'Igname de Chine*, in the *Revue Horticole*, 1st July and Dec. 1853 ; *Flore des Serres et Jardins*, x. pl. 971.

[2] *On the Study and Value*, etc., p. 12.

[3] Franchet and Savatier, *Enum. Plant. Japoniæ*, ii. p. 47.

[4] Blume, *Enum. Plant. Javæ*, p. 22.

he attributes pl. 51 in Rheede's *Hortus Malabaricus*, vol. viii., grows in damp places in the mountains of Java and of Malabar. In order to put faith in these assertions, it would be necessary to have carefully studied the question of species from authentic specimens.

The yam, which is most commonly cultivated in the Pacific Isles under the name *ubi*, is the *Dioscorea alata* of Linnæus. The authors of the seventeenth and eighteenth centuries speak of it as widely spread in Tahiti, in New Guinea, in the Moluccas, etc.[1] It is divided into several varieties, according to the shape of the rhizome. No one pretends to have found this species in a wild state, but the flora of the islands whence it probably came, in particular that of Celebes and of New Guinea, is as yet little known.

Passing to America, we find there also several species of this genus growing wild, in Brazil and Guiana, for instance, but it seems more probable that the cultivated varieties were introduced. Authors indicate but few cultivated species or varieties (Plumier one, Sloane two) and few common names. The most widely spread is *yam, igname*, or *inhame*, which is of African origin, according to Hughes, and so also is the plant cultivated in his time in Barbados.[2]

He says that the word *yam* means "to eat," in several negro dialects on the coast of Guinea. It is true that two travellers nearer to the date of the discovery of America, whom Humboldt quotes,[3] heard the word *igname* pronounced on the American continent : Vespucci in 1497, on the coast of Paria ; Cabral in 1500, in Brazil. According to the latter, the name was given to a root of which bread was made, which would better apply to the manioc, and leads me to think there must be some mistake, more especially since a passage from Vespucci, quoted elsewhere by Humboldt,[4] shows the

---

[1] Forster, *Plant. Esculent.*, p. 56 ; Rumphius, *Amboin*, vol. v., pl. 120, 121, etc.

[2] Hughes, *Hist. Nat. Barb.*, 1750, p. 226.

[3] Humboldt, *Nouvelle Espagne*, 2nd edit., vol. ii. p. 468.

[4] *Ibid.*, p. 403.

confusion he made between the manioc and the yam.
D. Cliffortiana, Lam., grows wild in Peru[1] and in
Brazil,[2] but it is not proved to be cultivated. Presl says
verosimiliter colitur, and the *Flora Brasiliensis* does
not mention cultivation.

The species chiefly cultivated in French Guiana,
according to Sagot,[3] is *Dioscorea triloba*, Lam., called
Indian yam, which is also common in Brazil and
the West India Islands. The common name argues a
native origin, whereas another species, *D. cayennensis*,
Kunth, also cultivated in Guiana, but under the name of
negro-country yam, was most likely brought from Africa,
an opinion the more probable that Sir W. Hooker likens
a yam cultivated in Africa on the banks of the Nun and
the Quorra,[4] to *D. cayennensis*. Lastly, the *free yam*
of Guiana is, according to Dr. Sagot, *D. alata* introduced
from the Malay Archipelago and Polynesia.

In Africa there are fewer indigenous *Dioscoreœ* than
in Asia and America, and the culture of yams is less
widely spread. On the west coast, according to Thonning,[5] only one or two species are cultivated; Lockhardt[6]
only saw one in Congo, and that only in one locality.
Bojer[7] mentions four cultivated species in Mauritius,
which are, he says, of Asiatic origin, and one, *D. bul-
bifera*, Lam., from India, if the name be correct. He
asserts that it came from Madagascar, and has spread
into the woods beyond the plantations. In Mauritius
it bears the name *Cambare marron*. Now, *cambare*
is something like the Hindu name *kam*, and *marron*
(marroon) indicates a plant escaped from cultivation.
The ancient Egyptians cultivated no yams, which argues
a cultivation less ancient in India than that of the colocasia. Forskal and Delile mention no yams cultivated
in Egypt at the present day.

To sum up : several *Dioscoreœ* wild in Asia (especially

---

[1] Hænke, in Presl, *Rel.*, p. 133.   [2] Martius, *Fl. Bras.*, v. p. 43.
[3] Sagot, *Bull. Soc. Bot. France*, 1871, p. 305.
[4] Hooker, *Fl. Nigrit*, p. 53.
[5] Schumacher and Thonning, *Besk. Guin*, p. 447.
[6] Brown, *Congo*, p. 49.   [7] Bojer, *Hortus Mauritianus*.

in the Asiatic Archipelago), and others less numerous
growing in America and in Africa, have been introduced
into cultivation as alimentary plants, probably more
recently than many other species. This last conjecture is
based on the absence of a Sanskrit name, on the limited
geographical range of cultivation, and on the date, which
appears to be not very ancient, of the inhabitants of the
Pacific Isles.

Arrowroot—*Maranta   arundinacea*, Linnæus. A
plant of the family of the *Scitamineæ*, allied to the genus
Canna, of which the underground suckers[1] produce the
excellent fecula called arrowroot.   It is cultivated in the
West India Islands and in several tropical countries of
continental America.   It has also been introduced into
the old world—on the coast of Guinea, for instance.[2]

Maranta arundinacea is certainly American.   According to Sloane,[3] it was brought from Dominica to
Barbados, and thence to Jamaica, which leads us to
suppose that it was not indigenous in the West Indies.
Körnicke, the last author who studied the genus Maranta,[4] saw several specimens which were gathered in
Guadaloupe, in St. Thomas, in Mexico, in Central
America, in Guiana, and in Brazil; but he did not con-
cern himself to discover whether they were taken from
wild, cultivated, or naturalized plants.   Collectors hardly
ever indicate this; and for the study of the American
continent (excepting the United States) we are unprovided with local floras, and especially with floras made
by botanists residing in the. country.   In published
works I find the species mentioned as cultivated[5] or
growing in plantations,[6] or without any explanation.   A
locality in Brazil, in the thinly peopled province of
Matto Grosso, mentioned by Körnicke, supposes an
absence of cultivation.   Seemann[7] mentions that the
species is found in sunny spots near Panama.

---

[1] See Tussac's description, *Flore des Antilles*, i. p. 183.
[2] Hooker, *Niger Flora*, p. 531.
[3] Sloane, *Jamaica*, 1707, vol. i. p. 254.
[4] In *Bull. Soc. des Natur. de Moscou*, 1822, vol. i. p. 34.
[5] Aublet, *Guyane*, i. p. 3.   [6] Meyer, *Flora Essequibo*, p. 11.
[7] Seemann, *Bot. of Herald.*, p. 213.

A species is also cultivated in the West Indies, *Maranta indica*, which, Tussac says, was brought from the East Indies. Körnicke believes that *M. ramosissima* of Wallich found at Sillet, in India, is the same species, and thinks it is a variety of *M. arundinacea*. Out of thirty-six more or less known species of the genus Maranta, thirty at least are of American origin. It is therefore unlikely that two or three others should be Asiatic. Until Sir Joseph Hooker's *Flora of British India* is completed, these questions on the species of the *Scitamineæ* and their origin will be very obscure.

Anglo-Indians obtain arrowroot from another plant of the same family, *Curcuma angustifolia*, Roxburgh, which grows in the forests of the Deccan and in Malabar.[1] I do not know whether it is cultivated.

[1] Roxburgh, *Fl. Ind.*, i. p. 31; Porter, *The Tropical Agriculturalist*, p. 241; Ainslie, *Materia Medica*, i. p. 19.

CHAPTER II.

PLANTS CULTIVATED FOR THEIR STEMS OR LEAVES.

Article I.—Vegetables.

Common Cabbage—*Brassica oleracea*, Linnæus.

The cabbage in its wild state, as it is represented in *Eng. Bot.*, t. 637, the *Flora Danica*, t. 2056, and elsewhere, is found on the rocks by the sea-shore : (1) in the Isle of Laland, in Denmark, the island of Heligoland, the south of England and Ireland, the Channel Isles, and the islands off the coast of Charente Inférieure;[1] (2) on the **north** coast of the Mediterranean, near Nice, Genoa, and Lucca.[2] A traveller of the last century, Sibthorp, said that he found it at Mount Athos, but this has not been confirmed by any modern botanist, and the species appears to be foreign in Greece, on the shores of the Caspian, as also in Siberia, where Pallas formerly said he had seen it, and in Persia.[3] Not only the numerous travellers who have explored these countries have not found the cabbage, but the winters of the east of Europe and of Siberia appear to be too severe for it. Its distribution into somewhat isolated places, and in two different regions of Europe, suggests the suspicion either that plants apparently indi-

[1] Fries, *Summa*, p. 29 ; Nylander, *Conspectus*, p. 46 ; Bentham, *Handb. Brit. Fl.*, edit. 4, p. 40 ; Mackay, *Fl. Hibern.*, p. 28 ; Brebisson, *Fl. de Normandie*, edit. 2, p. 18 ; Babbington, *Primitiæ Fl. Sarnicæ*, p. 8 ; Clavaud, *Flore de la Gironde*, i. p. 68.

[2] Bertoloni, *Fl. Ital.*, vii. p. 146 ; Nylander, *Conspectus.*

[3] Ledebour, *Fl. Ross.*; Griesbach, *Spiciligium Fl. Rumel.*; Boissier, *Flora Orientalis*, etc.

genous may in several cases be the result of self-sowing
from cultivation,[1] or that the species was formerly common, and is tending to disappear. Its presence in the
western islands of Europe favours the latter hypothesis,
but its absence in the islands of the Mediterranean is
opposed to it.[2]

Let us see whether historical and philological data
add anything to the facts of geographical botany.

In the first place, it is in Europe that the countless
varieties of cabbage have been formed,[3] principally since
the days of the ancient Greeks. Theophrastus distinguished three, Pliny double that number, Tournefort
twenty, De Candolle more than thirty. These modifications did not come from the East—another sign of an
ancient cultivation in Europe and of a European origin.

The common names are also numerous in European
languages, and rare or modern in those of Asia. Without
repeating a number of names I have given elsewhere,[4] I
shall mention the five or six distinct and ancient roots
from which the European names are derived.

Kap or *kab* in several Keltic and Slav names. The
French name *cabus* comes from it. Its origin is clearly
the same as that of *caput*, because of the head-shaped
form of the cabbage.

Caul, kohl, in several Latin (*caulis*, stem or cabbage),
German (*Chôli* in Old German, *Kohl* in modern German,
kaal in Danish), and Keltic languages (*kaol* and *kol* in
Breton, *cal* in Irish).[5]

Bresic, bresych, brassic, of the Keltic and Latin
(*brassica*) languages, whence, probably, *berza* and *verza* of
the Spaniards and Portuguese, *varza* of the Roumanians.[6]

---

[1] Watson, who is careful on these points, doubts whether the cabbage
is indigenous in England (*Compendium of the Cybele*, p. 103), but most
authors of British floras admit it to be so.

[2] *Br. balearica* and *Br. cretica* are perennial, almost woody, not
biennial; and botanists are agreed in separating them from *Br. oleracea*.

[3] Aug. Pyr. de Candolle has published a paper on the divisions and
subdivisions of *Br. oleracea* (*Transactions of the Hort. Soc.*, vol. v., translated into German and in French in the *Bibl. Univ. Agric.*, vol. viii.),
which is often quoted.

[4] Alph. de Candolle, *Géogr. Bot. Raisonnée*, p. 839.

[5] Ad. Pictet, *Les.Origines Indo-Européennes*, edit. 2, vol. i. p. 380.

[6] Brandza, *Prodr. Fl. Romane*, p. 122.

Aza of the Basques (Iberians), considered by de Charencey[1] as proper to the Euskarian tongue, but which differs little from the preceding.

Krambai, crambe, of the Greeks and Latins.

The variety of names in Keltic languages tends to show the existence of the species on the west coast of Europe. If the Aryan Kelts had brought the plant from Asia, they would probably not have invented names taken from three different sources. It is easy to admit, on the contrary, that the Aryan nations, seeing the cabbage wild, and perhaps already used in Europe by the Iberians or the Ligurians, either invented names or adopted those of the earlier inhabitants.

Philologists have connected the *krambai* of the Greeks with the Persian name *karamb, karam, kalam,* the Kurdish *kalam,* the Armenian *gaghamb ;*[2] others with a root of the supposed mother-tongue of the Aryans; but they do not agree in matters of detail. According to Fick,[3] *karambha,* in the primitive Indo-Germanic tongue, signifies " *Gemüsepflanze* (vegetable), *Kohl* (cabbage), *karambha* meaning stalk, like *caulis.*" He adds that *karambha,* in Sanskrit, is the name of two vegetables. Anglo-Indian writers do not mention this supposed Sanskrit name, but only a name from a modern Hindu dialect, *kopee.*[4] Pictet, on his side, speaks of the Sanskrit word *kalamba,* " vegetable stalk, applied to the cabbage."

I have considerable difficulty, I must own, in admitting these Eastern etymologies for the Greco-Latin word *crambe.* The meaning of the Sanskrit word (if it exists) is very doubtful, and as to the Persian word, we ought to know if it is ancient. I doubt it, for if the cabbage had existed in ancient Persia, the Hebrews would have known it.[5]

For all these reasons, the species appears to me of

---

[1] De Charencey, *Recherches sur les Noms Basques,* in *Actes de la Société Philologique,* 1st March, 1869.
[2] Ad. Pictet, *Les Origines Indo-Européennes,* edit. 2, vol. i. p. 380.
[3] Fick, *Vörterb. d. Indo-Germ. Sprachen,* p. 34.
[4] Piddington, *Index ;* Ainslie, *Mat. Med. Ind.*
[5] Rosenmüller, *Bibl. Alterth.,* mentions no name.

European origin. The date of its cultivation is probably very ancient, earlier than the Aryan invasions, but no doubt the wild plant was gathered before it was cultivated.

Garden-Cress—*Lepidium sativum*, Linnæus.

This little Crucifer, now used as a salad, was valued in ancient times for certain properties of the seeds. Some authors believe that it answers to a certain *cardamon* of Dioscorides; while others apply that name to *Erucaria aleppica*.[1] In the absence of sufficient description, as the modern common name is *cardamon*,[2] the first of these two suppositions is probably correct.

The cultivation of the species must date from ancient times and be widely diffused, for very different names exist: *reschad* in Arab, *turehtezuk*[3] in Persian, *diéges*[4] in Albanian, a language derived from the Pelasgic; without mentioning names drawn from the similarity of taste with that of the water-cress (*Nasturtium officinale*). There are very distinct names in Hindustani and Bengali, but none are known in Sanskrit.[5]

At the present day the plant is cultivated in Europe, in the north of Africa, in Eastern Asia, India, and elsewhere, but its origin is somewhat obscure. I possess several specimens gathered in India, where Sir Joseph Hooker[6] does not consider the species indigenous. Kotschy brought it back from Karrak, or Karek Island, in the Persian Gulf. The label does not say that it was a cultivated plant. Boissier[7] mentions it without comment, and he afterwards speaks of specimens from Ispahan and Egypt gathered in cultivated ground. Olivier is quoted as having found the cress in Persia, but it is not said whether it was growing wild.[8] It has been asserted that Sibthorp found it in Cyprus, but reference to his work shows it was in the fields.[9] Poech does not mention

---

[1] See Fraas, *Syn. Fl. Class.*, pp. 120, 124; Lenz, *Bot. der Alten*, p. 617.

[2] Sibthorp, *Prodr. Fl. Græc.*, ii. p. 6; Heldreich, *Nutzpfl. Griechenl.*, p. 47.

[3] Ainslie, *Mat. Med. Ind.*, i. p. 95.   [4] Heldreich, *Nutz. Gr.*

[5] Piddington, *Index;* Ainslie, *Mat. Med. Ind.*, i. p. 95.

[6] Hooker, *Fl. Brit. Ind.*, i. p. 160.   [7] Boissier, *Fl. Orient.*, vol. i.

[8] De Candolle, *Syst.*, ii. p. 533.

[9] Sibthorp and Smith, *Prodr. Fl. Græcœ*, ii. p. 6.

it in Cyprus.[1] Unger and Kotschy[2] do not consider it
to be wild in that island. According to Ledebour,[3] Koch
found it round the convent on Mount Ararat; Pallas
near Sarepta; Falk on the banks of the Oka, a tributary
of the Volga ; lastly, H. Martius mentions it in his flora of
Moscow ; but there is no proof that it was wild in these
various localities. Lindemann,[4] in 1860, did not reckon
the species among those of Russia, and he only indicates it
as cultivated in the Crimea.[5] According to Nyman,[6] the
botanist Schur found it wild in Transylvania, while the
Austro-Hungarian floras either do not mention the species,
or give it as cultivated, or growing in cultivated ground.

I am led to believe, by this assemblage of more or
less doubtful facts, that the plant is of Persian origin,
whence it may have spread, after the Sanskrit epoch,
into the gardens of India, Syria, Greece, and Egypt, and
even as far as Abyssinia.[7]

Purslane—*Portulaca oleracea*, Linnæus.

Purslane is one of the kitchen garden plants most
widely diffused throughout the old world from the earliest
times. It has been transported into America,[8] where it
spreads itself, as in Europe, in gardens, among rubbish,
by the wayside, etc. It is more or less used as a vegetable, a medicinal plant, and is excellent food for pigs.

A Sanskrit name for it is known, *lonica* or *lounia*,
which recurs in the modern languages of India.[9] The

---

[1] Poech, *Enum. Pl. Cypri*, 1842.

[2] Unger and Kotschy, *Inseln Cypern.*, p. 331.

[3] Ledebour, *Fl. Ross.*, i. p. 203.

[4] Lindemann, *Index Plant. in Ross., Bull. Soc. Nat. Mosc.* 1860, vol. xxxiii.

[5] Lindemann, *Prodr. Fl. Cherson*, p. 21.

[6] Nyman, *Conspectus Fl. Europ.*, 1878, p. 65.

[7] Schweinfurth, *Beitr. Fl. Æth.*, p. 270.

[8] In the United States purslane was believed to be of foreign origin
(Asa Gray, *Fl. of Northern States*, ed. 5 ; *Bot. of California*, i. p. 79), but
in a recent publication, Asa Gray and Trumbull give reasons for believing
that it is indigenous in America as in the old world. Columbus had
noticed it at San Salvador and at Cuba; Oviedo mentions it in St.
Domingo and De Lery in Brazil. This is not the testimony of botanists,
but Nuttall and others found it wild in the upper valley of the Missouri,
in Colorado, and Texas, where, however, from the date it might have
been introduced.—AUTHOR'S NOTE, 1884.

[9] Piddington, *Index to Indian Plants*.

Greek name *andrachne* and the Latin *portulaca* are very different, as also the group of names, *cholza* in Persian, *khursa* or *koursa* in Hindustani, *kourfa kara-or* in Arab and Tartar, which seem to be the origin of *kurza noka* in Polish, *kurj-noha* in Bohemian, *Kreusel* in German, without speaking of the Russian name *schrucha*, and some others of Eastern Asia.[1] One need not be a philologist to see certain derivations in these names showing that the Asiatic peoples in their migrations transported with them their names for the plant, but this does not prove that they transported the plant itself. They may have found it in the countries to which they came. On the other hand, the existence of three or four different roots shows that European peoples anterior to the Asiatic migrations had already names for the species, which is consequently very ancient in Europe as well as in Asia.

It is very difficult to discover in the case of a plant so widely diffused, and which propagates itself so easily by means of its enormous number of little seeds, whether a specimen is cultivated, naturalized by spreading from cultivation, or really wild.

It does not appear to be so ancient in the east as in the west of the Asiatic continent, and authors never say that it is a wild plant.[2] In India the case is very different. Sir Joseph Hooker says[3] that it grows in India to the height of five thousand feet in the Himalayas. He also mentions having found in the north-west of India the variety with upright stem, which is cultivated together with the common species in Europe. I find nothing positive about the localities in Persia, but so many are mentioned, and in countries so little cultivated, on the shores of the Caspian Sea, in the neighbourhood of the Caucasus, and even in the south of Russia,[4] that it is difficult not to admit that the plant is indigenous in that central region whence the Asiatic peoples overran

[1] Nemnich, *Polyglot. Lex. Naturgesch.*, ii. p. 1047.

[2] Loureiro, *Fl. Cochin.*, i. p. 359 ; Franchet and Savatier, *Enum. Pl. Japon.*, i. p. 53 ; Bentham, *Fl. Hongkong*, p. 127.

[3] Hooker, *Fl. Brit. Ind.*, i. p. 240.

[4] Ledebour, *Fl. Ross.*, ii. p. 145 ; Lindemann, in *Prodr. Fl. Chers.*, p. 74, says, " In desertis et arenosis inter Cherson et Berislaw, circa Odessam."

Europe. In Greece the plant is wild as well as cultivated.[1] Further to the west, in Italy, etc., we begin to
find it indicated in floras,. but only growing in fields,
gardens, rubbish-heaps, and other suspicious localities.[2]

Thus the evidence of philology and botany alike show
that the species is indigenous in the whole of the region
which extends from the western Himalayas to the south
of Russia and Greece.

New Zealand Spinach—*Tetragonia expansa,* Murray.

This plant was brought from New Zealand at the time
of Cook's famous voyage, and cultivated by Sir Joseph
Banks, and hence its name. It is a singular plant from a
double point of view. In the first place, it is the only
cultivated species which comes from New Zealand; and
secondly, it belongs to an order of usually fleshy plants,
the *Ficoideæ,* of which no other species is used. Horticulturists[3] recommend it as an annual vegetable, of
which the taste resembles that of spinach, but which
bears drought better, and is therefore a resource in
seasons when spinach fails.

Since Cook's voyage it has been found wild chiefly on
the sea coast, not only in New Zealand but also in Tasmania, in the south and west of Australia, in Japan, and
in South America.[4] It remains to be discovered whether
in the latter places it is not naturalized, for it is found
in the neighbourhood of towns in Japan and Chili.[5]

Garden Celery—*Apium graveolens,* Linnæus.

Like many Umbellifers which grow in damp places,
wild celery has a wide range. It extends from Sweden to
Algeria, Egypt, Abyssinia, and in Asia from the Caucasus
to Beluchistan, and the mountains of British India.[6]

---

[1] Lenz, *Bot. der Alten,* p. 632; Heldreich, *Fl. Attisch. Ebene.,* p. 483.

[2] Bertoloni, *Fl. It.,* vol. **v.**; Gussone, *Fl. Sic.,* vol. i.; Moris, *Fl. Sard.,*
vol. ii.; Willkomm and Lange, *Prodr. Fl. Hisp.,* vol. iii.

[3] *Botanical Magazine,* t. 2362; *Bon Jardinier,* 1880, p. 567.

[4] Sir J. Hooker, *Handbook of New Zealand Flora,* p. 84; Bentham,
Flora Australiensis, iii. p. 327; Franchet and Savatier, *Enum. Plant.
Japoniæ,* i. p. 177.

[5] Cl. Gay, *Flora Chilena,* ii. p. 468.

[6] Fries, *Summa Veget. Scand.;* Munby, *Catal. Alger.,* p. 11; Boissier,
Fl. Orient., vol. ii. p. 856; Schweinfurth and Ascherson, *Aufzählung,*
p. 272; Hooker, *Fl. Brit. Ind.,* ii. p. 679.

It is spoken of in the *Odyssey* under the name of *selinon*, and in Theophrastus; but later, Dioscorides and Pliny[1] distinguish between the wild and cultivated celery. In the latter the leaves are blanched, which greatly diminishes their bitterness. The long course of cultivation explains the numerous garden varieties. The one which differs more widely from the wild plant is that of which the fleshy root is eaten cooked.

Chervil—*Scandix cerefolium*, Linnæus; *Anthriscus cerefolium*, Hoffmann.

Not long ago the origin of this little Umbellifer, so common in our gardens, was unknown. Like many annuals, it sprang up on rubbish-heaps, in hedges, in waste places, and it was doubted whether it should be considered wild. In the west and south of Europe it seems to have been introduced, and more or less naturalized; but in the south-east of Russia and in western temperate Asia it appears to be indigenous. Steven[2] tells us that it is found "here and there in the woods of the Crimea." Boissier[3] received several specimens from the provinces to the south of the Caucasus, from Turcomania and the mountains of the north of Persia, localities of which the species is probably a native. It is wanting in the floras of India and the east of Asia.

Greek authors do not mention it. The first mention of the plant by ancient writers occurs in Columella and Pliny,[4] that is, at the beginning of the Christian era. It was then cultivated. Pliny calls it *cerefolium*. The species was probably introduced into the Greco-Roman world after the time of Theophrastus, that is in the course of the three centuries which preceded our era.

Parsley—*Petroselinum sativum*, Mœnch.

This biennial Umbellifer is wild in the south of Europe, from Spain to Turkey. It has also been found at Tlemcen in Algeria, and in Lebanon.[5]

---

[1] Dioscorides, *Mat. Med.*, l. 3, c. 67, 68; Pliny, *Hist.*, l. 19, c. 7, 8; Lenz, *Bot. der Alten Griechen und Römer*, p. 557.

[2] Steven, *Verzeichniss Taurischen Halbinseln*, p. 183.

[3] Boissier, *Fl. Orient.* ii. p. 913.

[4] Lenz, *Bot. d. Alt. Gr. und R.*, p. 572.

[5] Munby, *Catal. Alger.*, edit. 2, p. 22; Boissier, *Fl. Orient.*, ii. p. 857.

Dioscorides and Pliny speak of it under the names of *Petroselinon* and *Petroselinum*,[1] but only as a wild medicinal plant. Nothing proves that it was cultivated in their time. In the Middle Ages Charlemagne counted it among the plants which he ordered to be cultivated in his gardens.[2] Olivier de Serres in the sixteenth century cultivated parsley. English gardeners received it in 1548.[3] Although this cultivation is neither ancient nor important, it has already developed two varieties, which would be called species if they were found wild; the parsley with crinkled leaves, and that of which the fleshy root is edible.

Smyrnium, or **Alexanders** — *Smyrnium olus-atrum*, Linnæus.

Of all the Umbellifers used as vegetables, this was one of the commonest in gardens for nearly fifteen centuries, and it is now abandoned. We can trace its beginning and end. Theophrastus spoke of it as a medicinal plant under the name of *Ipposelinon*, but three centuries later Dioscorides [4] says that either the root or the leaves might be eaten, which implies cultivation. The Latins called it *olus-atrum*, Charlemagne *olisatum*, and commanded it to be sown in his farms.[5] The Italians made great use of it under the name *macerone*.[6] At the end of the eighteenth century the tradition existed in England that this plant had been formerly cultivated; later English and French horticulturists do not mention it.[7]

The *Smyrnium olus-atrum* is wild throughout Southern Europe, in Algeria, Syria, and Asia Minor.[8]

Corn Salad, or **Lamb's Lettuce**—*Valerianella olitoria*, Linnæus.

---

[1] Dioscorides, *Mat. Med.*, 1. 3, c. 70 ; Pliny, *Hist.*, 1. 20, ch. 12.
[2] The list of these plants may be found in Meyer, *Gesch. der Bot.*, iii. p. 401.
[3] Phillips, *Companion to the Kitchen Garden*, ii. p. 35.
[4] Theophrastus, *Hist.*, 1. 1, 9 ; 1. 2, 2 ; 1. 7, 6 ; Dioscorides, *Mat. Med.*, 1. 3, c. 71.
[5] E. Meyer, *Gesch. der Bot.*, iii. p. 401.
[6] Targioni, *Cenni Storici*, p. 58.
[7] *English Botany*, t. 230 ; Phillips, *Companion to the Kitchen Garden*; *Le Bon Jardinier*.
[8] Boissier, *Fl. Orient.*, ii. p. 927.

Frequently cultivated as a salad, this annual, of the Valerian family, is found wild throughout temperate Europe to about the sixtieth degree of latitude, in Southern Europe, in the Canary Isles, Madeira, and the Azores, in the north of Africa, Asia Minor, and the Caucasus.[1] It often grows in cultivated ground, near villages, etc., which renders it somewhat difficult to know where it grew before cultivation. It is mentioned, however, in Sardinia and Sicily, in the meadows and mountain pastures.[2] I suspect that it is indigenous only in these islands, and that everywhere else it is introduced or naturalized. The grounds for this opinion are the fact that no name which it seems possible to assign to this plant has been found in Greek or Latin authors. We cannot even name any botanist of the Middle Ages or of the sixteenth century who has spoken of it. Neither is it mentioned among the vegetables used in France in the seventeenth century, either by the *Jardinier Français* of 1651, or by Laurenberg's work, *Horticultura* (Frankfurt, 1632). The cultivation and even the use of this salad appear to be modern, a fact which has not been noticed.

Cardoon—*Cynara cardunculus*, Linnæus.

Artichoke—*Cynara scolymus*, Linnæus; *C. cardunculus*, var. *sativa*, Moris.

For a long time botanists have held the opinion that the artichoke is probably a form obtained by cultivation from the wild cardoon.[3] Careful observations have lately proved this hypothesis. Moris,[4] for instance, having cultivated, in the garden at Turin, the wild Sardinian plant side by side with the artichoke, affirmed that true characteristic distinctions no longer existed.

Willkomm and Lange,[5] who have carefully observed the plant in Spain, both wild and cultivated, share the

---

[1] Krok, *Monographie des Valerianella*, Stockholm, 1864, p. 88; Boissier, *Fl. Orient.*, iii. p. 104.

[2] Bertoloni, *Fl. Ital.*, i. p. 185; Moris, *Fl. Sard.*, ii. p. 314; Gussone, *Synopsis Fl. Siculæ*, edit. 2, vol. i. p. 30.

[3] Dodoens, *Hist. Plant.*, p. 724; Linnæus, *Species*, p. 1159; De Candolle, *Prodr.*, vi. p. 620.

[4] Moris, *Flora Sardoa*, ii. p. 61.

[5] Willkomm and Lange, *Prodr. Fl. Hisp.*, ii. p. 180.

same opinion. Moreover, the artichoke has not been found out of gardens ; and since the Mediterranean region, the home of all the *Cynaræ*, has been thoroughly explored, it may safely be asserted that it exists nowhere wild.

The cardoon, in which we must also include *C. horrida* of Sibthorp, is indigenous in Madeira and in the Canary Isles, in the mountains of Marocco near Mogador, in the south and east of the Iberian peninsula, the south of France, of Italy, of Greece, and in the islands of the Mediterranean Sea as far as Cyprus.[1] Munby [2] does not allow *C. cardunculus* to be wild in Algeria, but he does admit *Cynara humilis* of Linnæus, which is considered by a few authors as a variety.

The cultivated cardoon varies a good deal with regard to the division of the leaves, the number of spines, and the size—diversities which indicate long cultivation. The Romans eat the receptacle which bears the flowers, and the Italians also eat it, under the name of *girello*. Modern nations cultivate the cardoon for the fleshy part of the leaves, a custom which is not yet introduced into Greece.[3]

The artichoke offers fewer varieties, which bears out the opinion that it is a form derived from the cardoon. Targioni,[4] in an excellent article upon this plant, relates that the artichoke was brought from Naples to Florence in 1466, and he proves that ancient writers, even Athenæus, were not acquainted with the artichoke, but only with the wild and cultivated cardoons. I must mention, however, as a sign of its antiquity in the north of Africa, that the Berbers have two entirely distinct names for the two plants : *addad* for the cardoon, *taga* for the artichoke.[5]

[1] Webb, *Phyt. Canar.*, iii. sect. 2, p. 384 ; Ball, *Spicilegium Fl. Maroc.*, p. 524 ; Willkomm and Lange, *Pr. Fl. Hisp. ;* Bertoloni, *Fl. Ital.*, ix. p, 86 ; Boissier, *Fl. Orient.*, iii. p. 357 ; Unger and Kotschy, *Inseln Cypern.* p. 246.

[2] Munby, *Catal.*, edit. 2.

[3] Heldreich, *Nutzpflanzen Griechenlands*, p. 27.

[4] Targioni, *Cenni Storici*, p. 52.

[5] *Dictionnaire Français-Berbère*, published by the Government, 1 vol. in 8vo.

It is believed that the *kactos, kinara*, and *scolimos* of the Greeks, and the *carduus* of Roman horticulturists, were *Cynara cardunculus*,[1] although the most detailed description, that of Theophrastus, is sufficiently confused. "The plant," he said, "grows in Sicily "—as it does to this day—"and," he added, " not in Greece." It is, therefore, possible that the plants observed in our day in that country may have been naturalized from cultivation. According to Athenæus,[2] the Egyptian king Ptolemy Energetes, of the second century before Christ, had found in Libya a great quantity of wild *kinara*, by which his soldiers had profited.

Although the indigenous species was to be found at such a little distance, I am very doubtful whether the ancient Egyptians cultivated the cardoon or the artichoke. Pickering and Unger [3] believed they recognized it in some of the drawings on the monuments ; but the two figures which Unger considers the most admissible seem to me extremely doubtful. Moreover, no Hebrew name is known, and the Jews would probably have spoken of this vegetable had they seen it in Egypt. The diffusion of the species in Asia must have taken place somewhat late. There is an Arab name, *hirschuff* or *kerschouff*, and a Persian name, *kunghir*,[4] but no Sanskrit name, and the Hindus have taken the Persian word *kunjir*,[5] which shows that it was introduced at a late epoch. Chinese authors do not mention any *Cynara*.[6] The cultivation of the artichoke was only introduced into England in 1548.[7] One of the most curious facts in the history of *Cynara cardunculus* is its naturalization in the present century over a vast extent of the Pampas of Buenos Ayres, where its abundance is a hindrance to travellers.[8]

---

[1] Theophrastus, *Hist.*, 1. 6, c. 4 ; Pliny, *Hist.*, 1. 19, c. 8; Lenz, *Bot. der Alten Griechen and Römer*, p. 480.

[2] Athenæus, *Deipn.*, ii. 84.

[3] Pickering, *Chron. Arrangement*, p. 71 ; Unger, *Pflanzen der Alten Ægyptens*, p. 46, figs. 27 and 28.

[4] Ainslie, *Mat. Med. Ind.*, i. p. 22.   [5] Piddington, *Index.*

[6] Bretschneider, *Study*, etc., and Letters of 1881.

[7] Phillips, *Companion to the Kitchen Garden*, p. 22.

[8] Aug. de Saint Hilary, *Plantes Remarkables du Bresil*, Introd., p. 58; Darwin, *Animals and Plants under Domestication*, ii. p. 34.

It is becoming equally troublesome in Chili.[1] It is not asserted that the artichoke has anywhere been naturalized in this manner, and this is another sign of its artificial origin.

Lettuce—*Latuca Scariola*, var. *sativa.*

Botanists are agreed in considering the cultivated lettuce as a modification of the wild species called *Latuca Scariola*.[2] The latter grows in temperate and southern Europe, in the Canary Isles, Madeira,[3] Algeria,[4] Abyssinia,[5] and in the temperate regions of Eastern Asia. Boissier speaks of specimens from Arabia Petrea to Mesopotamia and the Caucasus.[6] He mentions a variety with crinkled leaves, similar therefore to some of our garden lettuces, which the traveller Hausknecht brought with him from the mountains of Kurdistan. I have a specimen from Siberia, found near the river Irtysch, and it is now known with certainty that the species grows in the north of India, in Kashmir, and in Nepal.[7] In all these countries it is often near cultivated ground or among rubbish, but often also in rocky ground, clearings, or meadows, as a really wild plant.

The cultivated lettuce often spreads from gardens, and sows itself in the open country. No one, as far as I know, has observed it in such a case for several generations, or has tried to cultivate the wild *L. Scariola*, to see whether the transition is easy from the one form to the other. It is possible that the original habitat of the species has been enlarged by the diffusion of cultivated lettuces reverting to the wild form. It is known that there has been a great increase in the number of cultivated varieties in the course of the last two thousand

---

[1] Cl. Gay, *Flora Chilena*, iv. p. 317.

[2] The author who has gone into this question most carefully is Bischoff, in his *Beiträge zur Flora Deutschlands und der Schweitz*, p. 184. See also Moris, *Flora Sardoa*, ii. p. 530.

[3] Webb, *Phytogr. Canariensis*, iii. p. 422 ; Lowe, *Flora of Madeira*, p. 544.

[4] Munby, *Catal.*, edit. 2, p. 22, under the name of *L. sylvestris*.

[5] Schweinfurth and Ascherson, *Aufzählung*, p. 285.

[6] Boissier, *Fl. Orient.*, iii. p. 809.

[7] Clarke, *Compos. Indicæ*, p. 263.

years. Theophrastus indicated three;[1] *le Bon Jardinier* of 1880 gives forty varieties existing in France.

The ancient Greeks and Romans cultivated the lettuce, especially as a salad. In the East its cultivation possibly dates from an earlier epoch. Nevertheless it does not appear, from the original common names both in Asia and Europe, that this plant was generally or very anciently cultivated. There is no Sanskrit nor Hebrew name known, nor any in the reconstructed Aryan tongue. A Greek name exists, *tridax;* Latin, *latuca;* Persian and Hindu, *kahn;* and the analogous Arabic form *chuss* or *chass*. The Latin form exists also, slightly modified, in the Slav and Germanic languages,[2] which may indicate either that the Western Aryans diffused the plant, or that its cultivation spread with its name at a later date from the south to the north of Europe.

Dr. Bretschneider has confirmed my supposition [3] that the lettuce is not very ancient in China, and that it was introduced there from the West. He says that the first work in which it is mentioned dates from A.D. 600 to A.D. 900.[4]

Wild Chicory—*Cichorium Intybus*, Linnæus.

The wild perennial chicory, which is cultivated as a salad, as a vegetable, as fodder, and for its roots, which are used to mix with coffee, grows throughout Europe, except in Lapland, in Marocco, and Algeria,[5] from Eastern Europe to Afghanistan and Beluchistan,[6] in the Punjab and Kashmir,[7] and from Russia to Lake Baikal in Siberia.[8] The plant is certainly wild in most of these countries; but as it often grows by the side of roads and fields, it is probable that it has been transported by man from its original home. This must be the case in India, for there is no known Sanskrit name.

The Greeks and Romans employed this species wild

[1] Theophrastus, 1. 7, c. 4.   [2] Nemnich, *Polygl. Lexicon.*
[3] A. de Candolle, *Géogr. Bot. Raisonnée,* p. 843.
[4] Bretschneider, *Study and Value of Chinese Botanical Works,* p. 17.
[5] Ball, *Spicilegium Fl. Marocc.,* p. 534; Munby, *Catal.,* edit. 2, p. 21.
[6] Boissier, *Fl. Orient.,* iii. p. 715.
[7] Clarke, *Compos. Ind.,* p. 250.
[8] Ledebour, *Fl. Ross.,* ii. p. 774.

and cultivated,[1] but their notices of it are too brief to be clear. According to Heldreich, the modern Greeks apply the general name of *lachana,* a vegetable or salad, to seventeen different chicories, of which he gives a list.[2] He says that the species commonly cultivated is *Cichorium divaricatum,* Schousboe (*C. pumilum,* Jacquin); but it is an annual, and the chicory of which Theophrastus speaks was perennial.

Endive—*Cichorium Endivia,* Linnæus.

The white chicories or endives of our gardens are distinguished from *Cichorium Intybus,* in that they are annuals, and less bitter to the taste. Moreover, the hairs of the pappus which crowns the seed are four times longer, and unequal instead of being equal. As long as this plant was compared with *C. Intybus,* it was difficult not to admit two species. The origin of *C. Endivia* is uncertain. When we received, forty years ago, specimens of an Indian *Cichorium,* which Hamilton named *C. cosmia,* they seemed to us so like the endive that we supposed the latter to have an Indian origin, as has been sometimes suggested ;[3] but Anglo-Indian botanists said, and continue to assert, that in India the plant only grows under cultivation.[4] The uncertainty persisted as to the geographical origin. After this, several botanists[5] conceived the idea of comparing the endive with an annual species, wild in the region of the Mediterranean, *Cichorium pumilum,* Jacquin (*C. divaricatum,* Schousboe), and the differences were found to be so slight that some have suspected, and others have affirmed, their specific identity. For my part, after having seen wild specimens from Sicily, and compared the good illustrations published by Reichenbach (*Icones,* vol. xix., pls. 1357, 1358), I am disposed to take the cultivated endives for varieties.

---

[1] Dioscorides, ii. c. 160 ; Pliny, xix. c. 8; Palladius, xi. c. 11.   See other authors quoted by Lenz, *Bot. d. Alten,* p. 483.

[2] Heldreich, *Die Nutzpflanzen Griechenlands,* pp. 28, 76.

[3] Aug. Pyr. de Candolle, *Prodr.,* vii. p. 84; Alph. de Candolle, *Géogr. Bot.,* p. 845.

[4] Clarke, *Compos. Ind.,* p. 250.

[5] De Viviani, *Flora Dalmat.,* ii. p. 97 ; Schultz in Webb, *Phyt. Canar.,* sect. ii. p. 391 ; Boissier, *Fl. Orient.,* iii. p. 716.

of the same species as *C. pumilum.* In this case the
oldest name being *C. Endivia,* it is the one which ought
to be retained, as has been done by Schultz. It resembles,
moreover, a popular name common to several languages.

The wild plant exists in the whole region, of which
the Mediterranean is the centre, from Madeira,[1] Marocco,[2]
and Algeria,[3] as far as Palestine,[4] the Caucasus, and
Turkestan.[5] It is very common in the islands of the
Mediterranean and in Greece. Towards the west, in
Spain and Madeira, for instance, it is probable that it has
become naturalized from cultivation, judging from the
positions it occupies in the fields and by the wayside.

No positive proof is found in ancient authors of the
use of this plant by the Greeks and Romans;[6] but it
is probable that they made use of it and several other
Cichoria. The common names tell us nothing, since they
may have been applied to two different species. These
names vary little,[7] and suggest a cultivation of Græco-
Roman origin. A Hindu name, *kasni,* and a Tamul one,
koschi,[8] are mentioned, but no Sanskrit name, and this
indicates that the cultivation of this plant was of late
origin in the east.

Spinach—*Spinacia oleracea,* Linnæus.

This vegetable was unknown to the Greeks and
Romans.[9] It was new to Europe in the sixteenth century,[10]
and it has been a matter of dispute whether it should be
called *spanacha,* as coming from Spain, or *spinacia,* from
its prickly fruit.[11] It was afterwards shown that the
name comes from the Arabic *isfânâdsch, esbanach,* or
sepanach, according to different authors.[12] The Persian

---

[1] Lowe, *Flora of Madeira,* p. 521.   [2] Ball, *Spicilegium,* p. 534.
[3] Munby, *Catal.,* edit. 2, p. 21.   [4] Boissier, *Fl. Orient.,* iii. p. 716.
[5] Bunge, *Beiträge zur Flora Russlands und Central Asiens,* p. 197.
[6] Lenz, *Bot. der Alten,* p. 483 ; Heldreich, *Die Nutzpflanzen Griechenlands,* p. 74.
[7] Nemnich, *Polygl. Lex.,* at the word *Cichorium Endivia.*
[8] Royle, *Ill. Himal.,* p. 247 ; Piddington, *Index.*
[9] J. Bauhin, *Hist.,* ii. p. 964; Fraas, *Syn. Fl. Class. ;* Lenz, *Bot. der
Alten.*
[10] Brassavola, p. 176.   [11] Mathioli, ed Valgr., p. 343.
[12] Ebn Baithar, ueberitz von Sondtheimer, i. p. 34; Forskal, *Egypt,*
p. 77; Delile, *Ill. Ægypt.,* p. 29.

name is *ispany,* or *ispanaj,*[1] and the Hindu *isfany,* or *palak,* according to Piddington, and also *pinnis,* according to the same and to Roxburgh. The absence of any Sanskrit name shows a cultivation of no great antiquity in these regions. Loureiro saw the spinach cultivated at Canton, and Maximowicz in Mantschuria;[2] but Bretschneider tells us that the Chinese name signifies *herb of Persia,* and that Western vegetables were commonly introduced into China a century before the Christian era.[3] It is therefore probable that the cultivation of this plant began in Persia from the time of the Græco-Roman civilization, or that it did not quickly spread either to the east or to the west of its Persian origin. No Hebrew name is known, so that the Arabs must have received both plant and name from the Persians. Nothing leads us to suppose that they carried this vegetable into Spain. Ebn Baithar, who was living in 1235, was of Malaga; but the Arabic works he quotes do not say where the plant was cultivated, except one of them, which says that its cultivation was common at Nineveh and Babylon. Herrera's work on Spanish agriculture does not mention the species, although it is inserted in a supplement of recent date, whence it is probable that the edition of 1513 did not speak of it; so that the European cultivation must have come from the East about the fifteenth century.

Some popular works repeat that spinach is a native of Northern Asia, but there is nothing to confirm this supposition. It evidently comes from the empire of the ancient Medes and Persians. According to Bosc,[4] the traveller, Olivier brought back some seeds of it, found in the East in the open country. This would be a positive proof, if the produce of these seeds had been examined by a botanist in order to ascertain the species and the variety. In the present state of our knowledge it must

---

[1] Roxburgh, *Fl. Ind.,* ed. 1832, v. iii. p. 771, applied to *Spinacia tetandra,* which seems to be the same species.

[2] Maximowicz, *Primitiæ Fl. Amur.,* p. 222.

[3] Bretschneider, *Study and Value of Chin. Bot. Works,* pp. 17, 15.

[4] *Dict. d'Agric.,* v. p. 906.

be owned that spinach has not yet been found in a wild state, unless it be a cultivated modification of *Spinacia tetandra*, Steven, which is wild to the south of the Caucasus, in Turkestan, in Persia, and in Afghanistan, and which is used as a vegetable under the name of *schamum*.[1]

Without entering here into a purely botanical discussion, I may say that, after reading the descriptions quoted by Boissier, and looking at Wight's[2] plate of *Spinacia tetandra*, Roxb., cultivated in India, and the specimens of several herbaria, I see no decided difference between this plant and the cultivated spinach with prickly fruit. The term *tetandra* implies that one of the plants has five and the other four stamens, but the number varies in our cultivated spinaches.[3]

If, as seems probable, the two plants are two varieties, the one cultivated, the other sometimes wild and sometimes cultivated, the oldest name, *S. oleracea*, ought to persist, especially as the two plants are found in the cultivated grounds of their original country.

The *Dutch* or *great spinach*, of which the fruit has no spines, is evidently a garden product. Tragus, or Bock was the first to mention it in the sixteenth century.[4]

Amaranth—*Amarantus gangeticus*, Linnæus.

Several annual amaranths are cultivated as a green vegetable in Mauritius, Bourbon, and the Seychelles Isles, under the name of *brède de Malabar*.[5] This appears to be the principal species. It is much cultivated in India. Anglo-Indian botanists mistook it for a time for *Amarantus oleraceus* of Linnæus, and Wight gives an illustration of it under this name,[6] but it is now acknowledged to be a different species, and belongs to *A. gangeticus*. Its numerous varieties, differing in size, colour, etc., are called in the Telinga dialect *tota kura*, with the occasional addition of an adjective for each.

[1] Boissier, *Fl. Orient.*, vi. p. 234.   [2] Wight, *Icones*, t. 818.

[3] Nees, *Gen. Plant. Fl. Germ.*, l. 7, pl. 15.

[4] Bauhin, *Hist.*, ii. p. 965.

[5] *A. gangeticus*, *A. tristis*, and *A. hybridis* of Linnæus, according to Baker, *Flora of Mauritius*, p. 266.

[6] Wight, *Icones*, p. 715.

There are other names in Bengali and Hindustani. The young shoots sometimes take the place of asparagus at the table of the English.[1] *A. melancholicus*, often grown as an ornamental plant in European gardens, is considered one of the forms of this species.

Its original home is perhaps India, but I cannot discover that the plant has ever been found there in a wild state; at least, this is not asserted by any author. All the species of the genus *Amarantus* spread themselves in cultivated ground, on rubbish-heaps by the wayside, and thus become half-naturalized in hot countries as well as in Europe. Hence the extreme difficulty in distinguishing the species, and above all in guessing or proving their origin. The species most nearly akin to A. *gangeticus* appear to be Asiatic.

A. gangeticus is said by trustworthy authorities to be wild in Egypt and Abyssinia;[2] but this is perhaps only the result of such naturalization as I spoke of just now. The existence of numerous varieties and of different names in India, render its Indian origin most probable.

The Japanese cultivate as vegetables *A. caudatus*, *A. mangostanus*, and A. *melancholicus* (or *gangeticus*) of Linnæus,[3] but there is no proof that any of them are indigenous. In Java *A. polystachyus*, Blume, is cultivated; it is very common among rubbish, by the wayside, etc.[4]

I shall speak presently of the species grown for the seed.

Leek—*Allium ampeloprasum*, var. *Porrum*.

According to the careful monograph by J. Gay,[5] the leek, as early writers[6] suspected, is only a cultivated variety of *Allium ampeloprasum* of Linnæus, so common in the East, and in the Mediterranean region,

---

[1] Roxburgh, *Flora Indica*, edit. 2, vol. iii. p. 606.
[2] Boissier, *Flora Orientalis*, iv. p. 990; Schweinfurth and Ascherson, *Aufzählung*, etc., p. 289.
[3] Franchet and Savatier, *Enum. Plant. Japoniæ*, i. p. 390.
[4] Hasskarl, *Plant. Javan. Rariores*, p. 431.
[5] Gay, *Ann. des Sc. Nat.*, 3rd series, vol. viii.
[6] Linnæus, *Species Pl.*; De Candolle, *Fl. Franç.*, iii. p. 219.

especially in Algeria, which in Central Europe sometimes becomes naturalized in vineyards and round ancient cultivations.[1] Gay seems to have mistrusted the indications of the floras of the south of Europe, for, contrary to his method with other species of which he gives the localities out of Algeria, he only quotes in the present case the Algerian localities; admitting, however, the identity of name in the authors for other countries.

The cultivated variety of *Porrum* has not been found wild. It is only mentioned in doubtful localities, such as vineyards, gardens, etc. Ledebour[2] indicates for *A. ampeloprasum* the borders of the Crimea, and the provinces to the south of the Caucasus. Wallich brought a specimen from Kamaon, in India,[3] but we cannot be sure that it was wild. The works on Cochin-China (Loureiro), China (Bretschneider), and Japan (Franchet and Savatier) make no mention of it.

Article II.—Fodder.

Lucern—*Medicago sativa*, Linnæus.

The lucern was known to the Greeks and Romans. They called it in Greek *medicai*, in Latin *medica*, or *herba medica*, because it had been brought from Media at the time of the Persian war, about 470 years before the Christian era.[4] The Romans often cultivated it, at any rate from the beginning of the first or second century. Cato does not speak of it,[5] but it is mentioned by Varro, Columella, and Virgil. De Gasparin[6] notices that Crescenz, in 1478, does not mention it in Italy, and that in 1711 Tull had not seen it beyond the Alps. Targioni, however, who could not be mistaken on this head, says that the cultivation of lucern was maintained in Italy, especially in Tuscany,

---

[1] Koch, *Synopsis Fl. Germ.*; Babington, *Man. of Brit. Bot.*; *English Bot.*, etc.
[2] Ledebour, *Flora Ross.*, iv. p. 163.
[3] Baker, *Journal of Bot.*, 1874, p. 295.
[4] Strabo, xii. p. 560; Pliny, bk. xviii. c. 16.
[5] Hehn, *Culturpflanzen*, etc., p. 355.
[6] Gasparin, *Cours d'Agric.*, iv. p. 424.

from ancient times.[1] It is rare in modern Greece.[2] French cultivators have often given to the lucern the name of *sainfoin*, which belongs properly to *Ono-brychis sativa;* and this transposition still exists, for instance in the neighbourhood of Geneva. The name *lucern* has been supposed to come from the valley of Luzerne, in Piedmont; but there is another and more probable origin. The Spaniards had an old name, *eruye*, mentioned by J. Bauhin,[3] and the Catalans call it *userdas*,[4] whence perhaps the patois name in the south of France, *laouzerdo*, nearly akin to *luzerne*. It was so commonly cultivated in Spain that the Italians have sometimea called it *herba spagna*.[5] The Spaniards have, besides the names already given, *mielga*, or *melga*, which appears to come from *Medica*, but they principally used names derived from the Arabic—*alfafa, alfasafat, alfalfa*. In the thirteenth century, the famous physician Ebn Baithar, who wrote at Malaga, uses the Arab word *fisfisat*, which he derives from the Persian *isfist*.[6] It will be seen that, if we are to trust to the common names, the origin of the plant would be either in Spain, Piedmont, or Persia. Fortunately botanists can furnish direct and possible proofs of the original home of the species.

It has been found wild, with every appearance of an indigenous plant, in several provinces of Anatolia, to the south of the Caucasus, in several parts of Persia, in Afghanistan, in Beluchistan,[7] and in Kashmir.[8] In the south of Russia, a locality mentioned by some authors, it is perhaps the result of cultivation as well as in the south of Europe. The Greeks may, therefore, have introduced the plant from Asia Minor as well as from India, which extended from the north of Persia.

This origin of the lucern, which is well established,

---

[1] Targioni-Tozzetti, *Cenni Storici*, p. 34.
[2] Fraas, *Synopsis Fl. Class.*, p. 63; Heldreich, *Die Nutzpflanzen Griechenlands*, p. 70. ·
[3] Bauhin, *Hist. Plant.*, ii. p. 381.   [4] Colmeiro, *Catal.*
[5] Tozzetti, *Dizion. Bot.*
[6] Ebn Baithar, *Heil und Nahrungsmittel*, translated from Arabic by Sontheimer, vol. ii. p. 257.·
[7] Boissier, *Fl. Orient.*, ii. p. 94.   [8] Royle, *Ill. Himal.*, p. 197.

makes me note as a singular fact that no Sanskrit name
is known.[1]   Clover and sainfoin have none either, which
leads us to suppose that the Aryans had no artificial
meadows.

Sainfoin—*Hedysarum Onobrychis*, Linnæus ; *Onobrychis sativa*, Lamarck.

This leguminous plant, of which the usefulness in the
dry and chalky soils of temperate regions is incontestable,
has not been long in cultivation.   The Greeks did not
grow it, and their descendants have not introduced it
into their agriculture to this day.[2]   The plant called
Onobrychis by Dioscorides and Pliny, is *Onobrychis
Caput-Galli* of modern botanists,[3] a species wild in Greece
and elsewhere, which is not cultivated.   The sainfoin, or
lupinella of the Italians, was highly esteemed as fodder
in the south of France in the time of Olivier de Serres,[4]
that is to say, in the sixteenth century ; but in Italy it
was only in the eighteenth century that this cultivation
spread, particularly in Tuscany.[5]

Sainfoin is a herbaceous plant, which grows wild in
the temperate parts of Europe, to the south of the
Caucasus, round the Caspian Sea,[6] and even beyond Lake
Baikal.[7]   In the south of Europe it grows only on the
hills.   Gussone does not reckon it among the wild species
of Sicily, nor Moris among those of Sardinia, nor Munby
among those of Algeria.

No Sanskrit, Persian, or Arabic names are known.
Everything tends to show that the cultivation of this
plant originated in the south of France as late perhaps
as the fifteenth century.

French Honeysuckle, or **Spanish Sainfoin**—*Hedysarum
coronarium*, Linnæus.

The cultivation of this leguminous plant, akin to the

[1] Piddington, *Index.*

[2] Heldreich, *Nutzpflanzen Griechenlands*, p. 72.

[3] Fraas, *Synopsis Fl. Class.*, p. 58 ; Lenz, *Bot. der Alten Gr. und
Röm.*, p. 731.

[4] O. de Serres, *Théâtre de l'Agric.*, p. 242.

[5] Targioni-Tozzetti, *Cenni Storici*, p. 34.

[6] Ledebour, *Fl. Ross.*, i. p. 708 ; Boissier, *Fl. Or.*, p. 532.

[7] Turczaninow, *Flora Baical. Dahur.*, i. p. 340.

sainfoin, and of which a good illustration may be found in the *Flora des Serres et des Jardins,* vol. xiii. pl. 1382, has been diffused in modern times through Italy, Sicily, Malta, and the Balearic Isles.[1] Marquis Grimaldi, who first pointed it out to cultivators in 1766, had seen it at Seminara, in Lower Calabria; De Gasparin[2] recommends it for Algeria, and it is probable that cultivators under similar conditions in Australia, at the Cape, in South America or Mexico, would do well to try it. In the neighbourhood of Orange, in Algeria, the plant did not survive the cold of 6° centigrade.

Hedysarum coronarium grows in Italy from Genoa to Sicily and Sardinia,[3] in the south of Spain[4] and in Algeria,[5] where it is rare. It is, therefore, a species of limited geographical area.

Purple Clover—*Trifolium pratense,* Linnæus.

Clover was not cultivated in ancient times, although the plant was doubtless known to nearly all the peoples of Europe and of temperate Western Asia. Its use was first introduced into Flanders in the sixteenth century, perhaps even earlier, and, according to Schwerz, the Protestants expelled by the Spaniards carried it into Germany, where they established themselves under the protection of the Elector Palatine. It was also from Flanders that the English received it in 1633, through the influence of Weston, Earl of Portland, then Lord Chancellor.[6]

Trifolium pratense is wild throughout Europe, in Algeria,[7] on the mountains of Anatolia, in Armenia and in Turkestan,[8] in Siberia towards the Altai Mountains,[9] and in Kashmir and Garwhall.[10]

[1] Targioni-Tozzetti, *Cenni Storici,* p. 35; Marès and Virgineix, *Catal des Baléares,* p. 100.

[2] De Gasparin, *Cours d'Agric.,* iv. p. 472.

[3] Bertoloni, *Flora Ital.,* viii. p. 6.

[4] Willkomm and Lange, *Prodr. Fl. Hisp.,* iii. p. 262.

[5] Munby, *Catal.,* edit. 2, p. 12.

[6] De Gasparin, *Cours d'Agric.,* iv. p. 445, according to Schwerz and A. Young.

[7] Munby, *Catal.,* edit. 2, p. 11.   [8] Boissier, *Fl. Orient.* i. p. 115.

[9] Ledebour, *Fl. Ross.,* i. p. 548.

[10] Baker, in Hooker's *Fl. of Brit. Ind.,* ii. p. 86.

The species existed, therefore, in Asia, in the land of the Aryan nations; but no Sanskrit name is known, whence it may be inferred that it was not cultivated.

Crimson or Italian Clover—*Trifolium incarnatum,* Linnæus.

An annual plant grown for fodder, whose cultivation, says Vilmorin, long confined to a few of the southern departments, becomes every day more common in France.[1] De Candolle, at the beginning of the present century, had only seen it in the department of Ariège.[2] It has existed for about sixty years in the neighbourhood of Geneva. Targioni does not think that it is of ancient date in Italy,[3] and the trivial name *trafoglio* strengthens his opinion.

The Catalan *fé, fench,*[4] and, in the patois of the south of France,[5] *farradje* (Roussillon), *farratage* (Languedoc), *feroutgé* (Gascony), whence the French name *farouch,* have, on the other hand, an original character, which indicates an ancient cultivation round the Pyrenees. The term which is sometimes used, " clover of Roussillon," also shows this.

The wild plant exists in Galicia, in Biscaya, and Catalonia,[6] but not in the Balearic Isles;[7] it is found in Sardinia[8] and in the province of Algiers.[9] It appears in several localities in France, Italy, and Dalmatia, in the valley of the Danube and Macedonia, but in many cases it is not known whether it may not have strayed from neighbouring cultivation. A singular locality in which it appears to be indigenous, according to English authors, is on the coast of Cornwall, near the Lizard. In this place, according to Bentham, it is the pale yellow variety, which is truly wild on the Continent, while the

---

[1] *Bon Jardinier,* 1880, pt. i. p. 618.

[2] De Candolle, *Fl. Franç.,* iv. p. 528.

[3] Targioni, *Cenni Storici,* p. 35.

[4] Costa, *Intro. Fl. di Catal.,* p. 60.

[5] Moritzi, *Dict. MS.,* compiled from floras published before the middle of the present century.

[6] Willkomm and Lange, *Prodr. Fl. Hisp.,* iii. p. 366.

[7] Marès and Virgineix, *Catal.,* 1880.

[8] Moris, *Fl. Sard.,* i. p. 467.   [9] Munby, *Catal.,* edit. 2.

crimson variety is only naturalized in England from cultivation.[1] I do not know to what degree this remark of Bentham's as to the wild nature of the sole variety of a yellow colour (var. *Molinerii,* Seringe) is confirmed in all the countries where the species grows. It is the only one indicated by Moris in Sardinia, and in Dalmatia by Viviani,[2] in the localities which appear natural (*in pascuis collinis, in montanis, in herbidis*). The authors of the *Bon Jardinier*[3] affirm with Bentham that *Trifolium Molinerii* is wild in the north of France, that with crimson flowers being introduced from the south; and while they admit the absence of a good specific distinction, they note that in cultivation the variety *Molinerii* is of slower growth, often biennial instead of annual.

Alexandrine or **Egyptian Clover**—*Trifolium Alexandrinum,* Linnæus.

This species is extensively cultivated in Egypt as fodder. Its Arab name is *bersym* or *berzun.*[4] There is nothing to show that it has been long in use; the name does not occur in Hebrew and Armenian botanical works. The species is not wild in Egypt, but it is certainly wild in Syria and Asia Minor.[5]

Ervilia—*Ervum Ervilia,* Linnæus; *Vicia Ervilia,* Willdenow.

Bertoloni[6] gives no less than ten common Italian names—*ervo, lero, zirlo,* etc. This is an indication of an ancient and general culture. Heldreich[7] says that the modern Greeks cultivate the plant in abundance as fodder. They call it *robai,* from the ancient Greek *orobos,* as *ervos* comes from the Latin *ervum.* The cultivation of the species is mentioned by ancient Greek and Latin authors.[8] The Greeks made use of the seed; for some has been

---

[1] Bentham, *Handbook Brit. Fl.,* edit. 4, p. 117.
[2] Moris, *Fl. Sard.,* i. p. 467; Viviani, *Fl. Dalmat.,* iii. p. 290.
[3] *Bon Jardinier,* 1880, p. 619.
[4] Forskal, *Fl. Egypt.,* p. 71; Delile, *Plant. Cult. en Egypt.,* p. 10; Wilkinson, *Manners and Customs of Ancient Egyptians,* ii. p. 398.
[5] Boissier, *Fl. Orient.,* ii. p. 127.   [6] Bertoloni, *Fl. It.,* vii. p. 500.
[7] *Nutzpflanzen Griechenlands,* p. 71.
[8] See Lenz, *Bot. d. Alten,* p. 727; Fraas, *Fl. Class.,* p. 54.

discovered in the excavations on the site of Troy.[1]   There
are a number of common names in Spain, some of them
Arabic,[2] but the species has not been so widely cultivated
there for several centuries.[3]   In France it is so little
grown that many modern works on agriculture do not
mention it.   It is unknown in British India.[4]

General botanical works indicate *Ervum Ervilia* as
growing in Southern Europe, but if we take severally the
best floras, it will be seen that it is in such localities as
fields, vineyards, or cultivated ground.   It is the same in
Western Asia, where Boissier[5] speaks of specimens from
Syria, Persia, and Afghanistan.   Sometimes, in abridged
catalogues,[6] the locality is not given, but nowhere do I
find it asserted that the plant has been seen wild in places
far from cultivation. The specimens in my own herbarium
furnish no further proof on this head.

In all likelihood the species was formerly wild in
Greece, Italy, and perhaps Spain and Algeria, but the
frequency of its cultivation in the very regions where it
existed prevent us from now finding the wild stocks.

Tare, or **Common Vetch**—*Vicia sativa*, Linnæus.

Vicia sativa is an annual leguminous plant wild
throughout Europe, except in Lapland.   It is also common
in Algeria,[7] and to the south of the Caucasus as far as the
province of Talysch.[8]   Roxburgh pronounces it to be
wild in the north-west provinces and in Bengal, but Sir
Joseph Hooker admits this only as far as the variety called
angustifolia[9] is concerned.   No Sanskrit name is known,
and in the modern languages of India only Hindu names.[10]
Targioni believes it to be the *ketsach* of the Hebrews.[11]

[1] Wittmack, *Sitzungsber Bot. Vereins Brandenburg*, Dec. 19, 1879.
[2] Willkomm and Lange, *Prodr. Fl. Hisp.*, iii. p. 308.
[3] Baker, in Hooker's *Fl. Brit. Ind.*
[4] Herrera, *Agricultura*, edit. 1819, iv. p. 72.
[5] Baker, in Hooker's *Fl. Brit. Ind.*
[6] For instance, Munby, *Catal. Plant Algeriæ*, edit. 2, p. 12.
[7] Munby, *Catal.*, edit. 2.
[8] Ledebour, *Fl. Ross.*, i. p. 666; Hohenacker, *Enum. Plant. Talysch*,
p. 113; C. A. Meyer, *Verzeichniss*, p. 147.
[9] Roxburgh, *Fl. Ind.*, edit. 1832, iii. p. 323; Hooker, *Fl. Brit. Ind.*,
ii p. 178.
[10] Piddington's *Index* gives four.   [11] Targioni, *Cenni Storici*, p. 30.

I have received specimens from the Cape and from California. The species is certainly not indigenous in the two last-named regions, but has escaped from cultivation.

The Romans sowed this plant both for the sake of the seed and as fodder as early as the time of Cato.[1] I have discovered no proof of a more ancient cultivation. The name *vik*, whence *vicia*, dates from a very remote epoch in Europe, for it exists in Albanian,[2] which is believed to be the language of the Pelasgians, and among the Slav, Swedish, and Germanic nations, with slight modifications. This does not prove that the species was cultivated. It is distinct enough and useful enough to herbivorous animals to have received common names from the earliest times.

Flat-podded Pea—*Lathyrus Cicera*, Linnæus.

An annual leguminous plant, esteemed as fodder, but whose seed, if used as food in any quantity, becomes dangerous.[3]

It is grown in Italy under the name of *mochi*.[4] Some authors suspect that it is the *cicera* of Columella and the *ervilia* of Varro,[5] but the common Italian name is very different to these. The species is not cultivated in Greece.[6] It is more or less grown in France and Spain, without anything to show that its use dates from ancient times. However, Wittmack[7] attributes to it, but doubtfully, some seeds brought by Virchow from the Trojan excavations.

According to the floras, it is evidently wild in dry places, beyond the limits of cultivation in Spain and Italy.[8] It is also wild in Lower Egypt, according to

---

[1] Cato, *De re Rustica*, edit. 1535, p. 34; Pliny, bk. xviii. c. 15.

[2] Heldreich, *Nutzpflanzen Griechenlands*, p. 71. In the earlier language than the Indo-Europeans, *vik* bears another meaning, that of "hamlet" (Fick, *Vorterb. Indo-Germ.*, p. 189).

[3] Vilmorin, *Bon Jardinier*, 1880, p. 603.

[4] Targioni, *Cenni Storici*, p. 31; Bertoloni, *Fl. Ital.*, vii. pp. 444, 447.

[5] Lenz, *Botanik. d. Alten*, p. 730.

[6] Fraas, *Fl. Class.*; Heldreich, *Nutzflanzen Griechenlands*.

[7] Wittmack, *Sitz. Ber. Bot. Vereins Brandenburg*, Dec. 19, 1879.

[8] Willkomm and Lange, *Prodr. Fl. Hisp.*, iii. p. 313; Bertoloni, *Fl. Ital.*

Schweinfurth and Ascherson ;[1] but there is no trace of ancient cultivation in this country or among the Hebrews. Towards the East its wild character becomes less certain. Boissier indicates the plant "in cultivated ground from Turkey in Europe, and Egypt as far as the south of the Caucasus and Babylon."[2]   It is not mentioned in India either as wild or cultivated, and has no Sanskrit name.[3]

The species is probably a native of the region comprised between Spain and Greece, perhaps also of Algeria,[4] and diffused by a cultivation, not of very ancient date, over Western Asia.

Chickling Vetch—*Lathyrus sativus*, Linnæus.

An annual leguminous plant, cultivated in the South of Europe, from a very early age, as fodder, and also for the seeds.   The Greeks called it *lathyros*[5] and the Latins *cicercula*.[6]   It is also cultivated in the temperate regions of Western Asia, and even in the north of India ;[7] but it has no Hebrew[8] nor Sanskrit name,[9] which argues a not very ancient cultivation in these regions.

Nearly all the floras of the south of Europe and of Algeria give the plant as cultivated and half-wild, rarely and only in a few localities as truly wild.   It is easy to understand the difficulty of recognizing the wild character of a species often mixed with cereals, and which persists and spreads itself after cultivation.   Heldreich does not allow that it is indigenous in Greece.[10]   This is a strong presumption that in the rest of Europe and in Algeria the plant has escaped from cultivation.

It is probable that this was not the case in Western Asia ; for authors cite sufficiently wild localities, where agriculture plays a less considerable part than in Europe.

[1] Schweinfurth and Ascherson, *Aufzählung*, etc., p. 257.
[2] Boissier, *Fl. Orient.*, ii. p. 605.
[3] J. Baker, in Hooker's *Fl. of Brit. Ind.*
[4] Munby, *Catal.*
[5] Theophrastus, *Hist. Plant.*, viii., c. 2, 10.
[6] Columella, *De rei rustica*, ii. c. 10 ; Pliny, xviii. c. 13, 32.
[7] Roxburgh, *Fl. Ind.* ; Hooker, *Fl. Brit. Ind.*, ii. p. 178.
[8] Rosenmüller, *Handb. Bibl. Alterth.*, vol. i.
[9] Piddington, *Index.*
[10] Heldreich, *Pflanz. d. Attisch. Ebene*, p. 476 ; *Nutzpf. Gr.*, p. 72.

Ledebour,[1] for instance, mentions specimens gathered in the desert, near the Caspian Sea, and in the province of Lenkoran. Meyer [2] confirms the assertion with respect to Lenkoran. Baker, in his flora of British India, after indicating the species as scattered here and there in the northern provinces, adds, "often cultivated," whence it may be inferred that he considers it as indigenous, at least in the north. Boissier asserts nothing with regard to the localities in Persia which he mentions in his Oriental flora.[3]

To sum up, I think it probable that the species was indigenous before cultivation in the region extending from the south of the Caucasus, or of the Caspian Sea, to the north of India, and that it spread towards Europe in the track of ancient cultivation, mixed perhaps with cereals.

Ochrus—*Pisum ochrus*, Linnæus ; *Lathyrus ochrus,* de Candolle.

Cultivated as an annual fodder in Catalonia, under the name of *tapisots*,[4] and in Greece, particularly in the island of Crete, under that of *ochros*,[5] mentioned by Theophrastus,[6] but without a word of description. Latin authors do not speak of it, which argues a rare and local cultivation in ancient times.

The species is certainly wild in Tuscany.[7] It appears to be wild also in Greece and Sardinia, where it is found in hedges,[8] and in Spain, where it grows in uncultivated ground ;[9] but as for the south of France, Algeria, and Sicily, authors are either silent as to the locality, or mention only fields and cultivated ground. The plant is unknown further east than Syria,[10] where probably it is not wild.

[1] Ledebour, *Fl. Ross.*, i. p. 681.
[2] C. A. Meyer, *Verzeichniss,* p. 148.
[3] Boissier, *Fl. Orient.*, ii. p. 606.
[4] Willkomm and Lange, *Prodr. Fl. Hisp.*, iii. p. 312.
[5] Lenz, *Bot. d. Alten,* p. 730; Heldreich, *Nutzpfl. Gr.,* p. 72.
[6] Lenz.
[7] Caruel, *Fl. Tosc.,* p. 193 ; Gussone, *Syn. Fl. Sic.,* edit. 2.
[8] Boissier, *Fl. Orient.*, ii. p. 602; Moris, *Fl. Sard.,* i. p. 582.
[9] Willkomm and Lange, *Prodr. Fl. Hisp.*   [10] Boissier, *Fl. Orient.*

The fine plate published by Sibthorp, *Flora Græca*, 589, suggests that the species is worthy of more general cultivation.

Trigonel, or **Fenugreek**—*Trigonella fœnum-græcum,* Linnæus.

The cultivation of this annual leguminous plant was common in ancient Greece and Italy,[1] either for spring forage, or for the medicinal properties of its seeds. Abandoned almost everywhere in Europe, and notably in Greece,[2] it is maintained in the East and in India,[3] where it is probably of very ancient date, and throughout the Nile Valley.[4] The species is wild in the Punjab, and in Kashmir,[5] in the deserts of Mesopotamia and of Persia,[6] and in Asia Minor,[7] where, however, the localities cited do not appear sufficiently distinct from the cultivated ground. It is also indicated [8] in several places in Southern Europe, such as Mount Hymettus and other localities in Greece, the hills above Bologna and Genoa, and a few waste places in Spain ; but the further west we go the more we find mentioned such localities as fields, cultivated ground, etc.; and careful authors do not fail to note that the species has probably escaped from cultivation.[9] I do not hesitate to say that if a plant of this nature were indigenous in Southern Europe, it would be far more common, and would not be wanting to the insular floras, such as those of Sicily, Ischia, and the Balearic Isles.[10]

The antiquity of the species and of its use in India is confirmed by the existence of several different names in

---

[1] Theophrastus, *Hist. Plant.*, viii. c. 8 ; Columella, *De rei rustica*, ii. c. 10 ; Pliny, *Hist.*, xviii. c. 16.

[2] Fraas, *Syn. Fl. Class.*, p. 63 ; Lenz, *Bot. der Alten*, p. 719.

[3] Baker, in Hooker's *Fl. Brit. Ind.*, ii. p. 57.

[4] Schweinfurth, *Beitr. z. Fl. Æthiop.*, p. 258.

[5] Baker, in Hooker's *Fl. Brit. Ind.*

[6] Boissier, *Fl. Orient.*, ii. p. 70.   [7] Boissier, *ibid.*

[8] Sibthorp, *Fl. Græca*, t. 766 ; Lenz, *Bot. der Alten*, Bertoloni, *Fl. Ital.*, viii. p. 250 ; Willkomm and Lange, *Prodr. Fl. Hisp.*, iii. p. 390.

[9] Caruel, *Fl. Tosc.*, p. 256 ; Willkomm and Lange.

[10] The plants which spread from one country to another introduce themselves into islands with more difficulty, as will be seen from the remarks I formerly published *Géogr. Bot. Raisonnée*, p. 706).

different dialects, and above all of a Sanskrit and modern Hindu name, *methi*.[1] There is a Persian name, *schemlit*, and an Arab name, *helbeh*;[2] but none is known in Hebrew.[3] One of the names of the plant in ancient Greek, *tailis* (τηλις), may, perhaps, be considered by philologists as akin to the Sanskrit name,[4] but of this I am no judge. The species may have been introduced by the Aryans, and the primitive name have left no trace in northern languages, since it can only live in the south of Europe.

Bird's Foot—*Ornithopus sativus,* Brotero ; *O. isthmocarpus,* Cosson.

The true bird's foot, wild and cultivated in Portugal, was described for the first time in 1804 by Brotero,[5] and Cosson has distinguished it more clearly from allied species.[6] Some authors had confounded it with *Ornithopus roseus* of Dufour, and agriculturists have sometimes given it the name of a very different species, *O. perpusillus,* which by reason of its small size is unsuited for cultivation. It is only necessary to see the pod of *Ornithopus sativus* to make certain of the species, for it is when ripe contracted at intervals and considerably bent. If there are in the fields plants of a similar appearance, but whose pods are straight and not contracted, they are the result of a cross with *O. roseus,* or, if the pod is curved but not contracted, with *O. compressus.* From the appearance of these plants, it seems that they might be grown in the same manner, and would present, I suppose, the same advantages.

The bird's foot is only suited to a dry and sandy soil. It is an annual which furnishes in Portugal a very early spring fodder. Its cultivation has been successfully introduced into Campine.[7]

---

[1] Piddington, *Index.*   [2] Ainslie, *Mat. Med. Ind.,* i. p. 130.

[3] Rosenmüller, *Bibl. Alterth.*

[4] As usual, Fick's dictionary of Indo-European languages does not mention the name of this plant, which the English say is Sanskrit.

[5] Brotero, *Flora Lusitanica,* ii. p. 160.

[6] Cosson, *Notes sur Quelques Plantes Nouvelles ou Critiques du Midi de l'Espagne,* p. 36.

[7] *Bon Jardinier,* 1880, p. 512.

O. sativus appears to be wild in several districts of
Portugal and the south of Spain. I have a specimen
from Tangier; and Cosson found it in Algeria. It is
often found in abandoned fields, and even elsewhere. It
is difficult to say whether the specimens are not from
plants escaped from cultivation, but localities are cited
where this seems improbable; for instance, a pine wood
near Chiclana, in the south of Spain (Willkomm).

Spergula, or Corn Spurry—*Spergula arvensis*, Linnæus.

This annual, belonging to the family of the Caryo-
phylaceæ, grows in sandy fields and similar places in
Europe, in North Africa and Abyssinia,[1] in Western Asia
as far as Hindustan,[2] and even in Java.[3] It is difficult to
know over what extent of the old world it was originally
indigenous. In many localities we do not know if it is
really wild or naturalized from cultivation. Sometimes
a recent introduction may be suspected. In India, for
instance, numerous specimens have been gathered in the
last few years; but Roxburgh, who was so diligent a
collector at the end of the last and the beginning of the
present century, does not mention the species. No
Sanskrit or modern Hindu name is known,[4] and it has
not been found in the countries between India and
Turkey.

The common names may tell us something with
regard to the origin of the species and to its cultivation.

No Greek or Latin name is known. *Spergula*, in
Italian *spergola*, seems to be a common name long in use
in Italy. Another Italian name, *erba renaiola*, indicates
only its growth in the sand (*rena*). The French (*spar-
goule*), Spanish (*esparcillas*), Portuguese (*espargata*), and
German (*Spark*), have all the same root. It seems that
throughout the south of Europe the species was taken
from country to country by the Romans, before the

---

[1] Boissier, *Fl. Orient.*, i. p. 731.
[2] Hooker, *Fl. Brit. Ind.*, i. p. 243, and several specimens from the
Nilgherries and Ceylon in my herbarium.
[3] Zollinger, No. 2556 in my herbarium.   [4] Piddington, *Index*.

division of the Latin languages. In the north the case is very different. There is a Russian name, *toritsa ;* [1] several Danish names, *humb* or *hum, girr* or *kirr ;* [2] and Swedish, *knutt, fryle, nägde, skorff.* [3] This great diversity shows that attention had long been drawn to this plant in this part of Europe, and argues an ancient cultivation. It was cultivated in the neighbourhood of Montbelliard in the sixteenth century, [4] and it is not stated that it was then of recent introduction. Probably it arose in the south of Europe during the Roman occupation, and perhaps earlier in the north. In any case, its original home must have been Europe.

Agriculturists distinguish a taller variety of spergula, [5] but botanists are not agreed with them in finding in it sufficient characteristics of a distinct species, and some do not even make it a variety.

Guinea Grass—*Panicum maximum,* Jacquin. [6]

This perennial grass has a great reputation in countries lying between the tropics as a nutritious fodder, easy of cultivation. With a little care a meadow of guinea grass will last for twenty years. [7]

Its cultivation appears to have begun in the West Indies. P. Browne speaks of it in his work on Jamaica, published in the middle of the last century, and it is subsequently mentioned by Swartz.

The former mentions the name guinea grass, without any remarks on the original home of the species. The latter says, "formerly brought from the coast of Africa to the Antilles." He probably trusted to the indication given by the common name ; but we know how fallacious

---

[1] Sobolewski, *Fl. Petrop.,* p. 109.

[2] Rafn, *Danmarks Flora,* ii. p. 799.

[3] Wahlenberg, quoted by Moritzi, *Dict. MS. ; Svensk Botanik,* t. 308.

[4] Bauhin, *Hist. Plant.,* iii. p. 722.

[5] *Spergula Maxima,* Böninghausen, an illustration published in Reichenbach's *Plantœ Crit.,* vi. p. 513.

[6] *Panicum maximum,* Jacq., Coll. 1, p. 71 (1786) ; Jacq., *Icones* 1, t. 13 ; Swartz, *Fl. Indiœ Occ.,* vii. p. 170 ; *P. polygamum,* Swartz, *Prodr.,* p. 24 (1788) ; *P. jumentorum,* Persoon Ench., i. p. 83 (1805); *P. altissimum* of some gardens and modern authors. According to the rule, the oldest name should be adopted.

[7] In Dominica according to Imray, in the *Kew Report* for 1879, p. 16.

such indications of origin sometimes are. Witness the so-called Turkey wheat, which comes from America.

Swartz, who is an excellent botanist, says that the plant grows in the dry cultivated pastures of the West Indies, where it is also wild, which may imply that it has become naturalized in places where it was formerly cultivated. I cannot find it anywhere asserted that it is really wild in the West Indies. It is otherwise in Brazil. From data collected by de Martius and studied by Nees,[1] data afterwards increased and more carefully studied by Dœll,[2] *Panicum maximum* grows in the clearings of the forests of the Amazon valley, near Santarem, in the provinces of Balria, Ceara, Rio de Janeiro, and Saint Paul. Although the plant is often cultivated in these countries, the localities given, by their number and nature, prove that it is indigenous. Dœll has also seen specimens from French Guiana and New Granada.

With respect to Africa, Sir William Hooker[3] mentioned specimens brought from Sierra Leone, from Aguapim, from the banks of the Quorra, and from the Island of St. Thomas, in Western Africa. Nees[4] indicates the species in several districts of Cape Colony, even in the bush and in mountainous country. Richard[5] mentions places in Abyssinia, which also seem to be beyond the limits of cultivation, but he owns to being not very sure of the species. Anderson, on the contrary, positively asserts that *Panicum maximum* was brought from the banks of the Mozambique and of the Zambesi rivers by the traveller Peters.[6]

The species is known to have been introduced into Mauritius by the Governour Labourdonnais,[7] and to have become naturalized from cultivation as in Rodriguez and the Seychelles Isles. Its introduction into Asia

---

[1] Nees, in Martius, *Fl. Brasil.*, in 8vo, vol. ii. p. 166.
[2] Dœll, in *Fl. Brasil.*, in fol., vol. ii. part 2.
[3] Sir W. Hooker, *Niger Fl.*, p. 560.
[4] Nees, *Floræ Africæ Austr. Gramineæ*, p. 36.
[5] A. Richard, *Abyssinie*, ii. p. 373.
[6] Peters, *Reise Botanik*, p. 546.
[7] Bojer, *Hortus Maurit.*, p. 565.

must be recent, for Roxburgh and Miquel do not mention the species. In Ceylon it is only cultivated.[1]

On the whole, it seems to me that the probabilities are in favour of an African origin, as its name indicates, and this is confirmed by the general, but insufficiently grounded opinion of authors.[2] However, as the plant spreads so rapidly, it is strange that it has not reached Egypt from the Mozambique or Abyssinia, and that it was introduced so late into the islands to the east of Africa. If the co-existence of phanerogamous species in Africa and America previous to cultivation were not extremely rare, it might be inferred in this case; but this is unlikely in the case of a cultivated plant of which the diffusion is evidently very easy.

Article III.—Various Uses of the Stem and Leaves.

Tea—*Thea sinensis*, Linnæus.

In the middle of the eighteenth century, when the shrub which produces tea was still very little known, Linnæus gave it the name of *Thea sinensis*. Soon afterwards, in the second edition of the *Species Plantatum*, he judged it better to distinguish two species, *Thea bohea* and *Thea viridis*, which he believed to correspond to the commercial distinction between black and green teas. It has since been proved that there is but one species, comprehending several varieties, from all of which either black or green tea may be obtained according to the process of manufacture. This question was settled, when another was raised, as to whether *Thea* really forms a genus by itself distinct from the genus *Camellia*. Some authors make *Thea* a section of the old genus *Camellia*; but from the characters indicated with great precision by Seemann,[3] it seems to me that we are justified in retaining the genus *Thea*, together with the old nomenclature of the principal species.

A Japanese legend, related by Kæmpfer,[4] is often

[1] Baker, *Fl. of Mauritius and Seychelles*, p. 436.
[2] Thwaites, *Enum. Pl. Zeylaniæ*.
[3] Seemann, *Tr. of the Linnæan Society*, xxii. p. 337, pl. 61.
[4] Kæmpfer, *Amœn. Japon.*

quoted. A priest who came from India into China in A.D. 519, having succumbed to sleep when he had wished to watch and pray, in a movement of anger cut off his two eyelids, which were changed into a shrub, the tea tree, whose leaves are eminently calculated to prevent sleep. Unfortunately for those people who readily admit legends in whole or in part, the Chinese have never heard of this story, although the event is said to have taken place in their country. Tea was known to them long before 519, and probably it was not brought from India. This is what Bretschneider tells us in his little work, rich in botanical and philological facts.[1] The *Pentsao*, he says, mentions tea 2700 B.C., the *Rye* 300 or 600 B.C.; and the commentator of the latter work, in the fourth century of our era, gave details about the plant and about the infusion of the leaves. Its use is, therefore, of very ancient date in China. It is perhaps more recent in Japan, and if it has been long known in Cochin-China, it is possible, but not proved, that it formerly spread thither from India; authors cite no Sanskrit name, nor even any name in modern Indian languages. This fact will appear strange when contrasted with what we have to say on the natural habitat of the species.

The seeds of the tea-plant often sow themselves beyond the limits of cultivation, thereby inspiring doubt among botanists as to the wild nature of plants encountered here and there. Thunberg believed the species to be wild in Japan, but Franchet and Savatier[2] absolutely deny this. Fortune,[3] who has so carefully examined the cultivation of tea in China, does not speak of the wild plant. Fontanier[4] says that the tea-plant grows wild abundantly in Mantschuria. It is probable that it exists in the mountainous districts of South-eastern China, where naturalists have not yet penetrated.

[1] Bretschneider, *On the Study and Value of Chin. Bot. Works*, pp. 13 and 45.
[2] Franchet and Savatier, *Enum. Pl. Jap.*, i. p. 61.
[3] Fortune, *Three Years' Wandering in China*, 1 vol. in 8vo.
[4] Fontanier, *Bulletin Soc. d'Acclim.*, 1870, p. 88.

Loureiro says that it is found both "cultivated and un-cultivated" in Cochin-China.[1] What is more certain is, that English travellers gathered specimens in Upper Assam[2] and in the province of Cachar.[3] So that the tea-plant must be wild in the mountainous region which separates the plains of India from those of China, but the use of the leaves was not formerly known in India.

The cultivation of tea, now introduced into several colonies, has produced admirable results in Assam. Not only is the product of a superior quality to that of average Chinese teas, but the quantity obtained increases rapidly. In 1870, three million pounds of tea were pro-duced in British India; in 1878, thirty-seven million pounds; and in 1880, a harvest of seventy million pounds was looked for.[4] Tea will not bear frost, and suffers from drought. As I have elsewhere stated,[5] the conditions which favour it are the opposite to those which suit the vine. On the other hand, it has been observed that tea flourishes in Azores, where good wine is made;[6] but it is possible to cultivate in gardens, or on a small scale, many plants which will not be profitable on a large scale. The vine grows in China, yet the manufacture of wine is unimportant. Conversely, no wine-growing country grows tea for exportation. After China, Japan, and Assam, it is in Java, Ceylon, and Brazil that tea is most largely grown, where, certainly, the vine is little culti-vated, or not at all; while the wines of dry regions, such as Australia and the Cape, are already known in the market.

Flax—*Linum usitatissimum*, Linnæus.

The question as to the origin of flax, or rather of the cultivated flax, is one of those which give rise to most interesting researches.

[1] Loureiro, *Fl. Cochin.*, p. 414.

[2] Griffith, *Reports;* Wallich, quoted by Hooker, *Fl. Brit. India,* i. p. 293.

[3] Anderson, quoted by Hooker.

[4] *The Colonies and India, Gardener's Chronicle,* 1880, i. p. 659.

[5] Speech at the Bot. Cong. of London in 1866.

[6] *Flora,* 1868, p. 64.

In order to understand the difficulties which it presents, we must first ascertain what nearly allied forms authors designate—sometimes as distinct species of the genus *Linum,* and sometimes as varieties of a single species.

The first important work on this subject was by Planchon, in 1848.[1] He clearly showed the differences between *Linum usitatissimum, L. humile,* and *L. angustifolium,* which were little known. Afterwards Heer,[2] when making profound researches into ancient cultivation, went again into the characters indicated, and by adding the study of two intermediate forms, as well as the comparison of a great number of specimens, he arrived at the conclusion that there was a single species, composed of several slightly different forms. I give a translation of his Latin summary of the characters, only adding a name for each distinct form, in accordance with the custom of botanical works.

Linum usitatissimum.

1. *Annuum* (annual). Root annual; stem single, upright; capsules 7 to 8 mm. long; seeds 4 to 6 mm., terminating in a point. *a. Vulgare* (common). Capsules 7 mm., not opening when ripe, and displaying glabrous partitions. German names, *Schliesslein, Dreschlein.* β. *Humile* (low). Capsules 8 mm., opening suddenly when ripe; the partitions hairy. *Linum humile,* Miller; *L. crepitans,* Böninghausen. German names, *Klanglein, Springlein.*

2. *Hyemale* (winter). Root annual or biennial; stems numerous, spreading at the base, and bent; capsules 7 mm., terminating in a point. *Linum hyemale romanum.* In German, *Winterlein.*

3. *Ambiguum* (doubtful). Root annual or perennial; stems numerous, leaves acuminate; capsules 7 mm., with partitions nearly free from hairs; seeds 4 mm., ending in a short point. *Linum ambiguum,* Jordan.

4. *Angustifolium* (narrow-leaved). Root annual or

[1] Planchon, in Hooker, *Journal of Botany,* vol. vii. p. 165.
[2] Heer, *Die Pflanzen der Pfahlbauten,* in 4to, Zürich, 1865, p. 35; *Ueber den Flachs und die Flachskultur,* in 4to, Zürich, 1872.

perennial; stems numerous, spreading at the base, and bent; capsules 6 mm., with hairy partitions; seeds 3 mm., slightly hooked at the top. *Linum angustifolium.*

It may be seen how easily one form passes into another. The quality of annual, biennial, or perennial, which Heer suspected to be uncertain, is vague, especially for the *angustifolium;* for Loret, who has observed this flax in the neighbourhood of Montpellier, says,[1] "In very hot countries it is nearly always an annual, and this is the case in Sicily according to Gussone; with us it is annual, biennial, or perennial, according to the nature of the soil in which it grows; and this may be ascertained by observing it on the shore, notably at Maguelone. There it may be seen that along the borders of trodden paths it lasts longer than on the sand, where the sun soon dries up the roots and the acidity of the soil prevents the plant from enduring more than a year."

When forms and physiological conditions pass from one into another, and are distinguished by characters which vary according to circumstances, we are led to consider the individuals as constituting a single species, although these forms and conditions possess a certain degree of heredity, and date perhaps from very early times. We are, however, forced to consider them separately in our researches into their origin. I shall first indicate in what country each variety has been discovered in a wild or half-wild state. I shall then speak of cultivation, and we shall see how far geographical and historical facts confirm the opinion of the unity of species.

The *common annual flax* has not yet been discovered, with absolute certainty, in a wild state. I possess several specimens of it from India, and Planchon saw others in the herbarium at Kew; but Anglo-Indian botanists do not admit that the plant is indigenous in British India. The recent flora of Sir Joseph Hooker speaks of it as a species cultivated principally for the oil extracted from the seeds; and Mr. C. B. Clarke, formerly director of the botanical gardens in Calcutta, writes to

[1] Loret, *Observations Critiques sur Plusieurs Plantes Montpelliéraines,* in the *Revue des Sc. Nat.,* 1875.

me that the specimens must have been cultivated, its
cultivation being very common in winter in the north of
India. Boissier [1] mentions *L. humile,* with narrow leaves,
which Kotschy gathered " near Schiraz in Persia, at the
foot of the mountain called Sabst Buchom." This is,
perhaps, a spot far removed from cultivation ; but I
cannot give satisfactory information on this head. Hohenacker found *L. usitatissimum* " half wild " in the province of *Talysch,* to the south of the Caucasus, towards the
Caspian Sea.[2] Steven is more positive with regard to
Southern Russia.[3] According to him, it " is found pretty
often on the barren hills to the south of the Crimea,
between Jalta and Nikita ; and Nordmann found it on
the eastern coast of the Black Sea." Advancing westward
in Southern Russia, or in the region of the Mediterranean,
the species is but rarely mentioned, and only as escaped
from cultivation, or half wild. In spite of doubts and of
the scanty data which we possess, I think it very possible that the annual flax, in one or other of these two
forms, may be wild in the district between the south of
Persia and the Crimea, at least in a few localities.

The *winter flax* is only known under cultivation in a
few provinces of Italy.[4]

The *Linum ambiguum* of Jordan grows on the coast
of Provence and of Languedoc in dry places.[5]

Lastly, *Linum angustifolium,* which hardly differs
from the preceding, has a well-defined and rather large
area. It grows wild, especially on hills throughout the
region of which the Mediterranean forms the centre ; that
is, in the Canaries and Madeira, in Marocco,[6] Algeria,[7]
and as far as the Cyrenaic ;[8] from the south of Europe,

[1] Boissier, *Flora Orient.,* i. p. 851. It is *L. usitatissimum* of Kotschy,
No. 164.

[2] Boissier, *ibid. ;* Hohenh., *Enum. Talysch.,* p. 168.

[3] Steven, *Verzeichniss der auf der taurischen Halbinseln wildwach-
senden Pflanzen,* Moscow, 1857, p. 91.

[4] Heer, *Ueb. d. Flachs,* pp. 17 and 22.

[5] Jordan, quoted by Walpers, *Annal.,* vol. ii., and by Heer, p. 22.

[6] Ball, *Spicilegium Fl. Marocc.,* p. 380.

[7] Munby, *Catal.,* edit. 2, p. 7.

[8] Rohlf, according to Cosson, *Bulle. Soc. Bot. de Fr.,* 1875, p. 46.

as far as England,[1] the Alps, and the Balkan Mountains; and lastly, in Asia from the south of the Caucasus[2] to Lebanon and Palestine.[3] I do not find it mentioned in the Crimea, nor beyond the Caspian Sea.

Let us now turn to the cultivation of flax, destined in most instances to furnish a textile substance, often also to yield oil, and cultivated among certain peoples for the nutritious properties of the seed. I first studied the question of its origin in 1855,[4] and with the following result :—

It was abundantly shown that the ancient Egyptians and the Hebrews made use of linen stuffs. Herodotus affirms this. Moreover, the plant may be seen figured in the ancient Egyptian drawings, and the microscope indubitably shows that the bandages which bind the mummies are of linen.[5] The culture of flax is of ancient date in Europe ; it was known to the Kelts, and in India according to history. Lastly, the widely different common names indicate likewise an ancient cultivation or long use in different countries. The Keltic name *lin*, and Greco-Latin *linon* or *linum*, has no analogy with the Hebrew *pischta*,[6] nor with the Sanskrit names *ooma*, *atasi*, *utasi*.[7] A few botanists mention the flax as "nearly wild" in the south-east of Russia, to the south of the Caucasus and to the east of Siberia, but it was not known to be truly wild. I then summed up the probabilities, saying, "The varying etymology of the names, the antiquity of cultivation in Egypt, in Europe, and in the north of India, the circumstance that in the latter district flax is cultivated for the yield of oil alone,

---

[1] Planchon, in Hooker's *Journal of Botany*, vol. 7 ; Bentham, *Handbk. of Brit. Flora*, edit. 4, p. 89.

[2] Planchon, *ibid.*   [3] Boissier, *Fl. Or.*, i. p. 861.

[4] A. de Candolle, *Géogr. Bot. Rais.*, p. 833.

[5] Thomson, *Annals of Philosophy*, June, 1834; Dutrochet, Larrey, and Costaz, *Comptes rendus de l'Acad. des. Sc.*, Paris, 1837, sem. i. p. 739 ; Unger, *Bot. Streifzüge*, iv. p. 62.

[6] Other Hebrew words are interpreted "flax," but this is the most certain. See Hamilton, *La Botanique de la Bible*, Nice, 1871, p. 58.

[7] Piddington, *Index Ind. Plants;* Roxburgh, *Fl. Ind.*, edit. 1832, ii. p. 110. The name *matusi* indicated by Piddington belongs to other plants, according to Ad. Pictet, *Origines Indo-Euro.*, edit. 2, vol. i. p. 396.

lead me to believe that two or three species of different origin, confounded by most authors under the name of *Linum usitatissimum*, were formerly cultivated in different countries, without imitation or communication the one with the other. . . . I am very doubtful whether the species cultivated by the ancient Egyptians was the species indigenous in Russia and in Siberia."

My conjectures were confirmed ten years later by a very curious discovery made by Oswald Heer. The lake-dwellers of Eastern Switzerland, at a time when they only used stone implements, and did not know the use of hemp, cultivated and wove a flax which is not our common annual flax, but the perennial flax called *Linum angustifolium*, which is wild south of the Alps. This is shown by the examination of the capsules, seeds, and especially of the lower part of a plant carefully extracted from the sediment at Robenhausen.[1] The illustration published by Heer shows distinctly a root surmounted by from two to four stems after the manner of perennial plants. The stems had been cut, whereas our common flax is plucked up by the roots, another proof of the persistent nature of the plant. With the remains of the Robenhausen flax some grains of *Silene cretica* were found, a species which is also foreign to Switzerland, and abundant in Italy in the fields of flax.[2] Hence Heer concluded that the Swiss lake-dwellers imported the seeds of the Italian flax. This was apparently the case, unless we suppose that the climate of Switzerland at that time differed from that of our own epoch, for the perennial flax would not at the present day survive the winters of Eastern Switzerland.[3] Heer's opinion is supported by the surprising fact that flax has not been found among the remains of the lake-dwellings of Laybach and Mondsee

---

[1] Heer, *Die Pflanzen der Pfahlbauten*, 8vo pamphlet, Zürich, 1865, p. 35; *Ueber den Flachs und die Flachskultur in Alterthum*, pamphlet in 8vo, Zurich, 1872.

[2] Bertoloni, *Fl. Ital.*, iv. p. 612.

[3] We have seen that flax is found towards the north-west of Europe, but not immediately north of the Alps. Perhaps the climate of Switzerland was formerly more equable than it is now, with more snow to shelter perennial plants.

of the Austrian States, where bronze has been discovered.[1] The late epoch of the introduction of flax into this region excludes the hypothesis that the inhabitants of Switzerland received it from Eastern Europe, from which, moreover, they were separated by immense forests.

Since the ingenious observations of the Zurich *savant*, a flax has been discovered which was employed by the prehistoric inhabitants of the peat-mosses of Lagozza, in Lombardy; and Sordelli has shown that it was the same as that of Robenhausen, *L. angustifolium*.[2] This ancient people was ignorant of the use of hemp and of metals, but they possessed the same cereals as the Swiss lake-dwellers of the stone age, and ate like them the acorns of *Quercus robur*, var. *sessiliflora*. There was, therefore, a civilization which had reached a certain development on both sides of the Alps, before metals, even bronze, were in common use, and before hemp and the domestic fowl were known.[3] It was probably before the arrival of the Aryans in Europe, or soon after that event.[4]

The common names of the flax in ancient European languages may throw some light on this question.

The name *lin, llin, linu, linon, linum, lein, lan,* exists in all the European languages of Aryan origin of the centre and south of Europe, Keltic, Slavonic, Greek, or Latin. This name is, however, not common to the Aryan languages of India; consequently, as Pictet[5] justly says, the cultivation must have been begun by the

---

[1] *Mittheil. Anthropol. Gesellschaft*, Wien, vol. vi. pp. 122, 161; *Abhandl., Wien Akad.*, 84, p. 488.

[2] Sordelli, *Sulle piante della torbiera e della stazione preistorica della Lagozza*, pp. 37, 51, printed at the conclusion of Castelfranco's *Notizie alla stazione lacustre della Lagozza*, in 8vo, *Atti della Soc. Ital. Sc. Nat.*, 1880.

[3] The fowl was introduced into Greece from Asia in the sixth century before Christ, according to Heer, *Ueb. d. Flachs*, p. 25.

[4] These discoveries in the peat-mosses of Lagozza and elsewhere in Italy show how far Hehn was mistaken in supposing that (*Kulturpfl.*, edit. 3, 1877, p. 524) the Swiss lake-dwellers were near the time of Cæsar. The men of the same civilization as they to the south of the Alps were evidently more ancient than the Roman republic, perhaps than the Ligurians.

[5] Ad. Pictet, *Origines Indo-Europ.*, edit. 2, vol. i. p. 396.

western Aryans, and before their arrival in Europe. Another idea occurred to me which led me into further researches, but they were unproductive. I thought that, since this flax was cultivated by the lake-dwellers of Switzerland and Italy before the arrival of the Aryan peoples, it was probably also grown by the Iberians, who then occupied Spain and Gaul; and perhaps some special name for it has remained among the Basques, the supposed descendants of the Iberians. Now, according to several dictionaries of their language,[1] *liho, lino,* or *li,* according to the dialects, signifies flax, which agrees with the name diffused throughout Southern Europe. The Basques seem, therefore, to have received flax from peoples of Aryan origin, or perhaps they have lost the ancient name and substituted that of the Kelts and Romans. The name *flachs* or *flax* of the Teutonic languages comes from the Old German *flahs.* There are also special names in the north-west of Europe—*pellawa, aiwina,* in Finnish;[2] *hor, härr, hor,* in Danish;[3] *hor* and *tone* in ancient Gothic.[4] *Haar* exists in the German of Salzburg.[5] This word may be in the ordinary sense of the German for thread or hair, as the name *li* may be connected with the same root as *ligare,* to bind, and as *hör,* in the plural *hörvar,* is connected by philologists[6] with *harva,* the German root for *Flachs;* but it is, nevertheless, a fact that in Scandinavian countries and in Finland terms have been used which differ from those employed throughout the south of Europe. This variety shows the antiquity of the cultivation, and agrees with the fact that the lake-dwellers of Switzerland and Italy cultivated a species of flax before the first invasion of the Aryans. It is possible, I might even say probable, that

---

[1] Van Eys, *Dict. Basque-Français,* 1876; Gèze, *Eléments de Grammaire Basque suivis d'un vocabulaire,* Bayonne, 1873; Salaberry, *Mots Basques Navarrais,* Bayonne, 1856; l'Ecluse, *Vocab. Franç.-Basque,* 1826.

[2] Nemnich, *Poly. Lex. d. Naturgesch.,* ii. p. 420; Rafn, *Danmark Flora,* ii. p. 390.

[3] Nemnich, *ibid.*   [4] *Ibid.*   [5] *Ibid.*

[6] Fick, *Vergl. Worterbuch. Ind. Germ.,* 2nd edit., i. p. 722. He also derives the name *Lina* from the Latin *linum ;* but this name is of earlier date, being common to several European Aryan languages.

the latter imported the name *li* rather than the plant or
its cultivation; but as there is no wild flax in the north
of Europe, an ancient people, the Finns, of Turanian
origin, introduced the flax into the north before the
Aryans. In this case they must have cultivated the
annual flax, for the perennial variety will not bear the
severity of the northern winters; while we know how
favourable the climate of Riga is in summer to the cultivation of the annual flax. Its first introduction into
Gaul, Switzerland, and Italy may have been from the
south, by the Iberians, and in Finland by the Finns; and
the Aryans may have afterwards diffused those names
which were commonest among themselves—that of *linum*
in the south, and of *flahs* in the north. Perhaps the
Aryans and Finns had brought the annual flax from
Asia, which would soon have been substituted for the
perennial variety, which is less productive and less
adapted to cold countries. It is not known precisely at
what epoch the cultivation of the annual flax in Italy
took the place of that of the perennial *linum angustifolium*, but it must have been before the Christian era;
for Latin authors speak of a well-established cultivation,
and Pliny says that the flax was sown in spring and
rooted up in the summer.[1] Metal implements were not
then wanting, and therefore the flax would have been
cut if it had been perennial. Moreover, the latter, if
sown in spring, would not have ripened till autumn.

For the same reasons the flax cultivated by the
ancient Egyptians must have been an annual. Hitherto
neither entire plants nor a great number of capsules have
been found in the catacombs of a nature to furnish direct
and incontestable proof. Unger[2] alone was able to examine a capsule taken from the bricks of a monument,
which Leipsius attributes to the thirteenth or fourteenth
century before Christ, and he found it more like those
of *L. usitatissimum* than of *L. angustifolium* Out of
three seeds which Braun[3] saw in the Berlin Museum,

[1] Pliny, bk. xix. c. 1: *Vere satum æstate vellitur.*
[2] Unger, *Botanische Streifzüge*, 1866, No. 7, p. 15.
[3] A. Braun, *Die Pflanzenreste des Ægyptischen Museums in Berlin*, in
8vo, 1877, p. 4.

mixed with those of other cultivated plants, one appeared to him to belong to *L. angustifolium*, and the other to *L. humile;* but it must be owned that a single seed without plant or capsule is not sufficient proof. Ancient Egyptian paintings show that flax was not reaped with a sickle like cereals, but uprooted.[1] In Egypt flax is cultivated in the winter, for the summer drought would no more allow of a perennial variety, than the cold of northern countries, where it is sown in spring, to be gathered in in summer. It may be added that the annual flax of the variety called *humile* is the only one now grown in Abyssinia, and also the only one that modern collectors have seen in Egypt.[2]

Heer suggests that the ancient Egyptians may have cultivated *L. angustifolium* of the Mediterranean region, sowing it as an annual plant.[3] I am more inclined to believe that they had previously imported or received their flax from Egypt, already in the form of the species *L. humile.* Their modes of cultivation, and the figures on the monuments, show that their knowledge of the plant dated from a remote antiquity. Now it is known that the Egyptians of the first dynasties before Cheops belonged to a proto-semitic race, which came into Egypt by the isthmus of Suez.[4] Flax has been found in a tomb of ancient Chaldea prior to the existence of Babylon,[5] and its use in this region is lost in the remotest antiquity. Thus the first Egyptians of white race may have imported the cultivated flax, or their immediate successors may have received it from Asia before the epoch of the Phœnician colonies in Greece, and before direct communication was established between Greece and Egypt under the fourteenth dynasty.[6]

[1] Rosellini, pls. 35 and 36, quoted by Unger, *Bot. Streifzüge*, No. 4, p. 62.

[2] W. Schimper, Ascherson, Boissier, Schweinfurth, quoted by Braun.

[3] Heer, *Ueb. d. Flachs*, p. 26.

[4] Maspero, *Histoire Ancienne des Peuples de l'Orient.*, edit. 3, Paris, 1878, p. 13.

[5] *Journal of the Royal Asiat. Soc.*, vol. xv. p. 271, quoted by Heer, *Ueb. den Fl.*

[6] Maspero, p. 213.

A very early introduction of the plant into Egypt from Asia does not prevent us from admitting that it was at different times taken from the East to the West at a later epoch than that of the first Egyptian dynasties. Thus the western Aryans and the Phœnicians may have introduced into Europe a flax more advantageous than *L. angustifolium* during the period from 2500 to 1200 years before our era.

The cultivation of the plant by the Aryans must have extended further north than that by the Phœnicians. In Greece, at the time of the Trojan war, fine linen stuffs were still imported from Colchis; that is to say, from that region at the foot of the Caucasus where the common annual flax has been found wild in modern times. It does not appear that the Greeks cultivated the plant at that epoch.[1] The Aryans had perhaps already introduced its cultivation into the valley of the Danube. However, I noticed just now that the lacustrine remains of Mondsee and Laybach show no trace of any flax. In the last centuries before the Christian era the Romans procured very fine linen from Spain, although the names of the plant in that country do not tend to show that the Phœnicians introduced it. There is not any Oriental name existing in Europe belonging either to antiquity or to the Middle Ages. The Arabic name *kattan, kettane,* or *kittane,* of Persian origin,[2] has spread westward only among the Kabyles of Algeria.[3]

The sum of facts and probabilities appear to me to lead to the following statements, which may be accepted until they are modified by further discoveries.

1. *Linum angustifolium,* usually perennial, rarely biennial or annual, which is found wild from the Canary Isles to Palestine and the Caucasus, was cultivated in Switzerland and the north of Italy by peoples more ancient than the conquerors of Aryan race. Its cultivation was replaced by that of the annual flax.

---

[1] The Greek texts are quoted in Lenz, *Bot. der Alt. Gr. und Röm.*, p. 672; and in Hehn, *Culturpfl. und Hausthiere*, edit. 3, p. 144.

[2] Ad. Pictet, *Origines Indo-Europ.*

[3] *Dictionnaire Franç.-Berbère*, 1 vol. in 8vo, 1844.

2. The annual flax (*L. usitatissimum*), cultivated for at least four thousand or five thousand years in Mesopotamia, Assyria, and Egypt, was and still is wild in the districts included between the Persian Gulf, the Caspian Sea, and the Black Sea.

3. This annual flax appears to have been introduced into the north of Europe by the Finns (of Turanian race), afterwards into the rest of Europe by the western Aryans, and perhaps here and there by the Phœnicians; lastly into Hindustan by the eastern Aryans, after their separation from the European Aryans.

4. These two principal forms or conditions of flax exist in cultivation, and have probably been wild in their modern areas for the last five thousand years at least. It is not possible to guess at their previous condition. Their transitions and varieties are so numerous that they may be considered as one species comprising two or three hereditary varieties, which are each again divided into subvarieties.

Jute—*Corchorus capsularis* and *Corchorus olitorius*, Linnæus.

The fibres of the jute, imported in great quantities in the last few years, especially into England, are taken from the stem of these two species of Corchorus, annuals of the family of the Tiliaceæ. The leaves are also used as a vegetable.

C. capsularis has a nearly spherical fruit, flattened at the top, and surrounded by longitudinal ridges. There is a good coloured illustration of it in the work of the younger Jacquin, *Eclogæ*, pl. 119. *C. olitorius*, on the contrary, has a long fruit, like the pod of a Crucifer. It is figured in the *Botanical Magazine*, fig. 2810, and in Lamarck, fig. 478.

The species of the genus are distributed nearly equally in the warm regions of Asia, Africa, and America; consequently the origin of each cannot be guessed. It must be sought in floras and herbaria, with the help of historical and other data.

Corchorus capsularis is commonly cultivated in the Sunda Islands, in Ceylon, in the peninsula of Hindustan, in Bengal, in Southern China, in the Philippine Islands,[1] generally in Southern Asia. Forster does not mention it in his work on the plants in use among the inhabitants of the Pacific, whence it may be inferred that at the time of Cook's voyages, a century ago, its cultivation had not spread in that direction. It may even be suspected from this fact that it does not date from a very remote epoch in the isles of the Indian Archipelago.

Blume says that *Corchorus capsularis* grows in the marshes of Java near Parang,[2] and I have two specimens from Java which are not given as cultivated.[3] Thwaites mentions it as "very common" in Ceylon.[4]

On the continent of Asia, authors speak more of it as a plant cultivated in Bengal and China. Wight, who gives a good illustration of the plant, does not mention its native place. Edgeworth,[5] who has studied on the spot the flora of the district of Banda, says that it is found in "the fields." In the *Flora of British India*, Masters, who drew up the article on the Tiliaceæ from the herbarium at Kew, says "in the hottest regions of India, cultivated in most tropical countries." [6] I have a specimen from Bengal which is not given as cultivated. Loureiro says "wild, and cultivated in the province of Canton in China,[7] which probably means wild in Cochin-China, and cultivated in Canton. In Japan the plant grows in cultivated soil.[8] In conclusion, I am not convinced that the species exists in a truly wild state north of Calcutta, although it may perhaps have spread from cultivation and have sown itself here and there.

C. capsularis has been introduced into various parts of tropical Africa and even of America, but it is only cultivated on a large scale for the production of jute thread in Southern Asia, and especially in Bengal.

[1] Rumphius, *Amboin*, vol. v. p. 212 ; Roxburgh, *Fl. Ind.*, ii. p. 581; Loureiro, *Fl. Cochinchine*, vi. p. 408.
[2] Blume, *Bijdragen*, i. p. 110.   [3] Zollinger, Nos. 1698 and 2761.
[4] Thwaites, *Enum. Pl. Zeylan.*, p. 31.
[5] Edgeworth, *Linnæan Soc. Journ.*, ix.
[6] Masters, in Hooker's *Fl. Brit. Ind.*, i. p. 397.
[7] Loureiro, *Fl. Cochin.*, i. p. 408.
[8] Franchet and Savatier, *Enum.*, i. p. 66.

C. olitorius is more used as a vegetable than for
its fibres. Out of Asia it is employed exclusively for
the leaves. It is one of the commonest of culinary
plants among the modern Egyptians and Syrians, who
call it in Arabic *melokych*, but it is not likely that they
had any knowledge of it in ancient times, as we know
of no Hebrew name.[1] The present inhabitants of Crete
cultivate it under the name of *mouchlia*,[2] evidently
derived from the Arabic, and the ancient Greeks were
not acquainted with it.

According to several authors[3] this species of Corchorus
is wild in several provinces of British India. Thwaites
says it is common in the hot districts of Ceylon ; but in
Java, Blume only mentions it as growing among rubbish
(*in ruderatis*). I cannot find it mentioned in Cochin-China
or Japan. Boissier saw specimens from Mesopotamia,
Afghanistan, Syria, and Anatolia, but gives as a general
indication, " culta, et in ruderatis subspontanea." No
Sanskrit name for the two cultivated species of Corchorus
is known.[4]

Touching the indigenous character of the plant in
Africa, Masters, in Oliver's *Flora of Tropical Africa* (i.
p. 262), says, " wild, or cultivated as a vegetable throughout tropical Africa." He attributes to the same species
two plants from Guinea which G. Don had described as
different, and as to whose wild nature he probably knew
nothing. I have a specimen from Kordofan gathered by
Kotschy, No. 45, "on the borders of the fields of sorghum."
Peters, as far as I know, is the only author who asserts
that the plant is wild. He found *C. olitorius* "in
dry places, and also in the meadows in the neighbourhood of Sena and Tette." Schweinfurth only gives it as
a cultivated plant in the whole Nile Valley.[5] This is
also the case in the flora of Senegambia by Guillemin,
Perrottet, and Richard.

[1] Rosenmüller, *Bibl. Naturgesch.*
[2] Von Heldreich, *Die Nützpfl. Griechenl.*, p. 53.
[3] Masters, in Hooker's *Fl. Brit. Ind.*, i. p. 397 ; Aitchison, *Catal.
Punjab*, p. 23 ; Roxburgh, *Fl. Ind.*, ii. p. 581.
[4] Piddington, *Index*.
[5] Schweinfurth, *Beitr. z. Fl. Æthiop.*, p. 264.

To sum up, *C. olitorius* seems to be wild in the moderately warm regions of Western India, of Kordofan, and probably of some intermediate countries. It must have spread from the coast of Timor, and as far as Northern Australia, into Africa and towards Anatolia, in the wake of a cultivation not perhaps ot earlier date than the Christian era, even at its origin.

In spite of the assertions made in various works, the cultivation of this plant is rarely indicated in America. I note, however, on Grisebach's authority,[1] that it has become naturalized in Jamaica from gardens, as often happens in the case of cultivated annuals.

Sumach.—*Rhus coriaria.*

This tree is cultivated in Spain and Italy [2] for the young shoots and leaves, which are dried and made into a powder for tanning. I recently saw a plantation in Sicily, of which the product was exported to America. As oak-bark becomes more rare and substances for tanning are more in demand, it is probable that this cultivation will spread; all the more that it is suitable to sandy, sterile regions. In Algeria, Australia, at the Cape, and in the Argentine Republic, it might be introduced with advantage.[3] Ancient peoples used the slightly acid fruits as a seasoning, and the custom has lingered here and there; but I find no proof that they cultivated the species.

It grows wild in the Canaries and in Madeira, in the Mediterranean region and in the neighbourhood of the Black Sea, preferring dry and stony ground. In Asia its area extends as far as the south of the Caucasus, the Caspian Sea, and Persia.[4] The species is so common that it may have been in use before it was cultivated.

---

[1] Grisebach, *Fl. of Brit. West Ind.*, p. 97.

[2] Bosc, *Dict. d'Agric.*, at the word "Sumac."

[3] The conditions and methods of the culture of the sumach are the subject of an important paper by Inzenga, translated in the *Bull. Soc. d'Acclim.*, Feb. 1877. In the *Trans. Bot. Soc. of Edinburgh*, ix. p. 341, may be seen an extract from an earlier paper by the author on the same subject.

[4] Ledebour, *Fl. Ross.*, i. p. 509; Boissier, *Fl. Orient.*, ii. p. 4.

Sumach is the Persian and Tartar name;[1] *rous, rhus,* the ancient name among the Greeks and Romans.[2] A proof of the persistence of certain common names is found in the French " Currier's *roux* or *roure.*"

Khât, or **Arab Tea**—*Catha edulis,* Forskal; *Celastrus edulis,* Vahl.

This shrub, belonging to the family of the *Celastraceæ,* is largely cultivated in Abyssinia, under the name of *tchut* or *tchat,* and in Arabia under that of *cat* or *gat.* Its leaves are chewed, when green, like those of the coca in America, and they have the same exciting and strengthening properties. Those of uncultivated plants have a stronger taste, and are even intoxicating. Botta saw that in Yemen as much importance is attributed to the cultivation of the *Catha* as to that of coffee, and he mentions that a sheik, who is obliged to receive many visits of ceremony, bought as much as a hundred francs' worth of leaves a day.[3] In Abyssinia an infusion is also made from the leaves.[4] In spite of the eagerness with which stimulants are sought, this species has not spread into the adjoining countries, such as Beluchistan, Southern India, etc., where it might succeed.

The *Catha* is wild in Abyssinia,[5] but has not yet been found wild in Arabia. It is true that the interior of the country is nearly unknown to botanists. It cannot be ascertained from Botta's account whether the wild plants he mentions are wild and indigenous, or escaped from cultivation and more or less naturalized. Perhaps the *Catha* was introduced from Abyssinia with the coffee plant, which likewise has not been discovered wild in Arabia.

Maté—*Ilex paraguariensis,* Saint-Hilaire.

The inhabitants of Brazil and of Paraguay have em-

---

[1] Nemnich, *Polygl. Lexicon,* ii. p. 1156; Ainslie, *Mat. Med. Ind.,* i. p. 414.

[2] Fraas, *Syn. Fl. Class.,* p. 85.

[3] Forskal, *Flora Ægypto-Arabica,* p. 65; Richard, *Tentamen Fl. Abyss.,* i. p. 134, pl. 30; Botta, *Archives du Muséum,* ii. p. 73.

[4] Hochstetter, *Flora,* 1841, p. 663.

[5] Schweinfurth and Ascherson, *Aufzählung,* p. 263; Oliver, *Fl. Trop. Afr.,* i. p. 364.

ployed from time immemorial the leaves of this shrub, as the Chinese have those of the tea plant. They gather them especially in the damp forests of the interior, between the degrees of 20 and 30 south latitude, and commerce transports them dried to great distances throughout the greater part of South America. These leaves contain, with aroma and tannin, a principle analogous to that of tea and coffee ; they are not, however, much liked in the countries where Chinese tea is known. The plantations of maté are not yet as important as the product of the wild shrub, but they may increase as the population increases. Moreover, the preparation is simpler than that of tea, as the leaves are not rolled.

Illustrations and descriptions of the species, with a number of details about its use and properties, may be found in the works of Saint-Hilaire, of Sir William Hooker, and of Martius.[1]

Coca.—*Erythroxylon Coca*, Lamarck.

The natives of Peru and of the neighbouring provinces, at least in the hot moist regions, cultivate this shrub, of which they chew the leaves, as the natives of India chew the leaves of the betel. It is a very ancient custom, which has spread even into elevated regions, where the species cannot live. Now that it is known how to extract the essential part of the coca, and its virtues are recognized as a tonic, which gives strength to endure fatigue without having the drawbacks of alcoholic liquors, it is probable that an attempt will be made to extend its cultivation in America and elsewhere. In Guiana, for instance, the Malay Archipelago, or the valleys of Sikkim and Assam, or in Hindustan, since both moisture and heat are requisite. Frost is very injurious to the species. The best sites are the slopes of hills where water cannot lie. An attempt made in the neighbourhood of Lima failed, because of the infrequency of rain and perhaps because of insufficient heat.[2]

---

[1] Aug. de Saint-Hilaire, *Mém. du Muséum,* ix. p. 351 ; *Ann. Sc. Nat.,* 3rd series, xiv. p. 52; Hooker, *London Journal of Botany,* i. p. 34; Martius, *Flora Brasiliensis,* vol. ii. part 1, p. 119.

[2] Martinet, *Bull. Soc. d'Acclim.,* 1874, p. 449.

I shall not repeat here what may be found in several excellent treatises on the coca;[1] I need only say that the original home of the species in America is not yet clearly ascertained. Gosse has shown that early authors, such as Joseph de Jussieu, Lamarck, and Cavanilles, had only seen cultivated specimens. Mathews gathered it in Peru, in the ravine (*quebrada*) of Chinchao,[2] which appears to be a place beyond the limits of cultivation. Some specimens from Cuchero, collected by Poeppig,[3] are said to be wild; but the traveller himself was not convinced of their wild nature.[4] D'Orbigny thinks he saw the wild coca on a hill in the eastern part of Bolivia.[5] Lastly, M. André has had the courtesy to send me the specimens of *Erythroxylon* in his herbarium, and I recognized the coca in several specimens from the valley of the river Cauca in New Granada, with the note " in abundance, wild or half-wild." Triana, however, does not admit that the species is wild in his country, New Granada.[6] Its extreme importance in Peru at the time of the Incas, compared to the rarity of its use in New Granada, seems to show that it has escaped from cultivation in places where it occurs in the latter country, and that the species is indigenous only in the east of Peru and Bolivia, according to the indications of the travellers mentioned above.

Dyer's Indigo.—*Indigofera tinctoria,* Linnæus.

The Sanskrit name is *nili.*[7] The Latin name, *indicum,* shows that the Romans knew that the indigo was a substance brought from India. As to the wild nature of the plant, Roxburgh says, "Native place unknown, for, though it is now common in a wild state in most of the provinces of India, it is seldom found far from the districts where it is now cultivated, or has been cultivated formerly." Wight and Royle, who have published illustrations of the species, tell us nothing on this head,

---

[1] Particularly in Gosse's *Monographie de l'Erythroxylon Coca,* in 8vo, 1861.

[2] Hooker, *Comp. to the Bot. Mag.,* ii. p. 25.

[3] Peyritsch, in the *Flora Brasil.,* fasc. 81, p. 156.

[4] Hooker, *Comp. to the Bot. Mag.*   [5] Gosse, *Monogr.,* p. 12.

[6] Triana and Planchon, *Ann. Sciences Nat.,* 4th series, vol. 18, p. 338

[7] Roxburgh, *Fl. Ind.,* iii. p. 379.

and more recent Indian floras mention the plant as cultivated.[1] Several other indigoes are wild in India.

This species has been found in the sands of Senegal,[2] but it is not mentioned in other African localities, and as it is often cultivated in Senegal, it seems probable that it is naturalized. The existence of a Sanskrit name renders its Asiatic origin most probable.

Silver Indigo—*Indigofera argentea.*

This species is certainly wild in Abyssinia, Nubia, Kordofan, and Senaar.[3] It is cultivated in Egypt and Arabia. Hence we might suppose that it was from this species that the ancient Egyptians extracted a blue dye;[4] but perhaps they imported their indigo from India, for its cultivation in Egypt is probably not of earlier date than the Middle Ages.[5]

A slightly different form, which Roxburgh gives as a separate species (*Indigofera cœrulea*), and which appears rather to be a variety, is wild in the plains of the peninsula of Hindustan and of Beluchistan.

American Indigoes.

There are probably one or two indigoes indigenous in America, but ill defined, and often intermixed in cultivation with the species of the old world, and naturalized beyond the limits of cultivation. This interchange makes the matter too uncertain for me to venture upon any researches into their original habitat. Some authors have thought that *I. Anil*, Linnæus, was one of these species. Linnæus, however, says that his plant came from India (*Mantissa*, p. 273). The blue dye of the ancient Mexicans was extracted from a plant which, according to Hernandez' account,[6] differs widely from the indigoes.

---

[1] Wight, *Icones*, t. 365 ; Royle, *Ill. Himal.*, t. 195 ; Baker, in *Flora of Brit. Ind.*, ii. p. 98 ; Brandis, *Forest Flora*, p. 136.

[2] Guillemin, Perrottet, and Richard, *Floræ Seneg. Tentamen*, p. 178.

[3] Richard, *Tentamen Fl. Abyss.*, i. 184 ; Oliver, *Fl. of Trop. Afr.*, ii. p. 97 ; Schweinfurth and Ascherson, *Aufzählung*, p. 256.

[4] Unger, *Pflanzen d. Alt. Ægyptens*, p. 66 ; Pickering, *Chronol. Arrang.*, p. 443.

[5] Reynier, *Economie des Juifs*, p. 439 ; *des Egyptiens*, p. 354.

[6] Hernandez, *Thes.*, p. 108.

Henna—*Lawsonia alba*, Lamarck (*Lawsonia inermis* and *L. spinosa* of different authors).

The custom among Eastern women of staining their nails red with the juice of henna-leaves dates from a remote antiquity, as ancient Egyptian paintings and mummies show.

It is difficult to know when and in what country this species was first cultivated to fulfil the requirements of a fashion as absurd as it is persistent, but it may be from a very early epoch, since the inhabitants of Babylon, Nineveh, and the towns of Egypt had gardens. It may be left to scholars to show whether the practice of staining the nails began in Egypt under this or that dynasty, before or after certain relations were established with Eastern nations. It is enough for our purpose to know that *Lawsonia*, a shrub belonging to the order of the Lythraceæ, is more or less wild in the warm regions of Western Asia and of Africa to the north of the equator.

I have in my possession specimens from India, Java, Timor, even from China[1] and Nubia, which are not said to be taken from cultivated plants, and others from Guiana and the West Indies, which are doubtless furnished by the imported species. Stocks found it indigenous in Beluchistan.[2] Roxburgh also considered it to be wild on the Coromandel[3] coast, and Thwaites[4] mentions it in Ceylon in a manner which seems to show that it is wild there. Clarke[5] says, "very common, and cultivated in India, perhaps wild in the eastern part." It is possible that it spread into India from its original home, as into Amboyna[6] in the seventeenth century, and perhaps more recently into the West Indies,[7] in the wake of cultivation; for the plant is valued for the scent of its flowers, as well as for the dye, and is easily propagated by seed.

---

[1] Fortune, No. 32.
[2] Aitchison, *Catal. of Pl. of Punjab and Sindh,* p. 60; Boissier, *Fl. Orient.,* ii. p. 744.
[3] Roxburgh, *Fl. Ind.,* ii. p. 258.
[4] Thwaites, *Enum. Pl. Zeyl.,* p. 122.
[5] Clarke, in Hooker's *Fl. Brit. Ind.,* ii. p. 273.
[6] Rumphius, *Amb.,* iv. p. 42.
[7] Grisebach, *Fl. Brit. W. Ind.,* i. p. 271.

There is the same doubt as to whether it is indigenous in Persia, Arabia, and Egypt (an essentially cultivated country), in Nubia, and even in Guinea, where specimens have been gathered.[1] It is even possible that the area of this shrub extends from India to Nubia. Such a wide geographical distribution is, however, always somewhat rare. The common names may furnish some indication.

A Sanskrit name, *sakachera*,[2] is attributed to the species, but as it has left no trace in the different modern languages of India, I am inclined to doubt its reality. The Persian name *hanna* is more widely diffused and retained than any other (*hina* of the Hindus, *henneh* and *alhenna* of the Arabs, *kinna* of the modern Greeks). That of *cypros*, used by the Syrians of the time of Dioscorides,[3] has not found so much favour. This fact supports the opinion that the species grew originally on the borders of Persia, and that its use as well as its cultivation spread from the East to the West, from Asia into Africa.

Tobacco—*Nicotiana Tabacum*, Linnæus ; and other species of *Nicotiana*.

At the time of the discovery of America, the custom of smoking, of snuff-taking, or of chewing tobacco was diffused over the greater part of this vast continent. The accounts of the earliest travellers, of which the famous anatomist Tiedemann [4] has made a very complete collection, show that the inhabitants of South America did not smoke, but chewed tobacco or took snuff, except in the district of La Plata, Uruguay, and Paraguay, where no form of tobacco was used. In North America, from the Isthmus of Panama and the West Indies as far as Canada and California, the custom of smoking was universal, and circumstances show that it was also very ancient. Pipes, in great numbers and of wonderful work-manship, have been discovered in the tombs of the Aztecs

[1] Oliver, *Fl. of Trop. Afr.*, ii. p. 483.
[2] Piddington, *Index*.
[3] Dioscorides, 1, c. 124 ; Lenz, *Bot. d. Alten*, p. 177.
[4] Tiedemann, *Geschichte des Tabaks*, in 8vo, 1854. For Brazil, see Martius, *Beitrage zur Ethnographie und Sprachkunde Amerikas*, i. p. 719.

in Mexico [1] and in the mounds of the United States; some of them represent animals foreign to North America.[2]

As the tobacco plant is an annual which gives a great quantity of seeds, it was easy to sow and to cultivate or naturalize them more or less in the neighbourhood of dwellings, but it must be noted that different species of the genus Nicotiana were employed in different parts of America, which shows that they had not all the same origin. *Nicotiana Tabacum,* commonly cultivated, was the most widely diffused, and sometimes the only one in use in South America and the West Indies. The use of tobacco was introduced into La Plata, Paraguay,[3] and Uruguay by the Spaniards, consequently we must look further to the north for the origin of the plant. De Martius does not think it was indigenous in Brazil,[4] and he adds that the ancient Brazilians smoked the leaves of a species belonging to their country known to botanists as *Nicotiana Langsdorfii.* When I went into the question in 1855,[5] I had not been able to discover any wild specimens of *Nicotiana Tabacum* except those sent by Blanchet from the province of Bahia, numbered 3223, *a.* No author, either before or since that time, has been more fortunate, and I see that Messrs. Flückiger and Hanbury, in their excellent work on vegetable drugs,[6] say positively, "The common tobacco is a native of the new world, though not now known in a wild state." I venture to gainsay this assertion, although the wild nature of a plant may always be disputed in the case of a plant which spreads so easily from cultivation.

We find in herbaria a number of specimens gathered in Peru without indication that they were cultivated or that they grew near plantations. Boissier's herbarium contains

---

[1] Tiedemann, p. 17, pl. 1.

[2] The drawings on these pipes are reproduced in Naidaillac's recent work, *Les Premiers Hommes et les Temps Préhistoriques,* vol. ii. pp. 45, 48.

[3] Tiedemann, pp. 38, 39.

[4] Martius, *Syst. Mat. Med. Bras.,* p. 120; *Fl. Bras.,* vol. x. p. 191.

[5] A. de Candolle, *Géogr. Bot. Raisonnée,* p. 849.

[6] Flückiger and Hanbury, *Pharmacographia,* p. 418.

two specimens collected by Pavon, from different localities.[1] Pavon says in his flora that the species grows in
the moist warm forests of the Peruvian Andes, and that it
is cultivated. But—and this is more significant—Edouard
André gathered specimens in the republic of Ecquador
at Saint Nicholas, on the western slope of the volcano of
Corazon in a virgin forest. These he was kind enough
to send me. They are evidently the tall variety (four to
six feet) of *N. Tabacum*, with the upper leaves narrow
and acuminate, as they are represented in the plates of
Hayne and Miller.[2] The lower leaves are wanting. The
flower, which gives the true characters of the species, is
certainly that of *N. Tabacum*, and it is well known that
the height of this plant and the breadth of the leaves
vary in cultivation.[3] It is very possible that its original
country extended north as far as Mexico, as far south as
Bolivia, and eastward to Venezuela.

Nicotiana rustica, Linnæus, a species with yellow
flowers, very different from *Tabacum*,[4] and which yields
a coarse kind of tobacco, was more often cultivated by
the Mexicans and the native tribes north of Mexico. I
have a specimen brought from California by Douglas in
1837, a time when colonists were still few; but American
authorities do not admit that the plant is wild, and Dr.
Asa Gray says that it sows itself in waste places.[5] This
was perhaps the case with the specimens in Boissier's
herbarium, gathered in Peru by Pavon, and which he
does not mention in the Peruvian flora. The species
grows in abundance about Cordova in the Argentine
Republic,[6] but from what epoch is unknown. From the

---

[1] One of these is classed under the name *Nicot. fruticosa*, which in
my opinion is the same species, tall, but not woody, as the name would
lead one to believe. *N. auriculata*, Bertero, is also *Tabacum*, according
to my authentic specimens.

[2] Hayne, *Arzneikunde Gewachse*, vol. xii. t. 41; Miller, *Figures of
Plants*, pl. 185, f. 1.

[3] The capsule is sometimes shorter and sometimes longer than the
calix, on the same plant, in André's specimens.

[4] See the figures of *N. rustica* in Plée, *Types de Familles Naturelles
de France, Solanées*; Bulliard, *Herbier de France*, t. 289.

[5] Asa Gray, *Syn. Flora of North Amer.* (1878), p. 241.

[6] Martin de Moussy, *Descr. de la Repub. Argent.*, i. p. 196.

ancient use of the plant and the home of the most analogous species, the probabilities are in favour of a Mexican, Texan, or Californian origin.

Several botanists, even Americans, have believed that the species came from the old world. This is certainly a mistake, although the plant has spread here and there even into our forests, and sometimes in abundance,[1] having escaped from cultivation. Authors of the sixteenth century spoke of it as a foreign plant introduced into gardens and sometimes spreading from them.[2] It occurs in some herbaria under the names of *N. tartarica, turcica,* or *sibirica ;* but these are garden-grown specimens, and no botanist has found the species in Asia, or on the borders of Asia, with any appearance of wildness.

This leads me to refute a widespread and more persistent error, in spite of what I proved in 1855, namely, that of regarding some species ill described from cultivated specimens as natives of the old world, of Asia in particular. The proofs of an American origin are so numerous and consistent that, without entering much into detail, I may sum them up as follows :—

A. Out of fifty species of the genus Nicotiana found in a wild state, two only are foreign to America ; namely, *N. suavolens* of New Holland, with which is joined *N. rotundifolia* of the same country, and that which Ventinat had wrongly styled *N. undulata ;* and *N. fragans,* Hooker, of the Isle of Pines, near New Caledonia, which differs very little from the preceding.

B. Though the Asiatic people are great lovers of tobacco, and have from a very early epoch sought the smoke of certain narcotic plants, none of them made use of tobacco before the discovery of America. Tiedemann has distinctly proved this fact by thorough researches into the writings of travellers in the Middle Ages.[3] He even quotes for a later epoch, not long after the discovery of America, between 1540 and 1603, the fact that

[1] Bulliard, *Herbier de France.*

[2] Cæsalpinus, lib. viii. cap. 44; Bauhin, *Hist.,* iii. p. 630.

[3] Tiedemann, *Geschichte des Tabaks* (1854), p. 208. Two years earlier, Volz, *Beitrage zur Culturgeschichte,* had collected a number of facts relative to the introduction of tobacco into different countries.

several travellers, some of whom were botanists, such as
Belon and Rauwolf, who travelled through the Turkish
and Persian empires, observing their customs with much
attention, have not once mentioned tobacco. It was
evidently introduced into Turkey at the beginning of the
seventeenth century, and the Persians soon received it
from the Turks. The first European who mentions the
smoking of tobacco in Persia is Thomas Herbert, in 1626.
No later travellers have omitted to notice the use of the
hookah as well established. Olearius describes this ap-
paratus, which he saw in 1633. The first mention of
tobacco in India is in 1605,[1] and it is probable that it
was of European introduction. It was first introduced
at Arracan and Pegu, in 1619, according to the traveller
Methold.[2] There are doubts about Java, because Rumphius, a very accurate observer, who wrote in the second
half of the seventeenth century, says [3] that, according
to the tradition of some old people, tobacco had been
employed as a medicine before the arrival of the Portuguese in 1496, and that only the practice of smoking it
had been communicated by the Europeans. Rumphius
adds, it is true, that the name *tabaco* or *tambuco*, which
is in use in all these places, is of foreign origin. Sir
Stamford Raffles,[4] in his numerous historical researches
on Java, gives, on the other hand, the year 1601 as the
date of the introduction of tobacco into Java. The
Portuguese had certainly discovered the coasts of Brazil
between 1500 and 1504, but Vasco di Gama and his
successors went to Asia round the Cape, or through the
Red Sea, so that they could hardly have established
frequent or direct communications between America and
Java. Nicot had seen the plant in Portugal in 1560, so
that the Portuguese probably introduced it into Asia
in the latter half of the sixteenth century. Thunberg
affirms [5] that the use of tobacco was introduced into

---

[1] According to an anonymous Indian author quoted by Tiedemann,
p. 229.
[2] Tiedemann, p. 234.   [3] Rumphius, *Herb. Amboin* v. p. 225.
[4] Raffles, *Descr. of Java*, p. 85.
[5] Thunberg, *Flora Japonica*, p. 91.

Japan by the Portuguese, and according to early travellers quoted by Tiedemann, this was at the beginning of the seventeenth century. Lastly, the Chinese have no original and ancient sign for tobacco ; their paintings on china in the Dresden collection often present, from the year 1700 and never before that date, details relating to tobacco,[1] and Chinese students are agreed that Chinese works do not mention the plant before the end of the sixteenth century.[2] If it be remembered with what rapidity the use of tobacco has spread wherever it has been introduced, these data about Asia have an incontestable force.

C. The common names of tobacco confirm its American origin. If there had been any indigenous species in the old world there would be a great number of different names; but, on the contrary, the Chinese, Japanese, Javanese, Indian, Persian, etc., names are derived from the American names, *petum,* or *tabak, tabok, tamboc,* slightly modified. It is true that Piddington gives Sanskrit names, *dhumrapatra* and *tamrakouta,*[3] but Adolphe Pictet informs me that the first of these names, which is not in Wilson's dictionary, means only leaf for smoking, and appears to be of modern composition; while the second is probably no older, and seems to be a modern modification of the American names. The Arabic word *docchan* simply means smoke.[4]

Lastly, we must inquire into the two so-called Asiatic *Nicotianæ.* The one, called by Lehmann *Nicotiana chinensis,* came from the Russian botanist Fischer, who said it was Chinese. Lehmann said he had seen it in a garden. Now, it is well known how often an erroneous origin is attributed to plants grown by horticulturists; and besides, from the description, it seems that it was simply *N. Tabacum,* of which the seeds had perhaps come from China.[5] The second species is *N. persica,*

[1] Klemm, quoted by Tiedemann, p. 256.
[2] Stanislas Julien, in de Candolle, *Géogr. Bot. Rais.,* p. 851; Bretschneider, *Study and Value,* etc., p. 17.
[3] Piddington, *Index.*   [4] Forskal, p. 63.
[5] Lehmann, *Historia Nicotinarum,* p. 18. The epithet *suffruticosa* is an exaggeration applied to the tobaccos, which are always annual. I have said already that *N. suffruticosa* of different authors is *N. Tabacum.*

Lindley, figured in the *Botanical Register* (pl. 1592), of which the seeds had been sent from Ispahan to the Horticultural Society of London, as those of the best tobacco cultivated in Persia, that of Schiraz. Lindley did not observe that it corresponded exactly to *N. alata*, drawn three years before by Link and Otto[1] from a plant in the gardens at Berlin. The latter was grown from seed sent by Sello from Southern Brazil. It is certainly a Brazilian species, with a white elongated corolla, allied to *N. suaveolens* of New Holland. Thus the tobacco cultivated sometimes in Persia along with the common species, is of American origin, as I declared in my *Geographical Botany* of 1855. I do not understand how this species was introduced into Persia. It must have been from seed taken from a garden, or brought by chance from America, and it is not likely that its cultivation is common in Persia, for Olivier and Bruguière, and other naturalists who have observed the tobacco plantations in that country, make no mention of it.

From all these reasons I conclude that no species of tobacco is a native of Asia. They are all American, except *N. suaveolens* of New Holland, and *N. fragrans* of the Isle of Pines to the south of New Caledonia.

Several *Nicotianæ*, besides *N. Tabacum* and *N. rustica*, have been cultivated here and there by savages, or as a curiosity by Europeans. It is strange that so little notice is taken of these attempts, by means of which very choice tobacco might be obtained. The species with white flowers would yield probably a light and perfumed tobacco, and as some smokers seek the strongest tobaccos and the most disagreeable to non-smokers, I would recommend to their notice *N. angustifolia* of Chili, which the natives call *tabaco del diablo*.[2]

---

[1] Link and Otto, *Icones Plant. Rar. Hort. Ber.*, in 4to, p. 63, t. 32. Sendtner, in *Flora Brasil*, vol. x. p. 167, describes the same plant as Sello, as it seems from the specimens collected by this traveller; and Grisebach, *Symbolæ Fl. Argent.*, p. 243, mentions *N. alata* in the province of *Entrerios* of the Argentine republic.

[2] Bertero, in De Cand., *Prodr.*, xii., sect. 1, p. 568.

Cinnamon—*Cinnamomum zeylanicum,* Breyn.

This little tree, belonging to the laurel tribe, of which the bark of the young branches forms the cinnamon of commerce, grows in great quantities in the forests of Ceylon. Certain varieties which grow wild on the continent of India were formerly considered to be so many distinct species, but Anglo-Indian botanists are agreed in connecting them with that of Ceylon.[1]

The bark of *C. zeylanicum,* and that of several uncultivated species of *Cinnamomum,* which produce the *cassia,* or *Chinese cassia,* have been an important article of commerce from a very early period. Flückiger and Hanbury[2] have treated of this historical question with so much learning and thoroughness, that we need only refer to their work, entitled *Pharmacographia, or History of the Principal Drugs of Vegetable Origin.* It is important from our point of view to note how modern the culture is of the cinnamon tree in comparison with the trade in its product. It was only between 1765 and 1770 that a Ceylon colonist, named de Koke, aided by Falck, the governor of the island, made some plantations which were wonderfully successful. They have diminished in Ceylon in the last few years, but others have been established in the tropical regions of the old and new worlds. The species becomes easily naturalized beyond the limits of cultivation,[3] as birds are fond of the fruit, and drop the seeds in the forests.

China Grass—*Boehmeria nivea,* Hooker and Arnott.

The cultivation of this valuable *Urticacea* has been introduced into the south of France and of the United States for about thirty years, but commerce had previously acquainted us with the great value of its fibres, more tenacious than hemp and in some cases flexible as silk. Interesting details on the manner of cultivating

---

[1] Thwaites, *Enum. Pl. Zelaniœ,* p. 252 ; Brandis, *Forest Flora of India,* p. 375.

[2] Flückiger and Hanbury, *Pharmacographia,* p. 467 ; Porter, *The Tropical Agriculturist,* p. 268.

[3] Brandis, *Forest Flora ;* Grisebach, *Flora of Brit. W. India Is.,* p. 179.

the plant and of extracting its fibres [1] may be found in several books; I shall confine myself here to defining as clearly as I can its geographical origin.

To attain this end we must not trust to the vague expressions of most authors, nor to the labels attached to the specimens in herbaria, since frequently no distinction has been made between cultivated, naturalized, or truly wild plants, and the two varieties of *Boehmeria nivea* (*Urtica nivea*, Linnæus), and *Boehmeria tenacissima*, Gaudichaud, or *B. candicans*, Hasskarl, have been confounded together; forms which appear to be varieties of the same species, because transitions between them have been observed by botanists. There is also a subvariety, with leaves green on both sides, cultivated by Americans and by M. de Malartic in the south of France.

The variety earliest known (*Urtica nivea*, L.), with leaves white on the under side, is said to grow in China and some neighbouring countries. Linnæus says it is found on walls in China, which would imply a plant naturalized on rubbish-heaps from cultivation. But Loureiro [2] says, "*habitat et abundanter colitur in Cochin-China et China*," and according to Bentham,[3] the collector Champion found it in abundance in the ravines of the island of Hongkong. According to Franchet and Savatier,[4] it exists in Japan in clearings and hedges (*in fruticetis umbrosis et sepibus*). Blanco [5] says it is common in the Philippine Isles. I find no proof that it is wild in Java, Sumatra, and other islands of the Malay Archipelago. Rumphius [6] knew it only as a cultivated plant. Roxburgh [7] believed it to be a native of Sumatra, but Miquel [8] does not confirm this belief. The other varieties

[1] De Malartic, *Journ. d'Agric. Pratique*, 1871, 1872, vol. ii. No. 31; de la Roque, *ibid.*, No. 29, *Bull. Soc. d'Acclim.*, 1872, p. 463; Vilmorin, *Bon Jardinier*, 1880, pt. 1, p. 700; Vetillart, *Études sur les Fibres Végétales Textiles*, p. 99, pl. 2.

[2] Loureiro, *Fl. Cochin.*, ii. p. 683.

[3] Bentham, *Fl. Hongkong*, p. 331.

[4] Franchet and Savatier, *Enum. Plant. Jap.*, i. p. 439.

[5] Blanco, *Flora de Filip.*, edit. 2, p. 484.

[6] Rumphius, *Amboin*, v. p. 214.

[7] Roxburgh, *Fl. Ind.*, iii. p. 590.

[8] Miquel, *Sumatra*, Germ. edit., p. 170.

have nowhere been found wild, which supports the theory that they are only the result of cultivation.

Hemp—*Cannabis sativa*, Linnæus.

Hemp is mentioned, in its two forms, male and female, in the most ancient Chinese works, particularly in the *Shu-King*, written 500 B.C.[1]

It has Sanskrit names, *bhanga* and *gangika*.[2] The root of these words, *ang* or *an*, recurs in all the Indo-European and modern Semitic languages : *bang* in Hindu and Persian, *ganga* in Bengali,[3] *hanf* in German, *hemp* in English, *chanvre* in French, *kanas* in Keltic and modern Breton,[4] *cannabis* in Greek and Latin, *cannab* in Arabic.[5]

According to Herodotus (born 484 B.C.), the Scythians used hemp, but in his time the Greeks were scarcely acquainted with it.[6]   Hiero II., King of Syracuse, bought the hemp used for the cordage of his vessels in Gaul, and Lucilius is the earliest Roman writer who speaks of the plant (100 B.C.).   Hebrew books do not mention hemp.[7] It was not used in the fabrics which enveloped the mummies of ancient Egypt.   Even at the end of the eighteenth century it was only cultivated in Egypt for the sake of an intoxicating liquid extracted from the plant.[8] The compilation of Jewish laws known as the Talmud, made under the Roman dominion, speaks of its textile properties as of a little-known fact.[9]   It seems probable that the Scythians transported this plant from Central Asia and from Russia when they migrated westward about 1500 B.C., a little before the Trojan war.   It may also have been introduced by the earlier incursions of the Aryans into Thrace and Western Europe ; yet in that case it would have been earlier known in Italy.   Hemp has

[1] Bretschneider, *On the Study and Value*, etc., pp. 5, 10, 48.

[2] Piddington, *Index ;* Roxburgh, *Fl. Ind.*, edit. 2, vol. iii. p. 772.

[3] Roxburgh, *ibid.*

[4] Reynier, *Économie des Celtes*, p. 448; Legonidec, *Dict. Bas-Breton.*

[5] J. Humbert, formerly professor of Arabic at Geneva, says the name is *kannab, kon-nab, hon-nab, hen-nab, kanedir*, according to the locality.

[6] Athenæus, quoted by Hehn, *Culturpflanzen*, p. 168.

[7] Rosenmüller, *Hand. Bibl. Alterth.*

[8] Forskal, *Flora ;* Delile, *Flore d'Egypte.*

[9] Reynier, *Économie des Arabes*, p. 434.

not been found in the lake-dwellings of Switzerland [1] and Northern Italy.[2]

The observations on the habitat of *Cannabis sativa* agree perfectly with the data furnished by history and philology. I have treated specially of this subject in a monograph in *Prodromus*, 1869.[3]

The species has been found wild, beyond a doubt, to the south of the Caspian Sea,[4] in Siberia, near the Irtysch, in the desert of the Kirghiz, beyond Lake Baikal, in Dahuria (government of Irkutsh). Authors mention it also throughout Southern and Central Russia, and to the south of the Caucasus,[5] but its wild nature is here less certain, seeing that these are populous countries, and that the seeds of the hemp are easily diffused from gardens. The antiquity of the cultivation of hemp in China leads me to believe that its area extends further to the east, although this has not yet been proved by botanists.[6] Boissier mentions the species as "almost wild in Persia." I doubt whether it is indigenous there, since in that case the Greeks and Hebrews would have known of it at an earlier period.

White Mulberry—*Morus alba*, Linnæus.

The mulberry tree, which is most commonly used in Europe for rearing silkworms, is *Morus alba*. Its very numerous varieties have been carefully described by Seringe,[7] and more recently by Bureau.[8] That most widely cultivated in India, *Morus indica*, Linnæus (*Morus alba*, var. *Indica*, Bureau), is wild in the Punjab and in Sikkim, according to Brandis, inspector-general of forests in British India.[9] Two other varieties, *serrata* and *cuspidata*, are also said to be wild in different pro-

---

[1] Heer, *Ueber d. Flachs*, p. 25.

[2] Sordelli, *Notizie sull. Staz. di Lagozza*, 1880.

[3] Vol. xvi. sect. 1, p. 30.

[4] De Bunge, *Bull. Soc. Bot. de Fr.*, 1860, p. 30.

[5] Ledebour, *Flora Rossica*, iii. p. 634.

[6] Bunge found hemp in the north of China, but among rubbish (*Enum.* No. 338).

[7] Seringe, *Description et Culture des Mûriers*.

[8] Bureau, in De Candolle, *Prodromus*, xvii. p. 238.

[9] Brandis, *Forest Flora of North-West and Central India*, 1874, p. 408. This variety has black fruit, like that of *Morus nigra*.

vinces of Northern India.[1]   The Abbé David found a
perfectly wild variety in Mongolia, described under the
name of *mongolica* by Bureau; and Dr. Bretschneider[2]
quotes a name *yen*, from ancient Chinese authors, for the
wild mulberry.

It is true he does not say whether this name applies
to the white mulberry, *pe-sang*, of the Chinese plantations.[3]   The antiquity of its culture in China,[4] and in
Japan, and the number of different varieties grown there,
lead us to believe that its original area extended eastward as far as Japan; but the indigenous flora of Southern
China is little known, and the most trustworthy authors
do not affirm that the plant is indigenous in Japan.
Franchet and Savatier[5] say that it is "cultivated from
time immemorial, and become wild here and there." It
is worthy of note also that the white mulberry appears
to thrive especially in mountainous and temperate countries, whence it may be argued that it was formerly
introduced from the north of China into the plains of
the south.   It is known that birds are fond of the fruit,
and bear the seeds to great distances and into uncultivated ground, and this makes it difficult to discover its
really original habitat.

This facility of naturalization doubtless explains the
presence in successive epochs of the white mulberry in
Western Asia and the south of Europe.   This must have
occurred especially after the monks brought the silkworm to Constantinople under Justinian in the sixth
century, and as the culture of silkworms was gradually
propagated westwards.   However, Targioni has proved
that only the black mulberry, *M. nigra*, was known in
Sicily and Italy when the manufacture of silk was introduced into Sicily in 1148, and two centuries later into

---

[1] Bureau, *ibid.*, from the specimens of several travellers.

[2] Bretschneider, *Study and Value*, etc., p. 12.

[3] This name occurs in the *Pent-sao*, according to Ritter, *Erdkunde*,
xvii. p. 489.

[4] Platt says (*Zeitschrift d. Gesellsch. Erdkunde*, 1871, p. 162) that
its cultivation dates from 4000 years B.C.

[5] Franchet and Savatier, *Enum. Plant. Jap.*, i. p. 433.

Tuscany.[1] According to the same author, the introduction of the white mulberry into Tuscany dates at the earliest from the year 1340. In like manner the manufacture of silk may have begun in China, because the silkworm is natural to that country; but it is very probable that the tree grew also in the north of India, where so many travellers have found it wild. In Persia, Armenia, and Asia Minor, I am inclined to believe that it was naturalized at a very early epoch, rather than to share Grisebach's opinion that it is indigenous in the basin of the Caspian Sea. Boissier does not give it as wild in that region.[2] Buhse[3] found it in Persia, near Erivan and Bashnaruschin, and he adds, "naturalized in abundance in Ghilan and Masenderan." Ledebour,[4] in his Russian flora, mentions numerous localities round the Caucasus, but he does not specify whether the species is wild or naturalized. In the Crimea, Greece, and Italy, it exists only in a cultivated state.[5] A variety, *tatarica*, often cultivated in the south of Russia, has become naturalized near the Volga.[6]

If the white mulberry did not originally exist in Persia and in the neighbourhood of the Caspian Sea, it must have penetrated there a long while ago. I may quote in proof of this the name *tut, tutti, tuta*, which is Persian, Arabic, Turkish, and Tartar. There is a Sanskrit name, *tula*,[7] which must be connected with the same root as the Persian name; but no Hebrew name is known, which is a confirmation of the theory of a successive extension towards the west of Asia.

I refer those of my readers who may desire more detailed information about the introduction of the mulberry and of silkworms to the able works of Targioni and

[1] Ant. Targioni, *Cenni Storici sull' Introduzione di Varie Piante nell' Agricoltura Toscana*, p. 188.

[2] Boissier, *Fl. Orient.*, iv. p. 1153.

[3] Buhse, *Aufzählung der Transcaucasien und Persien Pflanzen*, p. 203.

[4] Ledebour, *Fl. Ross.*, iii. p. 643.

[5] Steven, *Verseichniss d. Taurisch. Halbins*, p. 313; Heldreich, *Pflanzen des Attischen Ebene*, p. 508; Bertoloni, *Fl. Ital.*, x. p. 177; Caruel, *Fl. Toscana*, p. 171.

[6] Bureau, de Cand., *Prodr.*, xvii. p. 238.

[7] Roxburgh, *Fl. Ind.*; Piddington, *Index*.

Ritter, to which I have already referred. Recent discoveries made by various botanists have permitted me to add more precise data than those of Ritter on the question of origin, and if there are some apparent contradictions in our opinions on other points, it is because the famous geographer has considered a number of varieties as so many different species, whereas botanists, after a careful examination, have classed them together.

Black Mulberry—*Morus nigra*, Linnæus.

This tree is more valued for its fruit than for its leaves, and on that account I should have included it in the list of fruit trees; but its history can hardly be separated from that of the white mulberry. Moreover, its leaves are employed in many countries for the feeding of silkworms, although the silk produced is of inferior quality.

The black mulberry is distinguished from the white by several characters independently of the black colour of the fruit, which occurs also in a few varieties of the *M. alba*.[1] It has not a great number of varieties like the latter, which argues a less ancient and a less general cultivation and a narrower primitive area.

Greek and Latin authors, even the poets, have mentioned *Morus nigra*, which they compare to *Ficus sycomorus*, and which they even confounded originally with this Egyptian tree.

Commentators for the last two centuries have quoted a number of passages which leave no doubt on this head, but which are devoid of interest in themselves.[2] They furnish no proof touching the origin of the species, which is presumably Persian, unless we are to take seriously the fable of Pyramus and Thisbe, of which the scene was in Babylonia, according to Ovid.

Botanists have not yet furnished any certain proof that this species is indigenous in Persia. Boissier, who is the most learned in the floras of the East, contents

[1] Reichenbach gives good figures of both species in his *Icones Fl. Germ.*, 657, 658.

[2] Fraas, *Syn. Fl. Class.*, p. 236; Lenz, *Bot. der Alten Gr. und Röm.*, p. 419; Ritter, *Erdkunde*, xvii. p. 482; Hehn, *Culturpflanzen*, edit. 3, p. 336.

himself with quoting Hohenacker as the discoverer of *M. nigra* in the forests of Lenkoran, on the south coast of the Caspian Sea, and he adds, "probably wild in the north of Persia near the Caspian Sea." [1] Ledebour, in his Russian flora, had previously indicated, on the authority of different travellers, the Crimea and the provinces south of the Caucasus; [2] but Steven denies the existence of the species in the Crimea except in a cultivated state. [3] Tchihatcheff and Koch found the black mulberry in high wild districts of Armenia. It is very probable that in the region to the south of the Caucasus and of the Caspian Sea *Morus nigra* is wild and indigenous rather than naturalized. What leads me to this belief is (1) that it is not known, even in a cultivated state, in India, China, or Japan; (2) that it has no Sanskrit name; (3) that it was so early introduced into Greece, a country which had intercourse with Armenia at an early period. [4]

Morus nigra spread so little to the south of Persia, that no certain Hebrew name is known for it, nor even a Persian name distinct from that of *Morus alba*. It was widely cultivated in Italy until the superiority of the white mulberry for the rearing of silkworms was recognized. In Greece the black mulberry is still the most cultivated. [5] It has become naturalized here and there in these countries and in Spain. [6]

American Aloe—*Agave Americana*, Linnæus.

This ligneous plant, of the order of *Amaryllidaceæ*, has been cultivated from time immemorial in Mexico under the names *maguey* or *metl*, in order to extract from it, at the moment when the flower stem is developed, the wine known as *pulque*. Humboldt has given a full description of this culture, [7] and he tells us elsewhere [8] that the

[1] Boissier, *Fl. Orient.*, iv. p. 1153 (published 1879).

[2] Ledebour, *Fl. Ross.*, iii. p. 641.

[3] Steven, *Verseichniss d. Taur. Halb. Pflan.*, p. 313.

[4] Tchihatcheff, trans. of Grisebach's *Végétation du Globe*, i. 424.

[5] Heldreich, *Nutzpflanzen Griechenlands*, p. 19.

[6] Bertoloni, *Flora Ital.*, x. p. 179; Viviani, *Fl. Dalmat.*, i. p. 220; Willkomm and Lange, *Prodr. Fl. Hisp.*, i. p. 250.

[7] Humboldt, *Nouvelle Espagne*, ed. 2, p. 487.

[8] Humboldt, in Kunth, *Nova Genera*, i. p. 297.

species grows in the whole of South America as far as
five thousand feet of altitude. It is mentioned[1] in
Jamaica, Antigua, Dominica, and Cuba, but it must
be observed that it multiplies easily by suckers, and
that it is often planted far from dwellings to form
fences or to extract from it the fibre known as *pite*, and
this makes it difficult to ascertain its original habitat.
Transported long since into the countries which border
the Mediterranean, it occurs there with every appearance
of an indigenous species, although there is no doubt as
to its origin.[2] Probably, to judge from the various uses
made of it in Mexico before the arrival of the Europeans, it came originally from thence.

Sugar-Cane—*Saccharum officinarum*, Linnæus.

The origin of the sugar-cane, of its cultivation, and
of the manufacture of sugar, are the subject of a very
remarkable work by the geographer, Karl Ritter.[3] I need
not follow his purely agricultural and economical details;
but for that which interests us particularly, the primitive
habitat of the species, he is the best guide, and the facts
observed during the last forty years for the most part
support or confirm his opinions.

The sugar-cane is cultivated at the present day in all
the warm regions of the globe, but a number of historical
facts testify that it was first grown in Southern Asia,
whence it spread into Africa, and later into America.
The question is, therefore, to discover in what districts
of the continent, or in which of the southern islands of
Asia, the plant exists, or existed at the time it was first
employed.

Ritter has followed the best methods of arriving at a
solution. He notes first that all the species known in a

---

[1] Grisebach, *Fl. of Brit. W. Ind. Is.*, p. 582.

[2] Alph. de Candolle, *Géogr. Bot. Raisonnée*, p. 739; H. Hoffmann, in
Regel's *Gartenflora*, 1875, p. 70.

[3] K. Ritter, *Ueber die Geographische Verbreitung des Zuckerrohrs*,
in 4to, 108 pages (according to Pritzel, *Thes. Lit. Bot.*); *Die Cultur
des Zuckerrohrs, Saccharum, in Asien, Geogr. Verbreitung*, etc., etc., in
8vo, 64 pages, without date. This monograph is full of learning and
judgment, worthy of the best epoch of German science, when English
or French authors were quoted by all authors with as much care as
Germans.

wild state, and undoubtedly belonging to the genus *Saccharum*, grow in India, except one in Egypt.[1] Five species have since been described, growing in Java, New Guinea, Timor, and the Philippine Isles.[2] The probabilities are all in favour of an Asiatic origin, to judge from the data furnished by geographical botany.

Unfortunately no botanist had discovered at the time when Ritter wrote, or has since discovered, *Saccharum officinarum* wild in India, in the adjacent countries or in the archipelago to the south of Asia. All Anglo-Indian authors, Roxburgh, Wallich, Royle, etc., and more recently Aitchison,[3] only mention the plant as a cultivated one. Roxburgh, who was so long a collector in India, says expressly, "where wild I do not know." The family of the *Gramineæ* has not yet appeared in Sir Joseph Hooker's flora. For the island of Ceylon, Thwaites does not even mention the cultivated plant.[4] Rumphius, who has carefully described its cultivation in the Dutch colonies, says nothing about the home of the species. Miquel, Hasskarl, and Blanco mention no wild specimen in Sumatra, Java, or the Philippine Isles. Crawfurd tried to discover it, but failed to do so.[5] At the time of Cook's voyage Forster found the sugar-cane only as a cultivated plant in the small islands of the Pacific.[6] The natives of New Caledonia cultivate a number of varieties of the sugar-cane, and use it constantly, sucking the syrup from the cane; but Vieillard[7] takes care to say, "From the fact that isolated plants of *Saccharum officinarum* are often found in the middle of the bush and even on the mountains, it would be wrong to conclude that the plant is indigenous; for these specimens, poor and weak, only mark the site of old plantations, or

---

[1] Kunth, *Enum. Plant.* (1838), vol. i. p. 474. There is no more recent descriptive work on the family of the *Gramineæ*, nor the genus *Saccharum.*

[2] Miquel, *Floræ Indiæ Batavæ,* 1855, vol. iii. p. 511.

[3] Aitchison, *Catalogue of Punjab and Sindh Plants,* 1869, p. 173.

[4] Thwaites, *Enum. Pl. Zeylonicæ.*

[5] Crawfurd, *Indian Archip.,* i. p. 475.

[6] Forster, *De Plantis Esculentis.*

[7] Vieillard, *Annales des Sc. Nat.,* 4th series, vol. xvi. p. 32.

are sprung from fragments of cane left by the natives, who seldom travel without a piece of cane in the hand." In 1861, Bentham, who had access to the rich herbarium of Kew, says, in his *Flora of Hongkong*, "We have no authentic and certain proof of a locality where the common sugar-cane is wild."

I do not know, however, why Ritter and every one else has neglected an assertion of Loureiro, in his *Flora of Cochin-China*,[1] " Habitat, et colitur abundantissime in omnibus provinciis regni Cochin-Chinensis: simul in aliquibus imperii sinensis, sed minori copia." The word *habitat*, separated by a comma from the rest, is a distinct assertion. Loureiro could not have been mistaken about the *Saccharum officinarum*, which he saw cultivated all about him, and of which he enumerates the principal varieties. He must have seen plants wild, at least in appearance. They may have spread from some neighbouring plantation, but I know nothing which makes it unlikely that the plant should be indigenous in this warm moist district of the continent of Asia.

Forskal[2] mentions the species as wild in the mountains of Arabia, under a name which he believes to be Indian. If it came from Arabia, it would have spread into Egypt long ago, and the Hebrews would have known it.

Roxburgh had received in the botanical gardens of Calcutta in 1796, and had introduced into the plantations in Bengal, a *Saccharum* to which he gave the name of *S. sinense*, and of which he published an illustration in his great work *Plantœ Coromandelianœ*, vol. iii. pl. 232. It is perhaps only a form of *S. officinarum*, and moreover, as it is only known in a cultivated state, it tells nothing about the primitive country either of this or of any other variety.

A few botanists have asserted that the sugar-cane flowers more often in Asia than in America or Africa, and even that it produces seed[3] on the banks of the

[1] Loureiro, *Cochin-Ch.*, edit. 2, vol. i. p. 66.
[2] Forskal, *Fl. Ægypto-Arabica*, p. 103.
[3] Macfadyen, *On the Botanical Characters of the Sugar-Cane*, in Hooker's *Bot. Miscell.*, i. p. 101 ; Maycock, *Fl. Barbad.*, p. 50.

Ganges, which they regard as a proof that it is indigenous. Macfadyen says so without giving any proof. It was an assertion made to him in Jamaica by some traveller; but Sir W. Hooker adds in a note, " Dr. Roxburgh, in spite of his long residence on the banks of the Ganges, has never seen the seeds of the sugar-cane." It rarely flowers, and still more rarely bears fruit, as is commonly the case with plants propagated by buds or suckers, and if any variety of sugar-cane were disposed to seed, it would probably be less productive of sugar and would soon be abandoned. Rumphius, a better observer than many modern botanists, has given a good description of the cultivated cane in the Dutch colonies, and makes an interesting remark.[1] " It never produces flowers or fruit unless it has remained several years in a stony place." Neither he, nor any one else to my knowledge, has described or drawn the seed. The flower, on the contrary, has often been figured, and I have a fine specimen from Martinique.[2] Schacht is the only person who has given a good analysis of the flower, including the pistil; he had not seen the seed ripe.[3] De Tussac,[4] who gives a poor analysis, speaks of the seed, but he only saw it young in the ovary.

In default of precise information as to the native country of the species, accessory means, linguistic and historical, of proving an Asiatic origin, are of some interest. Ritter gives them carefully; I will content myself with an epitome. The Sanskrit name of the sugar-cane was *ikshu, ikshura,* or *ikshava,* but the sugar was called *sarkara,* or *sakkara,* and all its names in our European languages of Aryan origin, beginning with the ancient ones—Greek, for example—are clearly derived from this. This is an indication of Asiatic origin, and that the produce of the cane was of ancient use in the southern regions of Asia with which the ancient Sanskrit-speaking nation may have had commercial dealings. The two Sanskrit words have remained in Bengali under the

[1] Rumphius, *Amboin,* vol. v. p. 186.   [2] Hehn, No. 480.
[3] Schacht, *Madeira und Teneriffe,* tab. i.
[4] Tussac, *Flore des Antilles,* i. p. 153, pl. 23.

forms *ik* and *akh*.[1]   But in other languages beyond the
Indus, we find a singular variety of names, at least when
they are not akin to that of the Aryans; for instance:
panchadara in Telinga, *kyam* in Burmese, *mia* in the
dialect of Cochin-China, *kan* and *tche*, or *tsche*, in Chinese;
and further south, among the Malays, *tubu* or *tabu* for
the plant, and *gula* for the product.   This diversity
proves the great antiquity of its cultivation in those
regions of Asia in which botanical indications point out
the origin of the species.

The epoch of its introduction into different countries
agrees with the idea that its origin was in India, Cochin-
China, or the Malay Archipelago.

The Chinese were not acquainted with the sugar-cane
at a very remote period, and they received it from the
West.   Ritter contradicts those authors who speak of a
very ancient cultivation, and I find most positive confirmation of his opinion in Dr. Bretschneider's pamphlet,
drawn up at Pekin with the aid of all the resources of
Chinese literature.[2]   " I have not been able to discover,"
he says, "any allusion to the sugar-cane in the most
ancient Chinese books (the five classics)."   It appears to
have been mentioned for the first time by the authors of
the second century before Christ.   The first description
of it appears in the *Nan-fang-tsao-mu-chuang*, in the
fourth century : " The *chê chê, kan-chê (kan*, sweet, *chê,*
bamboo) grows," it says, " in Cochin-China.   It is several
inches in circumference, and resembles the bamboo.   The
stem, broken into pieces, is eatable and very sweet.   The
sap which is drawn from it is dried in the sun.   After a
few days it becomes sugar (here a compound Chinese
character), which melts in the mouth. . . . In the year
286 (of our era) the kingdom of Funan (in India, beyond
the Ganges) sent sugar as a tribute."   According to the
Pent-Sao, an 'emperor who reigned from 627 to 650 A.D.,
sent a man into the Indian province of Behar to learn
how to manufacture sugar.

There is nothing said in these works of the plant

[1] Piddington, *Index.*
[2] Bretschneider, *On the Study and Value,* etc., pp. 45–47.

growing wild in China; on the contrary, the origin in Cochin-China, indicated by Loureiro, finds an unexpected confirmation. It seems to me most probable that its primitive range extended from Bengal to Cochin-China. It may have included the Sunda Isles and the Moluccas, whose climate is very similar; but there are quite as many reasons for believing that it was early introduced into these from Cochin-China or the Malay peninsula.

The propagation of the sugar-cane from India westward is well known. The Greco-Roman world had a vague idea of the reed (*calamus*) which the Indians delighted to chew, and from which they obtained sugar.[1] On the other hand, the Hebrew writings do not mention sugar;[2] whence we may infer that the cultivation of the sugar-cane did not exist west of the Indus at the time of the Jewish captivity at Babylon. The Arabs in the Middle Ages introduced it into Egypt, Sicily, and the south of Spain,[3] where it flourished until the abundance of sugar in the colonies caused it to be abandoned. Don Henriquez transported the sugar-cane from Sicily to Madeira, whence it was taken to the Canaries in 1503.[4] Hence it was introduced into Brazil in the beginning of the sixteenth century.[5] It was taken to St. Domingo about 1520, and shortly afterwards to Mexico;[6] to Guadeloupe in 1644, to Martinique about 1650, to Bourbon when the colony was founded.[7] The variety known as *Otahiti*, which is not, however, wild in that island, and which is also called *Bourbon*, was introduced into the French and English colonies at the end of the last and the beginning of the present century.[8]

---

[1] See the quotations from Strabo, Dioscorides, Pliny, etc., in Lenz, *Botanik der Alten Griechen und Römer*, 1859, p. 267; Fingerhut, in *Flora*, 1839, vol. ii. p. 529; and many other authors.

[2] Rosenmüller, *Handbuch der Bibl. Alterth.*

[3] *Calendrier Rural de Harib*, written in the tenth century for Spain, translated by Dureau de la Malle in his *Climatologie de l'Italie et de l'Andalousie*, p. 71.

[4] Von Buch, *Canar. Ins.* [5] Piso, *Brésil*, p. 49.

[6] Humboldt, *Nouv. Espagne*, ed. 2, vol. iii. p. 34.

[7] *Not. Stat. sur les Col. Franc.*, i. pp. 207, 29, 83.

[8] Macfadyen, in Hooker, *Bot. Miscell.*, i. p. 101; Maycock, *Fl. Barbad.*, p. 50.

The processes of cultivation and preparation of the sugar are described in a number of works, among which the following may be recommended: de Tussac, *Flore des Antilles,* 3 vols., Paris; vol. i. pp. 151–182; and Macfadyen, in Hooker's *Botanical Miscellany,* 1830, vol. i. pp. 103–116.

CHAPTER III.

PLANTS CULTIVATED FOR THEIR FLOWERS, OR FOR THE ORGANS WHICH ENVELOP THEM.

Clove—*Caryophyllus aromaticus*, Linnæus.

The clove used for domestic purposes is the calix and flower-bud of a plant belonging to the order of Myrtaceæ. Although the plant has been often described and very well drawn from cultivated specimens, some doubt remains as to its nature when wild. I spoke of it in my *Geographical Botany* in 1855, but it does not appear that the question has made any further progress since then, which induces me to repeat here what I said then.

" The clove must have come originally from the Moluccas," as Rumphius asserts,[1] for its cultivation was limited two centuries ago to a few little islands in this archipelago. I cannot, however, find any proof that the true clove tree, with peduncles and aromatic buds, has been found in a wild state. Rumphius [2] considers that a plant of which he gives a description, and a drawing under the name *Caryophyllum sylvestre*, belongs to the same species, and this plant is wild throughout the Moluccas. A native told him that the cultivated clove trees degenerate into this form, and Rumphius himself found a plant of *C. sylvestre* in a deserted plantation of cultivated cloves. Nevertheless plate 3 differs from plate 1 of the cultivated clove in the shape of the leaves and of the teeth of the calix. I do not speak of plate 2, which appears to be an

[1] ii. p. 3. [2] ii. tab. 3.

abnormal form of the cultivated clove. Rumphius says that *C. sylvestre* has no aromatic properties; now, as a rule, the aromatic properties are more developed in the wild plants of a species than in the cultivated plants. Sonnerat [1] also publishes figures of the true clove and of a spurious clove found in a small island near the country of the Papuans. It is easy to see that his false clove differs completely by its blunt leaves from the true clove, and also from the two species of Rumphius. I cannot make up my mind to class all these different plants, wild and cultivated, together, as all authors have done. [2] It is especially necessary to exclude plate 120 of Sonnerat, which is admitted in the *Botanical Magazine*. An historical account of the cultivation of the clove, and of its introduction into different countries, will be found in the last-named work, in the *Dictionnaire d'Agriculture*, and in the dictionaries of natural history.

If it be true, as Roxburgh says, [3] that the Sanskrit language had a name, *luvunga*, for the clove, the trade in this spice must date from a very early epoch, even supposing the name to be more modern than the true Sanskrit. But I doubt its genuine character, for the Romans would have known of a substance so easily transported, and it does not appear that it was introduced into Europe before the discovery of the Moluccas by the Portuguese.

Hop—*Humulus Lupulus,* Linnæus.

The hop is wild in Europe from England and Sweden as far south as the mountains of the Mediterranean basin, and in Asia as far as Damascus, as the south of the Caspian Sea, and of Eastern Siberia, [4] but it is not found in India, the north of China, or the basin of the river Amur. [5]

[1] Sonnerat, *Voy. Nouv. Guin.*, tab. 119, 120.

[2] Thunberg, *Diss.*, ii. p. 326 ; De Candolle, *Prodr.*, iii. p. 262 ; Hooker, *Bot. Mag.*, tab. 2749 ; Hasskarl, *Cat. Hort. Bogor. Alt.*, p. 261.

[3] Roxburgh, *Flora Indica*, edit. 1832, vol. ii. p. 194.

[4] Alph. de Candolle, in *Prodromus*, vol. xvi., sect. 1, p. 29 ; Boissier, *Fl. Orient.*, iv. p. 1152 ; Hohenacker, *Enum. Plant. Talysch*, p. 30 ; Buhse *Aufzählung Transcaucasien*, p. 202.

[5] An erroneous transcription of what Asa Gray (*Botany of North. United States*, edit. 5) says of the hemp, wrongly attributed to the hop in *Prodromus*, and repeated in the French edition of this work, should

In spite of the entirely wild appearance of the hop in
Europe in districts far from cultivation, it has been sometimes asked if it is not of Asiatic origin.[1] I do not think
this can be proved, nor even that it is likely. The fact
that the Greeks and Latins have not spoken of the use
of the hop in making beer is easily explained, as they
were almost entirely unacquainted with this drink. If
the Greeks have not mentioned the plant, it is simply
perhaps because it is rare in their country. From the
Italian name *lupulo* it seems likely that Pliny speaks of
it with other vegetables under the name *lupus salictarius*.[2]
That the custom of brewing with hops only became
general in the Middle Ages proves nothing, except that
other plants were formerly employed, as is still the case
in some districts. The Kelts, the Germans, other peoples
of the north and even of the south who had the vine,
made beer [3] either of barley or of other fermented grain,
adding in certain cases different vegetable substances—the
bark of the oak or of the tamarisk, for instance, or the
fruits of *Myrica gale*.[4] It is very possible that they
did not soon discover the advantages of the hop, and that
even after these were recognized, they employed wild
hops before beginning to cultivate them. The first mention of hop-gardens occurs in an act of donation made by
Pepin, father of Charlemagne, in 768.[5] In the fourteenth
century it was an important object of culture in Germany,
but it began in England only under Henry VIII.[6]

The common names of the hop only furnish negative
indications as to its origin. There is no Sanskrit name,[7]

be corrected. *Humulus Lupulus* is indigenous in the east of the United
States, and also in the island of Yeso, according to a letter from
Maximowicz.—AUTHOR'S NOTE, 1884.

[1] Hehn, *Nutzpflanzen und Hausthiere in ihren Uebergang aus Asien*,
edit. 3, p. 415.

[2] Pliny, *Hist.*, bk. 21, c. 15. He mentions asparagus in this connection, and the young shoots of the hop are sometimes eaten in this
manner.

[3] Tacitus, *Germania*, cap. 25; Pliny, bk. 18, c. 7; Hehn, *Kulturpflanzen*, edit. 3, pp. 125–137.

[4] Volz, *Beitrage zur Culturgeschichte*, p. 149.   [5] *Ibid.*

[6] Beckmann, *Erfindungen*, quoted by Volz.

[7] Piddington, *Index;* Fick, *Wörterb. Indo-Germ. Sprachen*, i.; Ur-
sprache.

and this agrees with the absence of the species in the region
of the Himalayas, and shows that the early Aryan peoples
had not noticed and employed it.   I have quoted before [1]
some of the European names, showing their diversity,
although some few of them may be derived from a common stock.   Hehn, the philologist, has treated of their
etymology, and shown how obscure it is, but he has not
mentioned the names totally distinct from *humle, hopf* or
hop, and *chmeli* of the Scandinavian, Gothic, and Slav
races; for example, *Apini* in Lette, *Apwynis* in Lithuanian, *tap* in Esthonian, *blust* in Illyrian,[2] which have
evidently other roots.   This variety tends to confirm the
theory that the species existed in Europe before the
arrival of the Aryan nations.   Several different peoples
must have distinguished, known, and used this plant successively, which confirms its extension in Europe and in
Asia before it was used in brewing.

Carthamine—*Carthamus tinctorius*, Linnæus.

The composite annual which produces the dye called
carthamine is one of the most ancient cultivated species.
Its flowers are used for dyeing in red or yellow, and the
seeds yield oil.

The grave-cloths which wrap the ancient Egyptian
mummies are dyed with carthamine,[3] and quite recently
fragments of the plant have been found in the tombs
discovered at Deir el Bahari.[4]   Its cultivation must also
be ancient in India, since there are two Sanskrit names
for it, *cusumbha* and *kamalottara*, of which the first has
several derivatives in the modern languages of the
peninsula.[5]   The Chinese only received carthamine in
the second century B.C., when Chang-kien brought it
back from Bactriana.[6]   The Greeks and Latins were
probably not acquainted with it, for it is very doubtful
whether this is the plant which they knew as *cnikos* or
cnicus.[7]   At a later period the Arabs contributed largely

[1] A. de Candolle, *Géogr. Bot. Rais.*, p. 857.
[2] *Dict. MS.*, compiled from floras, Moritzi.
[3] Unger, *Die Pflanzen des Alten Ægyptens*, p. 47.
[4] Schweinfurth, in a letter to M. Boissier, 1882.   [5] Piddington, *Index*.
[6] Bretschneider, *Study and Value*, etc., p. 15.
[7] See Targioni, *Cenni Storici*, p. 108.

to diffuse the cultivation of carthamine, which they named *qorton, kurtum,* whence *carthamine,* or *usfur,* or *ihridh,* or *morabu,*[1] a diversity indicating an ancient existence in several countries of Western Asia or of Africa. The progress of chemistry threatens to do away with the cultivation of this plant as of many others, but it still subsists in the south of Europe, in the East, and throughout the valley of the Nile.[2]

No botanist has found the carthamine in a really wild state. Authors doubtfully assign to it an origin in India or Africa, in Abyssinia in particular, but they have never seen it except in a cultivated state, or with every appearance of having escaped from cultivation.[3]

Mr. Clarke,[4] formerly director of the Botanical Gardens in Calcutta, who has lately studied the *Compositæ* of India, includes the species only as a cultivated one. The summary of our modern knowledge of the plants of the Nile region, including Abyssinia, by Schweinfurth and Ascherson,[5] only indicates it as a cultivated species, nor does the list of the plants observed by Rohlfs on his recent journey mention a wild carthamine.[6]

As the species has not been found wild either in India or in Africa, and as it has been cultivated for thousands of years in both countries, the idea occurred to me of seeking its origin in the intermediate region; a method which had been successful in other cases.

Unfortunately, the interior of Arabia is almost unknown. Forskal, who has visited the coasts of Yemen, has learnt nothing about the carthamine; nor is it mentioned among the plants of Botta and of Bové. But an Arab, Abu Anifa, quoted by Ebn Baithar, a thirteenth-century writer, expressed himself as follows:[7]—" *Usfur,* this plant furnishes a substance used as a dye; there are two kinds, one cultivated and one wild, which both grow

---

[1] Forskal, *Fl. Ægypt.,* p. 73; Ebn Baithar, Germ. trans., ii. pp. 196, 293; i. p. 18.
[2] See Gasparin, *Cours d'Agric.,* iv. p. 217.
[3] Boissier, *Fl. Orient.,* iii. p. 710; Oliver, *Flora of Trop. Afr.,* iii. p. 439.
[4] Clarke, *Compositæ Indic e,* 1876, p. 244.
[5] Schweinfurth and Ascherson, *Aufzählung,* p. 283.
[6] Rohlfs, *Kufra,* in 8vo, 1881.   [7] Ebn Baithar, ii. p. 196.

in Arabia, of which the seeds are called *elkurthum.*"
Abu Anifa was very likely right.

Saffron—*Crocus sativus,* Linnæus.

The saffron was cultivated in very early times in the
west of Asia. The Romans praised the saffron of Cilicia,
which they preferred to that grown in Italy.[1] Asia Minor,
Persia, and Kashmir have been for a long time the
countries which export the most. India gets it from
Kashmir[2] at the present day. Roxburgh and Wallich
do not mention it in their works. The two Sanskrit
names mentioned by Piddington[3] probably applied to the
substance saffron brought from the West, for the name
kasmirajamma appears to indicate its origin in Kashmir.
The other name is *kunkuma.* The Hebrew word *karkom*
is commonly translated saffron, but it more probably
applies to carthamine, to judge from the name of the
latter in Arabic.[4] Besides, the saffron is not cultivated
in Egypt or in Arabia. The Greek name is *krokos.*[5]
Saffron, which recurs in all modern European languages,
comes from the Arabic *sahafaran,*[6] *zafran.*[7] The
Spaniards, nearer to the Arabs, call it *azafran.* The
Arabic name itself comes from *assfar,* yellow.

Trustworthy authors say that *C. sativus* is wild
in Greece[8] and in the Abruzzi mountains in Italy.[9]
Maw, who is preparing a monograph of the genus Crocus,
based on a long series of observations in gardens and
in herbaria, connects with *C. sativus* six forms which
are found wild in mountainous districts from Italy to
Kurdistan. None of these, he says,[10] are identical with
the cultivated variety; but certain forms described
under other names (*C. Orisnii, C. Cartwrightianus, C.
Thomasii*), hardly differ from it. These are from Italy
and Greece.

---

[1] Pliny, bk. xxi. c. 6.   [2] Royle, *Ill. Himal.,* p. 372.
[3] *Index,* p. 25.
[4] According to Forskal, Delile, Reynier, Schweinfurth, and Ascherson.
[5] Theophrastus, *Hist.,* 1. 6, c. 6.
[6] J. Bauhin, *Hist.,* ii. p. 637.   [7] Royle, *Ill. Himal.*
[8] Sibthorp, *Prodr.;* Fraas, *Syn. Fl. Class.,* p. 292.
[9] J. Gay, quoted by Babington, *Man. Brit. Fl.*
[10] Maw, in the *Gardener's Chron.,* 1881, vol. xvi.

The cultivation of saffron, of which the conditions are given in the *Cours d'Agriculture* by Gasparin, and in the *Bulletin de la Société d'Acclimatation* for 1870, is becoming more and more rare in Europe and Asia.[1] It has sometimes had the effect of naturalizing the species for a few years at least in localities where it appears to be wild.

[1] Jacquemont, *Voyage*, vol. iii. p. 238.

CHAPTER IV.

PLANTS CULTIVATED FOR THEIR FRUITS.[1]

Sweet Sop, Sugar Apple [2]—*Anona squamosa*, Linnæus. (In British India, **Custard Apple**; but this is the name of *Anona muricata* in America.)

The original home of this and other cultivated Anonaceæ has been the subject of doubts, which make it an interesting problem. I attempted to resolve them in 1855. The opinion at which I then arrived has been confirmed by the subsequent observations of travellers, and as it is useful to show how far probabilities based upon sound methods lead to true assertions, I will transcribe what I then said,[3] mentioning afterwards the more recent discoveries.

"Robert Brown proved in 1818 that all the species of the genus Anona, excepting *Anona senegalensis*, belong to America, and none to Asia. Aug. de Saint-Hilaire says that, according to Vellozo, *A. squamosa* was introduced into Brazil, that it is known there under the name of *pinha*, from its resemblance to a fir-cone, and of *ata*, evidently borrowed from the names *attoa* and *atis*, which are those of the same plant in Asia, and which belong to Eastern languages. Therefore, adds de

---

[1] The word fruit is here employed in the vulgar sense, for any fleshy part which enlarges after the flowering. In the strictly botanical sense, the Anonaceæ, strawberries, cashews, pine-apples, and breadfruit are not fruits.

[2] *A. squamosa* is figured in Descourtilz, *Flore des Antilles*, ii. pl. 83; Hooker's *Bot. Mag.*, 3095; and Tussac, *Flore des Antilles*, iii. pl. 4.

[3] A. de Candolle, *Géogr. Bot. Rais.*, p. 859.

Saint-Hilaire,[1] the Portuguese transported *A. squamosa* from their Indian to their American possessions, etc."

Having made in 1832 a review of the family of the Anonaceæ,[2] I noticed how Mr. Brown's botanical argument was ever growing stronger; for in spite of the considerable increase in the number of described Anonaceæ, no Anona, nor even any species of Anonaceæ with united ovaries, had been found to be a native of Asia. I admitted[3] the probability that the species came from the West Indies or from the neighbouring part of the American continent; but I inadvertently attributed this opinion to Mr. Brown, who had merely indicated an American origin in general.[4]

Facts of different kinds have since confirmed this view.

"*Anona squamosa* has been found wild in Asia, apparently as a naturalized plant; in Africa, and especially in America, with all the conditions of an indigenous plant. In fact, according to Dr. Royle,[5] the species has been naturalized in several parts of India; but he only saw it apparently growing wild on the side of the mountain near the fort of Adjeegurh in Bundlecund, among teak trees. When so remarkable a tree, in a country so thoroughly explored by botanists, has only been discovered in a single locality beyond the limits of cultivation, it is most probable that it is not indigenous in the country. Sir Joseph Hooker found it in the isle of St. Iago, of the Cape Verde group, forming woods on the hills which overlook the valley of St. Domingo.[6] Since *A. squamosa* is only known as a cultivated plant on the neighbouring continent;[7] as it is not even indicated in Guinea by Thonning,[8] nor in Congo,[9] nor in Senegambia,[10] nor in

[1] Aug. de Saint-Hilaire, *Plantes usuelles des Brésiliens*, bk. vi. p. 5.
[2] Alph. de Candolle, *Mem. Soc. Phys. et d'Hist. Nat. de Genève.*
[3] *Ibid.*, p. 19 of *Mem.* printed separately.
[4] See *Botany of Congo*, and the German translation of Brown's works, which has alphabetical tables.
[5] Royle, *Ill. Himal.*, p. 60.
[6] Webb, in *Fl. Nigr.*, p. 97.   [7] *Ibid.*, p. 204.
[8] Thonning, *Pl. Guin.*   [9] Brown, *Congo*, p. 6.
[10] Guillemin, Perrottet, and Richard, *Tentamen Fl. Seneg.*

Abyssinia and Egypt, which proves a recent introduction into Africa ; lastly, as the Cape Verde Isles have lost a great part of their primitive forests, I believe that this is a case of naturalization from seed escaped from gardens. Authors are agreed in considering the species wild in Jamaica. Formerly the assertions of Sloane[1] and Brown[2] might have been disregarded, but they are confirmed by Macfadyen.[3] Martius found the species wild in the virgin forests of Para.[4] He even says, ' *Sylvescentem in nemoribus paraensibus inveni,*' whence it may be inferred that these trees alone formed a forest. Splitgerber[5] found it in the forests of Surinam, but he says, ' *An spontanea ?* ' The number of localities in this part of America is significant. I need not remind my readers that no tree growing elsewhere than on the coast has been found truly indigenous at once in tropical Asia, Africa, and America.[6] The result of my researches renders such a fact almost impossible, and if a tree were robust enough to extend over such an area, it would be extremely common in all tropical countries.

"Moreover, historical and philological facts tend also to confirm the theory of an American origin. The details given by Rumphius[7] show that *Anona squamosa* was a plant newly cultivated in most of the islands of the Malay Archipelago. Forster does not mention the cultivation of any Anonacea in the small islands of the Pacific.[8] Rheede[9] says that *A. squamosa* is an exotic in Malabar, but was brought to India, first by the Chinese and the Arabs, afterwards by the Portuguese. It is certainly cultivated in China and in Cochin-China,[10] and in the Philippine Isles,[11] but we do not know from what epoch. It is doubtful whether the Arabs cultivate it.[12]

[1] Sloane, *Jam.*, ii. p. 168.   [2] P. Brown, *Jam.*, p. 257.
[3] Macfadyen, *Fl. Jam.*, p. 9.   [4] Martius, *Fl. Bras.*, fasc. ii. p. 15.
[5] Splitgerber, *Nederl. Kruidk. Arch.*, ii. p. 230.
[6] A. de Candolle, *Géogr. Bot. Rais.*, chap. x.
[7] Rumphius, i. p. 139.   [8] Forster, *Plantœ Esculentœ.*
[9] Rheede, *Malabar*, iii. p. 22.   [10] Loureiro, *Fl. Cochin.*, p. 427.
[11] Blanco, *Fl. Filip.*
[12] This depends upon the opinion formed with respect to *A. glabra,* Forskal (*A. Asiatica*, B. Dun. *Anon.*, p. 71 ; *A. Forskalii*, D. C. *Syst.*, i. p. 472), which was sometimes cultivated in gardens in Egypt when

It was cultivated in India in Roxburgh's day;[1] he had not seen the wild plant, and only mentions one common name in a modern language, the Bengali *ata*, which is already in Rheede. Later the name *gunda-gatra*[2] was believed to be Sanskrit, but Dr. Royle[3] having consulted Wilson, the famous author of the Sanskrit dictionary, touching the antiquity of this name, he replied that it was taken from the *Sabda Chanrika*, a comparatively modern compilation. The names of *ata, ati*, are found in Rheede and Rumphius.[4] This is doubtless the foundation of Saint-Hilaire's argument; but a nearly similar name is given to *Anona squamosa* in Mexico. This name is *ate, ahate di Panucho*, found in Hernandez[5] with two similar and rather poor figures which may be attributed either to *A. squamosa*, as Dunal[6] thinks, or to *A. cherimolia*, according to Martius.[7] Oviedo uses the name *anon*.[8] It is very possible that the name *ata* was introduced into Brazil from Mexico and the neighbouring countries. It may also, I confess, have come from the Portuguese colonies in the East Indies. Martius says, however, that the species was imported from the West India Islands.[9] I do not know whether he had any proof of this, or whether he speaks on the authority of Oviedo's work, which he quotes and which I cannot consult. Oviedo's article, translated by Marcgraf,[10] describes *A. squamosa* without speaking of its origin.

Forskal visited that country; it was called *keschta*, that is, coagulated milk. The rarity of its cultivation and the silence of ancient authors shows that it was of modern introduction into Egypt. Ebn Baithar (Sondtheimer's German translation, in 2 vols., 1840), an Arabian physician of the thirteenth century, mentions no Anonacea, nor the name *keschta*. I do not see that Forskal's description and illustration (*Descr.*, p. 102. ic. tab. 15) differ from *A. squamosa*. Coquebert's specimen, mentioned in the *Systema*, agrees with Forskal's plate; but as it is in flower while the plate shows the fruit, its identity cannot be proved.

[1] Roxburgh, *Fl. Ind.*, edit. 1832, v. ii. p. 657.
[2] Piddington, *Index*, p. 6.   [3] Royle, *Ill. Him.*, p. 60.
[4] Rheede and Rumphius, i. p. 139.
[5] Hernandez, pp. 348, 454.   [6] Dunal, *Mem. Anon.*, p. 70.
[7] Martius, *Fl. Bras.*, fasc. ii . p. 15.
[8] Hence the generic name *Anona*, which Linnæus changed to *Annona* (provision), because he did not wish to have any savage name, and did not mind a pun.
[9] Martius, *Fl. Bras.*, fasc. ii. p. 15.   [10] Marcgraf, *Brazil*, p. 94.

"The sum total of the facts is altogether in favour of
an American origin. The locality where the species
usually appears wild is in the forests of Para. Its cultivation is ancient in America, since Oviedo is one of the
first authors (1535) who has written about this country.
No doubt its cultivation is of ancient date in Asia likewise, and this renders the problem curious. It is not
proved, however, that it was anterior to the discovery
of America, and it seems to me that a tree of which the
fruit is so agreeable would have been more widely diffused
in the old world if it had always existed there. Moreover, it would be difficult to explain its cultivation in
America in the beginning of the sixteenth century, on the
hypothesis of an origin in the old world."

Since I wrote the above, I find the following facts
published by different authors :—

1. The argument drawn from the fact that there is no
Asiatic species of the genus Anona is stronger than ever.
A. Asiatica, Linnæus, was based upon errors (see my
note in the *Géogr. Bot.*, p. 862). *A. obtusifolia* (Tussac,
Fl. des Antilles, i. p. 191, pl. 28), cultivated formerly
in St. Domingo as of Asiatic origin, is also perhaps
founded upon a mistake. I suspect that the drawing
represents the flower of one species (*A. muricata*) and
the fruit of another (*A. squamosa*). No Anona has been
discovered in Asia, but four or five are now known in
Africa instead of only one or two,[1] and a larger number
than formerly in America.

2. The authors of recent Asiatic floras do not hesitate to consider the Anonæ, particularly *A. squamosa*,
which is here and there found apparently wild, as
naturalized in the neighbourhood of cultivated ground
and of European settlements.[2]

[1] See Baker, *Flora of Mauritius*, p. 3. The identity admitted by
Oliver, *Fl. Trop. Afr.*, i. p. 16, of the *Anona palustris* of America with
that of Senegambia, appears to me very extraordinary, although it is a
species which grows in marshes ; that is, having perhaps a very wide
area.

[2] Hooker, *Fl. of Brit. Ind.*, i. p. 78 ; Miquel, *Fl. Indo-Batava*, i. part 2,
p. 33 ; Kurz, *Forest Flora of Brit. Burm.*, i. p. 46 ; Stewart and Brandis,
Forests of India, p. 6.

3. In the new African floras already quoted, *A. squamosa* and the others of which I shall speak presently are always mentioned as cultivated species.

4. McNab, the horticulturist, found *A. squamosa* in the dry plains of Jamaica,[1] which confirms the assertions of previous authors. Eggers says[2] that the species is common in the thickets of Santa Cruz and Virgin Islands. I do not find that it has been discovered wild in Cuba.

5. On the American continent it is given as cultivated.[3] However, M. André sent me a specimen from a stony district in the Magdalena valley, which appears to belong to this species and to be wild. The fruit is wanting, which renders the matter doubtful. From the note on the ticket, it is a delicious fruit like that of *A. squamosa*. Warming[4] mentions the species as cultivated at Lagoa Santa in Brazil. It appears, therefore, to be cultivated or naturalized from cultivation in Para, Guiana, and New Granada.

In fine, it can hardly be doubted, in my opinion, that its original country is America, and in especial the West India Islands.

Sour Sop—*Anona muricata*, Linnæus.

This fruit-tree,[5] introduced into all the colonies in tropical countries is wild in the West Indies; at least, its existence has been proved in the islands of Cuba, St. Domingo, Jamaica, and several of the smaller islands.[6] It is sometimes naturalized on the continent of South America near dwellings.[7] André brought specimens from the district of Cauca in New Granada,

---

[1] Grisebach, *Fl. of Brit. W. I. Isles*, p. 5.

[2] Eggers, *Flora of St. Croix and Virgin Isles*, p. 23.

[3] Triana and Planchon, *Prodr. Fl. Novo-Granatensis*, p. 29; Sagot, *Journ. Soc. d'Hortic.*, 1872.

[4] Warming, *Symbolæ ad. Fl. Bras.*, xvi. p. 434.

[5] Figured in Descourtilz, *Fl. Med. des. Antilles*, ii. pl. 87, and in Tussac, *Fl. des Antilles*, ii. p. 24.

[6] Richard, *Plantes Vasculaires de Cuba*, p. 29; Swartz, *Obs.*, p. 221; P. Brown, *Jamaica*, p. 255; Macfadyen, *Fl. of Jam.*, p. 7; Eggers, *Fl. of St. Croix*, p. 23; Grisebach, *Fl. Brit. W. I.*, p. 4.

[7] Martius, *Fl. Brasil*, fasc. ii. p. 4; Splitgerber, *Pl. de Surinam*, in *Nederl. Kruidk. Arch.*, i. p. 226.

but he does not say they were wild, and I see that Triana (*Prodr. Fl. Granat.*) only mentions it as cultivated.

Custard Apple in the West Indies, **Bullock's Heart** in the East Indies—*Anona reticulata*, Linnæus.

This Anona, figured in Descourtilz, *Flore Médicale des Antilles*, ii. pl. 82, and in the *Botanical Magazine*, pl. 2912, is wild in Cuba, Jamaica, St. Vincent, Guadeloupe, Santa Cruz, and Barbados,[1] and also in the island of Tobago in the Bay of Panama,[2] and in the province of Antioquia in New Granada.[3] If it is wild in the last-named localities as well as in the West Indies, its area probably extends into several states of Central America and of New Granada.

Although the bullock's heart is not much esteemed as a fruit, the species has been introduced into most tropical colonies. Rheede and Rumphius found it in plantations in Southern Asia. According to Welwitsch, it has naturalized itself from cultivation in Angola, in Western Africa,[4] and this has also taken place in British India.[5]

Chirimoya—*Anona Cherimolia*, Lamarck.

The chirimoya is not so generally cultivated in the colonies as the preceding species, although the fruit is excellent. This is probably the reason that there is no illustration of the fruit better than that of Feuillée (*Obs.*, iii. pl. 17), while the flower is well represented in pl. 2011 of the *Botanical Magazine*, under the name of *A. tripetala*.

In 1855, I wrote as follows, touching the origin of the species:[6] "The chirimoya is mentioned by Lamarck and Dunal as growing in Peru; but Feuillée, who was the first to speak of it,[7] says that it is cultivated. Mac-

---

[1] Richard, Macfadyen, Grisebach, Eggers, Swartz, Maycock, *Fl. Barbad.*, p. 233.

[2] Seemann, *Bot. of the Herald*, p. 75.

[3] Triana and Planchon, *Prodr. Fl. Novo-Granat.*, p. 29.

[4] Oliver, *Fl. Trop. Afr.*, i. p. 15.

[5] Sir J. Hooker, *Fl. Brit. Ind.*, i. p. 78.

[6] De Candolle, *Géogr. Bot. Rais.*, p. 863.

[7] Feuillée, *Obs.*, iii. p. 23, t. 17.

fadyen[1] says it abounds in the Port Royal Mountains, Jamaica; but he adds that it came originally from Peru, and must have been introduced long ago, whence it appears that the species is cultivated in the higher plantations, rather than wild. Sloane does not mention it. Humboldt and Bonpland saw it cultivated in Venezuela and New Granada; Martius in Brazil,[2] where the seeds had been introduced from Peru. The species is cultivated in the Cape Verde Islands, and on the coast of Guinea,[3] but it does not appear to have been introduced into Asia. Its American origin is evident. I might even go further, and assert that it is a native of Peru, rather than of New Granada or Mexico. It will probably be found wild in one of these countries. Meyen has not brought it from Peru."[4]

My doubts are now lessened, thanks to a kind communication from M. Ed. André. I may mention first, that I have seen specimens from Mexico gathered by Botteri and Bourgeau, and that authors often speak of finding the species in this region, in the West Indies, in Central America, and New Granada. It is true, they do not say that it is wild. On the contrary, they remark that it is cultivated, or that it has escaped from gardens and become naturalized.[5] Grisebach asserts that it is wild from Peru to Mexico, but he gives no proof. André gathered, in a valley in the south-west of Ecuador, specimens which certainly belong to the species as far as it can be asserted without seeing the fruit. He says nothing as to its wild nature, but the care with which he points out in other cases plants cultivated or perhaps escaped from cultivation, leads me to think that he regards these specimens as wild. Claude Gay says that the species has been cultivated in Chili from time immemorial.[6] However, Molina, who mentions several fruit-

---

[1] Macfadyen, *Fl. Jam.*, p. 10. [2] Martius, *Fl. Bras.*, fasc. iii. p. 15.
[3] Hooker, *Fl. Nigr.*, p. 205. [4] *Nov. Act. Nat. Cur.*, xix. suppl. 1.
[5] Richard, *Plant. Vasc. de Cuba;* Grisebach, *Fl. Brit. W. Ind. Is.;* Hemsley, *Biologia Centr. Am.*, p. 118; Kunth, in Humboldt and Bonpland, *Nova Gen.*, v. p. 57; Triana and Planchon, *Prodr. Fl. Novo-Granat.*, p. 28.
[6] Gay, *Flora Chil.*, i. p. 66.

trees in the ancient plantations of the country, does not speak of it.[1]

In conclusion, I consider it most probable that the species is indigenous in Ecuador, and perhaps in the neighbouring part of Peru.

Oranges and Lemons—*Citrus*, Linnæus.

The different varieties of citrons, lemons, oranges, shaddocks, etc., cultivated in gardens have been the subject of remarkable works by several horticulturists, among which Gallesio and Risso[2] hold the first rank. The difficulty of observing and classifying so many varieties was very great. Fair results have been obtained, but it must be owned that the method was wrong from the beginning, since the plants from which the observations were taken were all cultivated, that is to say, more or less artificial, and perhaps in some cases hybrids. Botanists are now more fortunate. Thanks to the discoveries of travellers in British India, they are able to distinguish the wild and therefore the true and natural species. According to Sir Joseph Hooker,[3] who was himself a collector in India, the work of Brandis[4] is the best on the *Citrus* of this region, and he follows it in his flora. I shall do likewise in default of a monograph of the genus, remarking also that the multitude of garden varieties which have been described and figured for centuries, ought to be identified as far as possible with the wild species.[5]

The same species, and perhaps others also, probably grow wild in Cochin-China and in China; but this has not been proved in the country itself, nor by means of specimens examined by botanists. Perhaps the important works of Pierre, now in course of publication, will

---

[1] Molina, French trans.

[2] Gallesio, *Traité du Citrus*, in 8vo, Paris, 1811; Risso and Poiteau, *Histoire Naturelle des Orangers*, 1818, in folio, 109 plates.

[3] Hooker, *Fl. of Brit. Ind.*, i. p. 515.

[4] Brandis, *Forest Flora*, p. 50.

[5] For a work of this nature, the first step would be to publish good figures of wild species, showing particularly the fruit, which is not seen in herbaria. It would then be seen which forms represented in the plates of Risso, Duhamel, and others, are nearest to the wild types.

give information on this head for Cochin-China. With regard to China, I will quote the following passage from Dr. Bretschneider,[1] which is interesting from the special knowledge of the writer:—" Oranges, of which there are a great variety in China, are counted by the Chinese among their wild fruits. It cannot be doubted that most of them are indigenous, and have been cultivated from very early times. The proof of this is that each species or variety bears a distinct name, besides being in most cases represented by a particular character, and is mentioned in the *Shu-king, Rh-ya,* and other ancient works."

Men and birds disperse the seeds of Aurantiaceæ, whence results the extension of its area, and its naturalization in all the warm regions of the two worlds. It was observed [2] in America from the first century after the conquest, and now groves of orange trees have sprung up even in the south of the United States.

Shaddock—*Citrus decumana,* Willdenow.

I take this species first, because its botanical character is more marked than that of the others. It is a larger tree, and this species alone has down on the young shoots and the under sides of the leaves. The fruit is spherical, or nearly spherical, larger than an orange, sometimes even as large as a man's head. The juice is slightly acid, the rind remarkably thick. Good illustrations of the fruit may be seen in Duhamel, *Traité des Arbres,* edit. 2, vii. pl. 42, and in Tussac, *Flore des Antilles,* iii. pls. 17, 18. The number of varieties in the Malay Archipelago indicates an ancient cultivation. Its original country is not yet accurately known, because the trees which appear indigenous may be the result of naturalization, following frequent cultivation. Roxburgh says that the species was brought to Calcutta from Java,[3] and Rumphius [4] believed it to be a native of Southern China.

[1] Bretschneider, *On the Study and Value of Chinese Botanical Works,* p. 55.

[2] Acosta, *Hist. Nat. des Indes,* Fr. trans. 1598, p. 187.

[3] Roxburgh, *Flora Indica,* edit. 1832 iii. p. 393.

[4] Rumphius, *Hortus Amboinensis,* ii. p. 98.

Neither he nor modern botanists saw it wild in the Malay Archipelago.[1] In China the species has a simple name, *yu;* but its written character [2] appears too complicated for a truly indigenous plant. According to Loureiro, the tree is common in China and Cochin-China, but this does not imply that it is wild.[3] It is in the islands to the east of the Malay Archipelago that the clearest indications of a wild existence are found. Forster [4] formerly said of this species, "very common in the Friendly Isles." Seemann [5] is yet more positive about the Fiji Isles. "Extremely common," he says, "and covering the banks of the rivers."

It would be strange if a tree, so much cultivated in the south of Asia, should have become naturalized to such a degree in certain islands of the Pacific, while it has scarcely been seen elsewhere. It is probably indigenous to them, and may perhaps yet be discovered wild in some islands nearer to Java.

The French name, *pompelmouse,* is from the Dutch *pompelmoes.* Shaddock was the name of a captain who first introduced the species into the West Indies.[6]

Citron, Lemon—*Citrus medica,* Linnæus.

This tree, like the common orange, is glabrous in all its parts. Its fruit, longer than it is wide, is surmounted in most of its varieties by a sort of nipple. The juice is more or less acid. The young shoots and the petals are frequently tinted red. The rind of the fruit is often rough, and very thick in some subvarieties.[7]

Brandis and Sir Joseph Hooker distinguish four cultivated varieties :—

1. *Citrus medica proper* (*citron* in English, *cedratier* in French, *cedro* in Italian), with large, not

---

[1] Miquel, *Flora Indo-Batava,* i. pt. 2, p. 526.

[2] Bretschneider, *Study and Value,* etc.

[3] Loureiro, *Fl. Cochin.,* ii. p. 572. For another species of the genus, he says that it is cultivated and non-cultivated, p. 569.

[4] Forster, *De Plantis Esculentis Oceani Australis,* p. 35.

[5] Seemann, *Flora Vitiensis,* p. 33.

[6] Plukenet, *Almagestes,* p. 239; Sloane, *Jamaica,* i. p. 41.

[7] *Cedrat à gros fruit* of Duhamel, *Traité des Arbres,* edit. 2, vii. p. 68, pl. 22.

spherical fruit, whose highly aromatic rind is covered with lumps, and of which the juice is neither abundant nor very acid. According to Brandis, it was called *vijapûra* in Sanskrit.

2. *Citrus medica Limonum* (*citronnier* in French, *lemon* in English). Fruit of average size, not spherical, and abundant acid juice.

3. *Citrus medica acida* (*C. acida*, Roxburgh). Lime in English. Small flowers, fruit small and variable in shape, juice very acid. According to Brandis, the Sanskrit name was *jambira*.

4. *Citrus medica Limetta* (*C. Limetta* and *C. Lumia* of Risso), with flowers like those of the preceding variety, but with spherical fruit and sweet, non-aromatic juice. In India it is called the *sweet lime*.

The botanist Wight affirms that this last variety is wild in the Nilgherry Hills. Other forms, which answer more or less exactly to the three other varieties, have been found wild by several Anglo-Indian botanists [1] in the warm districts at the foot of the Himalayas, from Garwal to Sikkim, in the south-east at Chittagong and in Burmah, and in the south-west in the western Ghauts and the Satpura Mountains. From this it cannot be doubted that the species is indigenous in India, and even under different forms of prehistoric antiquity.

I doubt whether its area includes China or the Malay Archipelago. Loureiro mentions *Citrus medica* in Cochin-China only as a cultivated plant, and Bretschneider tells us that the lemon has Chinese names which do not exist in the ancient writings, and for which the written characters are complicated, indications of a foreign species. It may, he says, have been introduced. In Japan the species is only a cultivated one.[2] Lastly, several of Rumphius' illustrations show varieties cultivated in the Sunda Islands, but none of these are considered by the author as really wild and indigenous to the country. To indicate the locality, he sometimes used

---

[1] Royle, *Ill. Himal.*, p. 129; Brandis, *Forest Flora*, p. 52; Hooker, *Fl. of Brit. Ind.*, i. p. 514.

[2] Franchet and Savatier, *Enum. Plant. Jap.*, p. 129.

the expression "*in hortis sylvestribus*," which might be translated shrubberies.   Speaking of his *Lemon sussu* (vol. ii. pl. 25), which is a *Citrus medica* with ellipsoidal acid fruit, he says it has been introduced into Amboyna, but that it is commoner in Java, "usually in forests." This may be the result of an accidental naturalization from cultivation.   Miquel, in his modern flora of the Dutch Indies,[1] does not hesitate to say that *Citrus medica* and *C. Limonum* are only cultivated in the archipelago.

The cultivation of more or less acid varieties spread into Western Asia at an early date, at least into Mesopotamia and Media.   This can hardly be doubted, for two varieties had Sanskrit names; and, moreover, the Greeks knew the fruit through the Medes, whence the name *Citrus medica*.   Theophrastus[2] was the first to speak of it under the name of apple of Media and of Persia, in a phrase often repeated and commented on in the last two centuries.[3]   It evidently applies to *Citrus medica ;* but while he explains how the seed is first sown in vases, to be afterwards transplanted, the author does not say whether this was the Greek custom, or whether he was describing the practice of the Medes.   Probably the citron was not then cultivated in Greece, for the Romans did not grow it in their gardens at the beginning of the Christian era.

Dioscorides,[4] born in Cilicia, and who wrote in the first century, speaks of it in almost the same terms as Theophrastus.   It is supposed that the species was, after many attempts,[5] cultivated in Italy in the third or fourth century.   Palladius, in the fifth century, speaks of it as well established.

The ignorance of the Romans of the classic period touching foreign plants has caused them to confound, under the name of *lignum citreum*, the wood of *Citrus*, with that of *Cedrus*, of which fine tables were made, and

[1] Miquel, *Flora Indo-Batava*, i. pt. 2, p. 528.
[2] Theophrastus, 1. 4, c. 4.
[3] Bodæus, in Theophrastus, edit. 1644, pp. 322, 343 ; Risso, *Traité du Citrus*, p. 198 ; Targioni, *Cenni Storici*, p. 196.
[4] Dioscorides, i. p 166.   [5] Targioni, *Cenni Storici*.

which was a cedar, or a *Thuya*, of the totally different family of Coniferæ.

The Hebrews must have known the citron before the Romans, because of their frequent relations with Persia, Media and the adjacent countries. The custom of the modern Jews of presenting themselves at the synagogue on the day of the Feast of Tabernacles, with a citron in their hand, gave rise to the belief that the word *hadar* in Leviticus signified lemon or citron; but Risso has shown, by comparing the ancient texts, that it signifies a fine fruit, or the fruit of a fine tree. He even thinks that the Hebrews did not know the citron or lemon at the beginning of our era, because the Septuagint Version translates *hadar* by fruit of a fine tree. Nevertheless, as the Greeks had seen the citron in Media and in Persia in the time of Theophrastus, three centuries before Christ, it would be strange if the Hebrews had not become acquainted with it at the time of the Babylonish Captivity. Besides, the historian Josephus says that in his time the Jews bore Persian apples, *malum persicum*, at their feasts, one of the Greek names for the citron.

The varieties with very acid fruit, like *Limonum* and *acida*, did not perhaps attract attention so early as the citron, however the strongly aromatic odour mentioned by Dioscorides and Theophrastus appears to indicate them. The Arabs extended the cultivation of the lemon in Africa and Europe. According to Gallesio, they transported it, in the tenth century of our era, from the gardens of Oman into Palestine and Egypt. Jacques de Vitry, in the thirteenth century, well described the lemon which he had seen in Palestine. An author named Falcando mentions in 1260 some very acid " *lumias* " which were cultivated near Palermo, and Tuscany had them also towards the same period.[1]

Orange—*Citrus Aurantium*, Linnæus (excl. var. γ); *Citrus Aurantium*, Risso.

Oranges are distinguished from shaddocks (*C. decumana*) by the complete absence of down on the young shoots and leaves, by their smaller fruit, always spherical,

[1] Targioni, p. 217.

and by a thinner rind. They differ from lemons and citrons in their pure white flowers; in the fruit, which is never elongated, and without a nipple on the summit; in the rind, smooth or nearly so, and adhering but lightly to the pulp.

Neither Risso, in his excellent monograph of *Citrus*, nor modern authors, as Brandis and Sir Joseph Hooker, have been able to discover any other character than the taste to distinguish the sweet orange from more or less bitter fruits. This difference appeared to me of such slight importance from the botanical point of view, when I studied the question of origin in 1855, that I was inclined, with Risso, to consider these two sorts of orange as simple varieties. Modern Anglo-Indian authors do the same. They add a third variety, which they call *Bergamia*, for the bergamot orange, of which the flower is smaller, and the fruit spherical or pyriform, and smaller than the common orange, aromatic and slightly acid. This last form has not been found wild, and appears to me to be rather a product of cultivation.

It is often asked whether the seeds of sweet oranges yield sweet oranges, and of bitter, bitter oranges. It matters little from the point of view of the distinction into species or varieties, for we know that both in the animal and vegetable kingdoms all characters are more or less hereditary, that certain varieties are habitually so, to such a degree that they should be called races, and that the distinction into species must consequently be founded upon other considerations, such as the absence of intermediate forms, or the failure of crossed fertilization to produce fertile hybrids. However, the question is not devoid of interest in the present case, and I must answer that experiments have given results which are at times contradictory.

Gallesio, an excellent observer, expresses himself as follows:—" I have during a long series of years sown pips of sweet oranges, taken sometimes from the natural tree, sometimes from oranges grafted on bitter orange trees or lemon trees. The result has always been trees bearing sweet fruit; and the same has been observed for more than sixty years by all the gardeners of Finale. There

is no instance of a bitter orange tree from seed of sweet
oranges, nor of a sweet orange tree from the seed of
bitter oranges. . . . In 1709, the orange trees of Finale
having been killed by frost, the practice of raising sweet
orange trees from seed was introduced, and every one
of these plants produced the sweet-juiced fruit." [1]

Macfadyen,[2] on the contrary, in his *Flora of Jamaica*,
says, " It is a well-established fact, familiar to every one
who has been any length of time in this island, that the
seed of the sweet orange very frequently grows up into
a tree bearing the bitter fruit, numerous well-attested
instances of which have come to my own knowledge.   I
am not aware, however, that the seed of the bitter orange
has ever grown up into the sweet-fruited variety. . . .
We may therefore conclude," the author judiciously goes
on to say, " that the bitter orange was the original stock."
He asserts that in calcareous soil the sweet orange may
be raised from seed, but that in other soils it produces
fruits more or less sour or bitter.   Duchassaing says that
in Guadeloupe the seeds of sweet oranges often yield
bitter fruit,[3] while, according to Dr. Ernst, at Caracas
they sometimes yield sour but not bitter fruit.[4]   Brandis
relates that at Khasia, in India, as far as he can verify
the fact, the extensive plantations of sweet oranges are
from seed.   These differences show the variable degree of
heredity, and confirm the opinion that these two kinds
of orange should be considered as two varieties, not two
species.

I am, however, obliged to take them in succession,
to explain their origin and the extent of their cultivation
at different epochs.

Bitter Orange—*Arancio forte* in Italian, *bigaradier* in
French, *pomeranze* in German.   *Citrus vulgaris*, Risso ;
C. aurantium (var. *bigaradia*), Brandis and Hooker.

It was unknown to the Greeks and Romans, as well
as the sweet orange.   As they had had communication

[1] Gallesio, *Traité du Citrus*, pp. 32, 67, 355, 357.
[2] Macfadyen, *Flora of Jamaica*, p. 129.
[3] Quoted in Grisebach's *Veget. Karaiben*, p. 34.
[4] Ernst, in Seemann, *Journ. of Bot.*, 1867, p. 272.

with India and Ceylon, Gallesio supposed that these
trees were not cultivated in their time in the west of
India. He had studied from this point of view, ancient
travellers and geographers, such as Diodorus Siculus,
Nearchus, Arianus, and he finds no mention of the orange
in them. However, there was a Sanskrit name for the
orange—*nagarunga, nagrunga*.[1] It is from this that the
word orange came, for the Hindus turned it into *narun-
gee* (pron. *naroudji*), according to Royle, *nerunga* according to Piddington; the Arabs into *narunj*, according to
Gallesio, the Italians into *naranzi, arangi*, and in the
mediæval Latin it was *arancium, arangium*, afterwards
aurantium.[2] But did the Sanskrit name apply to the
bitter or to the sweet orange? The philologist Adolphe
Pictet formerly gave me some curious information on
this head. He had sought in Sanskrit works the descriptive names given to the orange or to the tree, and
had found seventeen, which all allude to the colour, the
odour, its acid nature (*danta catha*, harmful to the
teeth), the place of growth, etc., never to a sweet or
agreeable taste. This multitude of names similar to
epithets show that the fruit had long been known, but
that its taste was very different to that of the sweet
orange. Besides, the Arabs, who carried the orange tree
with them towards the West, were first acquainted with
the bitter orange, and gave it the name *narunj*,[3] and
their physicians from the tenth century prescribed the
bitter juice of this fruit.[4] The exhaustive researches of
Gallesio show that after the fall of the Empire the species
advanced from the coast of the Persian Gulf, and by the
end of the ninth century had reached Arabia, through
Oman, Bassora, Irak, and Syria, according to the Arabian
author Massoudi. The Crusaders saw the bitter orange
tree in Palestine. It was cultivated in Sicily from the
year 1002, probably a result of the incursions of the

---

[1] Roxburgh, *Fl. Indica*, edit. 1832, vol. ii. p. 392; Piddington, *Index*.
[2] Gallesio, p. 122.
[3] In the modern languages of India the Sanskrit name has been
applied to the sweet orange, so says Brandis, by one of those transposi-
tions which are so common in popular language.
[4] Gallesio, pp. 122, 247, 248.

Arabs. It was they who introduced it into Spain, and most likely also into the east of Africa. The Portuguese found it on that coast when they doubled the Cape in 1498.[1] There is no ground for supposing that either the bitter or the sweet orange existed in Africa before the Middle Ages, for the myth of the garden of Hesperides may refer to any species of the order *Aurantiaceæ*, and its site is altogether arbitrary, since the imagination of the ancients was wonderfully fertile.

The early Anglo-Indian botanists, such as Roxburgh, Royle, Griffith, Wight, had not come across the bitter orange wild; but there is every probability that the eastern region of India was its original country. Wallich mentions Silhet,[2] but without asserting that the species was wild in this locality. Later, Sir Joseph Hooker[3] saw the bitter orange certainly wild in several districts to the south of the Himalayas, from Garwal and Sikkim as far as Khasia. The fruit was spherical or slightly flattened, two inches in diameter, bright in colour, and uneatable, of mawkish and bitter taste (" if I remember right," says the author). *Citrus fusca*, Loureiro,[4] similar, he says, to pl. 23 of Rumphius, and wild in Cochin-China and China, may very likely be the bitter orange whose area extends to the east.

Sweet Orange — Italian, *Arancio dolce;* German, *Apfelsine. Citrus Aurantium sinense*, Gallesio.

Royle[5] says that sweet oranges grow wild at Silhet and in the Nilgherry Hills, but his assertion is not accompanied with sufficient detail to give it importance. According to the same author, Turner's expedition gathered "delicious" wild oranges at Buxedwar, a locality to the north-east of Rungpoor, in the province of Bengal. On the other hand, Brandis and Sir Joseph Hooker do not mention the sweet orange as wild in

[1] Gallesio, p. 240. Goeze, *Beitrag zur Kenntniss der Orangengewächse,* 1874, p. 13, quotes early Portuguese travellers on this head.

[2] Wallich, *Catalogue,* No. 6384.

[3] Hooker, *Fl. of Brit. Ind.,* i. p. 515.

[4] Loureiro, *Fl. Cochin.,* p. 571.

[5] Royle, *Illustr. of Himal.,* p. 129. He quotes Turner, *Journey to Thibet,* pp. 20, 387.

British India; they only give it as cultivated. Kurz
does not mention it in his forest flora of British Burmah.
Further east, in Cochin-China, Loureiro [1] describes a *C.
Aurantium*, with bitter-sweet (*acido-dulcis*) pulp, which
appears to be the sweet orange, and which is found both
wild and cultivated in China and Cochin-China. Chinese
authors consider orange trees in general as natives of
their country, but precise information about each species
and variety is wanting on this head.

From the collected facts, it seems that the sweet
orange is a native of Southern China and of Cochin-
China, with a doubtful and accidental extension of area
by seed into India.

By seeking in what country it was first cultivated,
and how it was propagated, some light may be thrown
upon the origin, and upon the distinction between the
bitter and sweet orange. So large a fruit, and one so
agreeable to the palate as the sweet orange, can hardly
have existed in any district, without some attempts
having been made to cultivate it. It is easily raised
from seed, and nearly always produces the wished-for
quality. Neither can ancient travellers and historians
have neglected to notice the introduction of so remarkable a fruit tree. On this historical point Gallesio's
study of ancient authors has produced extremely interesting results.

He first proves that the orange trees brought from
India by the Arabs into Palestine, Egypt, the south of
Europe, and the east coast of Africa, were not the sweet-
fruited tree. Up to the fifteenth century, Arab books
and chronicles only mention bitter, or sour oranges.
However, when the Portuguese arrived in the islands of
Southern Asia, they found the sweet orange, and apparently it had not previously been unknown to them.
The Florentine who accompanied Vasco de Gama, and
who published an account of the voyage, says, "*Sonvi
melarancie assai, ma tutte dolci*" (there are plenty of
oranges, but all sweet.) Neither this writer nor subsequent
travellers expressed surprise at the pleasant taste of the

---

[1] Loureiro, *Fl. Cochin.*, p. 569.

fruit, Hence Gallesio infers that the Portuguese were
not the first to bring the sweet orange from India, which
they reached in 1498, nor from China, which they
reached in 1518.   Besides, a number of writers in the
beginning of the sixteenth century speak of the sweet
orange as a fruit already cultivated in Spain and Italy.
There are several testimonies for the years 1523, and
1525.   Gallesio goes no further than the idea that the
sweet orange was introduced into Europe towards the
beginning of the fifteenth century ; [1] but Targioni quotes
from Valeriani a statute of Fermo, of the fourteenth
century, referring to citrons, sweet oranges, etc. ; [2] and
the information recently collected from early authors by
Goeze,[3] about the introduction into Spain and Portugal,
agrees with this date.   It therefore appears to me probable that the oranges imported later from China by the
Portuguese were only of better quality than those
already known in Europe, and that the common expressions, Portugal and Lisbon oranges, are due to this circumstance.

If the sweet orange had been cultivated at a very
early date in India, it would have had a special name
in Sanskrit; the Greeks would have known it after
Alexander's expedition, and the Hebrews would have
early received it through Mesopotamia. This fruit would
certainly have been valued, cultivated, and propagated
in the Roman empire, in preference to the lemon, citron,
and bitter orange. Its existence in India must, therefore, be less ancient.

In the Malay Archipelago the sweet orange was
believed to come from China.[4]   It was but little diffused
in the Pacific Isles at the time of Cook's voyages.[5]

We come back thus by all sorts of ways to the idea
that the sweet variety of the orange came from China

---

[1] Gallesio, p. 321.

[2] The date of this *statuto* is given by Targioni, on p. 205 of the *Cenni
Storici*, as 1379, and on p. 213 as 1309.   The *errata* do not notice this
discrepancy.

[3] Goeze, *Ein Beitrag zur Kenntniss der Orangengewächse.* Hamburg,
1874, p. 26.

[4] Rumphius, *Amboin.*, ii. c. 42.   Forster, *Plantis Esculentis*, p. 35.

and Cochin-China, and that it spread into India perhaps towards the beginning of the Christian era. It may have become naturalized from cultivation in many parts of India and in all tropical countries, but we have seen that the seed does not always yield trees bearing sweet fruit. This defect in heredity in certain cases is in support of the theory that the sweet orange was derived from the bitter, at some remote epoch, in China or Cochin-China, and has since been carefully propagated on account of its horticultural value.

Mandarin—*Citrus nobilis*, Loureiro.

This species, characterized by its smaller fruit, uneven on the surface, spherical, but flattened at the top, and of a peculiar flavour, is now prized in Europe as it has been from the earliest times in China and Cochin-China. The Chinese call it *kan*.[1] Rumphius had seen it cultivated in all the Sunda Islands,[2] and says that it was introduced thither from China, but it had not spread into India. Roxburgh and Sir Joseph Hooker do not mention it, but Clarke informs me that its culture has been greatly extended in the district of Khasia. It was new to European gardens at the beginning of the present century, when Andrews published a good illustration of it in the *Botanist's Repository* (pl. 608).

According to Loureiro,[3] this tree, of average size, grows in Cochin-China, and also, he adds, in China, although he had not seen it in Canton. This is not very precise information as to its wild character, but no other origin can be supposed. According to Kurz,[4] the species is only cultivated in British Burmah. If this is confirmed, its area would be restricted to Cochin-China and a few provinces in China.

Mangosteen—*Garcinia mangostana*, Linnæus.

There is a good illustration in the *Botanical Magazine*, pl. 4847, of this tree, belonging to the order Guttiferæ, of which the fruit is considered one of the best in existence.

[1] Bretschneider, *On the Study and Value*, etc., p. 11.
[2] Rumphius, *Amboin.*, ii. pls. 34, 35, where, however, the form of the fruit is not that of our mandarin.
[3] Loureiro, *Fl. Cochin.*, p. 570.   [4] Kurz, *Forest Fl. of Brit. Bur.*

It demands a very hot climate, for Roxburgh could not make it grow north of twenty-three and a half degrees of latitude in India,[1] and, transported to Jamaica, it bears but poor fruit.[2]   It is cultivated in the Sunda Islands, in the Malay Peninsula, and in Ceylon.

The species is certainly wild in the forests of the Sunda Islands [3] and of the Malay Peninsula.[4]   Among cultivated plants it is one of the most local, both in its origin, habitation, and in cultivation.   It belongs, it is true, to one of those families in which the mean area of the species is most restricted.

Mamey, or **Mammee Apple** — *Mammea Americana,* Jacquin.

This tree, of the order Guttiferæ, requires, like the mangosteen, great heat.   Although much cultivated in the West Indies and in the hottest parts of Venezuela,[5] its culture has seldom been attempted, or has met with but little success, in Asia and Africa, if we are to judge by the silence of most authors.

It is certainly indigenous in the forests of most of the West Indies.[6]   Jacquin mentions it also for the neighbouring continent, but I do not find this confirmed by modern authors.   The best illustration is that in Tussac's *Flore des Antilles,* iii. pl. 7, and this author gives a number of details respecting the use of the fruit.

Ochro, or **Gombo**—*Hibiscus esculentus,* Linnæus.

The young fruits of this annual, of the order of Malvaceæ, form one of the most delicate of tropical vegetables.   Tussac's *Flore des Antilles* contains a fine plate of the species, and gives all the details a *gourmet* could desire on the manner of preparing the *caloulou,* so much esteemed by the creoles of the French colonies.

---

[1] Royle, *Ill. Himal.,* p. 133, and Roxburgh, *Fl. Ind.,* ii. p. 618.

[2] Macfadyen, *Flora of Jamaica,* p. 134.

[3] Rumphius, *Amboin.,* i. p. 133; Miquel, *Plantæ Junghun.,* i. p. 290; *Flora Indo-Batava,* i. pt. 2, p. 506.

[4] Hooker, *Flora of Brit. Ind.,* i. p. 260.

[5] Ernst in Seemann, *Journal of Botany,* 1867, p. 273; Triana and Planchon, *Prodr. Fl. Novo-Granat.,* p. 285.

[6] Sloane, *Jamaica,* i. p. 123; Jacquin, *Amer.,* p. 268; Grisebach, *Fl. of Brit. W. Ind. Isles,* p. 118.

When I formerly[1] tried to discover whence this plant, cultivated in the old and new worlds, came originally, the absence of a Sanskrit name, and the frct that the first writers on the Indian flora had not seen it wild, led me to put aside the hypothesis of an Asiatic origin. However, as the modern flora of British India[2] mentions it as "probably of native origin," I was constrained to make further researches.

Although Southern Asia has been thoroughly explored during the last thirty years, no locality is mentioned where the *Gombo* is wild or half wild. There is no indication, even, of an ancient cultivation in Asia. The doubt, therefore, lies between Africa and America. The plant has been seen wild in the West Indies by a good observer,[3] but I can discover no similar assertion on the part of any other botanist, either with respect to the islands or to the American continent. The earliest writer on Jamaica, Sloane, had only seen the species in a state of cultivation. Marcgraf[4] had observed it in Brazilian plantations, and as he mentions a name from the Congo and Angola country, *quillobo*, which the Portuguese corrupted into *quingombo*, the African origin is hereby indicated.

Schweinfurth and Ascherson[5] saw the plant wild in the Nile Valley in Nubia, Kordofan, Senaar, Abyssinia, and in the Baar-el-Abiad, where, indeed, it is cultivated. Other travellers are mentioned as having gathered specimens in Africa, but it is not specified whether these plants were cultivated or wild at a distance from habitations. We should still be in doubt if Flückiger and Hanbury[6] had not made a bibliographical discovery which settles the question. The Arabs call the fruit *bamyah*, or *bâmiat*, and Abul-Abas-Elnabati, who visited Egypt long before the discovery of America, in 1216, has

---

[1] A. de Candolle, *Géogr. Bot. Rais.*, p. 768.
[2] *Flora of Brit. Ind.*, i. p. 343.
[3] Jacquin, *Observationes*, iii. p. 11.
[4] Marcgraf, *Hist. Plant.*, p. 32, with illustrations.
[5] Schweinfurth and Ascherson, *Aufzählung*, p. 265, under the name *abelmoschus*.
[6] Flückiger and Hanbury, *Pharmacographia*, p. 86. The description is in Ebn Baithar, Sondtheimer's trans., i. p. 118.

distinctly described the *gombo* then cultivated by the Egyptians.

In spite of its undoubtedly African origin, it does not appear that the species was cultivated in Lower Egypt before the Arab rule. No proof has been found in ancient monuments, although Rosellini thought he recognized the plant in a drawing, which differs widely from it according to Unger.[1] The existence of one name in modern Indian languages, according to Piddington, confirms the idea of its propagation towards the East after the beginning of the Christian era.

Vine—*Vitis vinifera*, Linnæus.

The vine grows wild in the temperate regions of Western Asia, Southern Europe, Algeria, and Marocco.[2] It is especially in the Pontus, in Armenia, to the south of the Caucasus and of the Caspian Sea, that it grows with the luxuriant wildness of a tropical creeper, clinging to tall trees and producing abundant fruit without pruning or cultivation. Its vigorous growth is mentioned in ancient Bactriana, Cabul, Kashmir, and even in Badakkhan to the north of the Hindu Koosh.[3] Of course, it is a question whether the plants found there, as elsewhere, are not sprung from seeds carried from vineyards by birds. I notice, however, that the most trustworthy botanists, those who have most thoroughly explored the Transcaucasian provinces of Russia, do not hesitate to say that the plant is wild and indigenous in this region. It is as we advance towards India and Arabia, Europe and the north of Africa, that we frequently find in floras the expression that the vine is " subspontaneous," perhaps wild, or become wild (*verwildert* is the expressive German term).

The dissemination by birds must have begun very early, as soon as the fruit existed, before cultivation, before the migration of the most ancient Asiatic peoples,

[1] Unger, *Die Pflanzen des Alten Ægyptens*, p. 50.
[2] Grisebach, *Végét. du Globe*, French trans. by Tchihatcheff, i. pp. 162, 163, 442; Munby, *Catal. Alger;* Ball, *Fl. Maroc. Spicel*, p. 392.
[3] Adolphe Pictet, *Origines Indo-Europ.* edit. 2, vol. 1, p. 295, quotes several travellers for these regions, among others Wood's *Journey to the Sources of the Oxus.*

perhaps before the existence of man in Europe or even in Asia. Nevertheless, the frequency of cultivation, and the multitude of forms of the cultivated grape, may have extended naturalization and introduced among wild vines varieties which originated in cultivation. In fact, natural agents, such as birds, winds, and currents, have always widened the area of species, independently of man, as far as the limits imposed in each age by geographical and physical conditions, together with the hostile action of other plants and animals, allow. An absolutely primitive habitation is more or less mythical, but habitations successively extended or restricted are in accordance with the nature of things. They constitute areas more or less ancient and real, provided that the species has maintained itself wild without the constant addition of fresh seed.

Concerning the vine, we have proofs of its great antiquity in Europe as in Asia. Seeds of the grape have been found in the lake-dwellings of Castione, near Parma, which date from the age of bronze,[1] in a prehistoric settlement of Lake Varese,[2] and in the lake-dwellings of Wangen, Switzerland, but in the latter instance at an uncertain depth.[3] And, what is more, vine-leaves have been found in the tufa round Montpellier, where they were probably deposited before the historical epoch, and in the tufa of Meyrargue in Provence, which is certainly prehistoric,[4] though later than the tertiary epoch of geologists.[5]

A Russian botanist, Kolenati,[6] has made some very interesting observations on the different varieties of the vine, both wild and cultivated, in the country which may be called the central, and perhaps the most ancient home of the species, the south of the Caucasus. I consider his opinion the more important that the author has based

---

[1] These are figured in Heer's *Pflanzen der Pfahlbauten*, p. 24, fig. 11.

[2] Ragazzoni, *Rivista Arch. della Prov. di Como*, 1880, fasc. 17, p. 30.

[3] Heer, *ibid*.

[4] Planchon, *Étude sur les Tufs de Montpellier*, 1864, p. 63.

[5] De Saporta, *La Flore des Tufs Quaternaires de Provence*, 1867, pp. 15, 27.

[6] Kolenati, *Bulletin de la Société Impériale des Naturalistes de Moscou*, 1846, p. 279.

his classification of varieties with reference to the downy character and veining of the leaves, points absolutely indifferent to cultivators, and which consequently must far better represent the natural conditions of the plant. He says that the wild vines, of which he had seen an immense quantity between the Black and Caspian Seas, may be grouped into two subspecies which he describes, and declares are recognizable at a distance, and which are the point of departure of cultivated vines, at least in Armenia and the neighbourhood. He recognized them near Mount Ararat, at an altitude where the vine is not cultivated, where, indeed, it could not be cultivated. Other characters—for instance, the shape and colour of the grapes—vary in each of the subspecies. We cannot enter here into the purely botanical details of Kolenati's paper, any more than into those of Regel's more recent work on the genus *Vitis ;* [1] but it is well to note that a species cultivated from a very remote epoch, and which has perhaps two thousand described varieties, presents in the district where it is most ancient, and probably presented before all cultivation, at least two principal forms, with others of minor importance. If the wild vines of Persia and Kashmir, of Lebanon and Greece, were observed with the same care, perhaps other subspecies of prehistoric antiquity might be found. The idea of collecting the juice of the grape and of allowing it to ferment may have occurred to different peoples, principally in Western Asia, where the vine abounds and thrives. Adolphe Pictet,[2] who has, in common with numerous authors, but in a more scientific manner, considered the historical, philological, and even mythological questions relating to the vine among ancient peoples,

---

[1] Regel, *Acta Horti Imp. Petrop.*, 1873. In this short review of the genus, M. Regel gives it as his opinion that *Vitis vinifera* is a hybrid between two wild species, *V. vulpina* and *V. labrusca*, modified by cultivation ; but he gives no proof, and his characters of the two wild species are altogether unsatisfactory. It is much to be desired that the wild and cultivated vines of Europe and Asia should be compared with regard to their seeds, which furnish excellent distinctions, according to Englemann's observations on the American vines.

[2] Ad. Pictet, *Origines Indo-Eur.*, 2nd edit., vol. i. pp. 298–321.

admits that both Semitic and Aryan nations knew the use of wine, so that they may have introduced it into all the countries into which they migrated, into India and Egypt and Europe. This they were the better able to do, since they found the vine wild in several of these regions.

The records of the cultivation of the grape and of the making of wine in Egypt go back five or six thousand years.[1] In the West the propagation of its culture by the Phenicians, Greeks, and Romans is pretty well known, but to the east of Asia it took place at a late period. The Chinese who now cultivate the vine in their northern provinces did not possess it earlier than the year 122 B.C.[2]

It is known that several wild vines exist in the north of China, but I cannot agree with M. Regel in considering *Vitis Amurensis*, Ruprecht, the one most analogous to our vine, as identical in species. The seeds drawn in the *Gartenflora*, 1861, pl. 33, differ too widely. If the fruit of these vines of Eastern Asia had any value, the Chinese would certainly have turned them to account.

Common Jujube—*Zizyphus vulgaris*, Lamarck.

According to Pliny,[3] the jujube tree was brought from Syria to Rome by the consul Sextus Papinius, towards the end of the reign of Augustus. Botanists, however, have observed that the species is common in rocky places in Italy,[4] and that, moreover, it has not yet been found wild in Syria, although it is cultivated there, as in the whole region extending from the Mediterranean to China and Japan.[5]

The result of the search for the origin of the jujube tree as a wild plant bears out Pliny's assertion, in spite

[1] M. Delchevalerie, in *l'Illustration Horticole*, 1881, p. 28. He mentions in particular the tomb of Phtah-Hotep, who lived at Memphis 4000 B.C.

[2] Bretschneider, *Study and Value*, etc., p. 16.

[3] Pliny, *Hist.*, lib. 15, c. 14.

[4] Bertoloni, *Fl. Ital.*, ii. p. 665; Gussone, *Syn. Fl. Sicul.*, ii. p. 276.

[5] Willkomm and Lange, *Prod. Fl. Hisp.*, iii. p. 480; Desfontaines, *Fl. Atlant.*, i. p. 200; Boissier, *Fl. Orient.*, ii. p. 12; J. Hooker, *Fl. Brit. Ind.*, i. p. 633; Bunge, *Enum. Pl. Chin.*, p. 14; Franchet and Savatier, *Enum. Pl. Jap.*, i. p. 81.

of the objections I have just mentioned. According to plant collectors and authors of floras, the species appears to be more wild and more anciently cultivated in the east than in the west of its present wide area. Thus, in the north of China, de Bunge says it is "very common and very troublesome (on account of its thorns) in mountainous places." He had seen the thornless variety in gardens. Bretschneider[1] mentions the jujube as one of the fruits most prized by the Chinese, who give it the simple name *tsao*. He also mentions the two varieties, with and without thorns, the former wild.[2] The species does not grow in the south of China and in India proper, because of the heat and moisture of the climate. It is found again wild in the Punjab, in Persia, and Armenia.

Brandis[3] gives seven different names for the jujube tree (or for its varieties) in modern Indian languages, but no Sanskrit name is known. The species was therefore probably introduced into India from China, at no very distant epoch, and it must have escaped from cultivation and have become wild in the dry provinces of the west. The Persian name is *anob*, the Arabic *unab*. No Hebrew name is known, a further sign that the species is not very ancient in the west of Asia.

The ancient Greeks do not mention the common jujube, but only another species, *Zizyphus lotus*. At least, such is the opinion of the critic and modern botanist, Lenz.[4] It must be confessed that the modern Greek name *pritzuphuia* has no connection with the names formerly attributed in Theophrastus and Dioscorides to some Zizyphus, but is allied to the Latin name *zizyphus* (fruit *zizyphum*) of Pliny, which does not occur in earlier authors, and seems to be rather of an Oriental than of a Latin character. Heldreich[5] does not admit that the jujube tree is wild in Greece, and others say "naturalized, half-wild," which confirms the hypothesis of a

[1] Bretschneider, *Study and Value*, etc., p. 11.
[2] *Zizyphus chinensis* of some authors is the same species.
[3] Brandis, *Forest Flora of British India*, p. 84.
[4] Lenz, *Botanik der Alten*, p. 651.
[5] Heldreich, *Nutzpflanzen Griechenlands*, p. 57.

recent introduction. The same arguments apply to Italy. The species may have become naturalized there after the introduction into gardens mentioned by Pliny.

In Algeria the jujube is only cultivated or half-wild.[1] So also in Spain. It is not mentioned in Marocco, nor in the Canary Isles, which argues no very ancient existence in the Mediterranean basin.

It appears to me probable, therefore, that the species is a native of the north of China; that it was introduced and became naturalized in the west of Asia after the epoch of the Sanskrit language, perhaps two thousand five hundred or three thousand years ago; that the Greeks and Romans became acquainted with it at the beginning of our era, and that the latter carried it into Barbary and Spain, where it became partially naturalized by the effect of cultivation.

Lotus Jujube—*Zizyphus lotus*, Desfontaines.

The fruit of this jujube is not worthy of attention except from an historical point of view. It is said to have been the food of the lotus-eater, a people of the Lybian coast, of whom Herod and Herodotos [2] have given a more or less accurate account. The inhabitants of this country must have been very poor or very temperate, for a berry the size of a small cherry, tasteless, or slightly sweet, would not satisfy ordinary men. There is no proof that the lotus-eaters cultivated this little tree or shrub. They doubtless gathered the fruit in the open country, for the species is common in the north of Africa. One edition of Theophrastus [3] asserts, however, that there were some species of lotus without stones, which would imply cultivation. They were planted in gardens, as is still done in modern Egypt,[4] but it does not seem to have been a common custom even among the ancients.

For the rest, widely different opinions have been held

---

[1] Munby, *Catal.*, edit. 2, p. 9.

[2] *Odyssey*, bk. 1, v. 84; Herodotos, l. 4, p. 177, trans. in Lenz, *Bot. der Alt.*, p. 653.

[3] Theophrastus, *Hist.*, l. 4, c. 4, edit. 1644. The edition of 1613 does not contain the words which refer to this detail.

[4] Schweinfurth and Ascherson, *Beitr. zur Fl. Æthiop.*, p. 263.

touching the lotus of the lotus-eaters,[1] and it is needless to insist upon a point so obscure, in which so much must be allowed for the imagination of a poet and for popular ignorance.

The jujube tree is now wild in dry places from Egypt to Marocco, in the south of Spain, Terracina, and the neighbourhood of Palermo.[2] In isolated Italian localities it has probably escaped from cultivation.

Indian Jujube[3]—*Zizyphus jujube,* Lamarck; *ber* among the Hindus and Anglo-Indians, *masson* in the Mauritius.

This jujube is cultivated further south than the common kind, but its area is equally extensive. The fruit is sometimes like an unripe cherry, sometimes like an olive, as is shown in the plate published by Bouton in Hooker's *Journal of Botany,* i. pl. 140. The great number of known varieties indicates an ancient cultivation. It extends at the present day from Southern China, the Malay Archipelago, and Queensland, through Arabia and Egypt as far as Marocco, and even to Senegal, Guinea, and Angola.[4] It grows also in Mauritius, but it does not appear to have been introduced into America as yet, unless perhaps into Brazil, as it seems from a specimen in my herbarium.[5] The fruit is preferable to the common jujube, according to some writers.

It is not easy to know what was the habitation of the species before all cultivation, because the stones sow themselves readily and the plant becomes naturalized outside gardens.[6] If we are guided by its abundance in a wild state, it would seem that Burmah and British India are its original abode. I have in my herbarium several specimens gathered by Wallich in the kingdom of Burmah,

---

[1] See the article on the carob tree.

[2] Desfontaines, *Fl. Atlant.,* i. p. 200; Munby, *Catal. Alger.,* edit. 2, p. 9; Ball, *Spicilegium, Fl. Maroc.,* p. 301; Willkomm and Lange, *Prodr. Fl. Hisp.,* iii. p. 481; Bertoloni, *Fl. Ital.,* ii. p. 664.

[3] This name, which is little used, occurs in Bauhin, as *Jujuba Indica.*

[4] Sir J. Hooker, *Fl. Brit. Ind.,* i. p. 632; Brandis, *Forest Fl.,* i. 87; Bentham, *Fl. Austral.,* i. p. 412; Boissier, *Fl. Orient.,* ii. p. 13; Oliver, *Fl. of Trop. Afr.,* i. p. 379.

[5] Received from Martius, No. 1070, from the *Cabo frio.*

[6] Bouton, in Hooker's *Journ. of Bot.;* Baker, *Fl. of Mauritius,* p. 61; Brandis.

and Kurz has often seen it in the dry forests of that country, near Ava and Prome.[1]   Beddone admits the species to be wild in the forests of British India, but Brandis had only found it in the neighbourhood of native settlements.[2]   In the seventeenth century Rheede[3] described this tree as wild on the Malabar coast, and botanists of the sixteenth century had received it from Bengal.   In support of an Indian origin, I may mention the existence of three Sanskrit names, and of eleven other names in modern Indian languages.[4]

It had been recently introduced into the eastern islands of the Amboyna group when Rumphius was living there,[5] and he says himself that it is an Indian species.   It was perhaps originally in Sumatra and in other islands near to the Malay Peninsula.   Ancient Chinese authors do not mention it ; at least Bretschneider did not know of it.   Its extension and naturalization to the east of the continent of India appear, therefore, to have been recent.

Its introduction into Arabia and Egypt appears to be of yet later date.   Not only no ancient name is known, but Forskal, a hundred years ago, and Delile at the beginning of the present century, had not seen the species, of which Schweinfurth has recently spoken as cultivated.   It must have spread to Zanzibar from Asia, and by degrees across Africa or in European vessels as far as the west coast.   This must have been quite recently, as Robert Brown (*Bot. of Congo*) and Thonning did not see the species in Guinea.[6]

Cashew—*Anacardium occidentale*, Linnæus.

The most erroneous assertions about the origin of this species were formerly made,[7] and in spite of what

[1] Kurz, *Forest Flora of Burmah*, i. p. 266.

[2] Beddone, *Forest Flora of India*, i. pl. 149 (representing the wild fruit, which is smaller than that of the cultivated plant) ; Brandis.

[3] Rheede, iv. pl. 141.

[4] Piddington, *Index*.

[5] Rumphius, *Amboyna*, ii. pl. 36.

[6] *Zizyphus abyssinicus*, Hochst, seems to be a different species.

[7] Tussac, *Flore des Antilles*, iii. p. 55 (where there is an excellent figure, pl. 13).   He says that it is an East Indian species, thus aggravating Linnæus' mistake, who believed it to be Asiatic and American.

I said on the subject in 1855,[1] I find them occasionally reproduced.

The French name *Pommier d'acajou* (mahogany apple tree) is as absurd as it is possible to be. It is a tree belonging to the order of *Terebintaceæ* or *Anacardiaceæ*, very different from the Rosaceæ and the Meliaceæ, to which the apple and the mahogany belong. The edible part is more like a pear than an apple, and botanically speaking is not a fruit, but the receptacle or support of the fruit, which resembles a large bean. The two names, French and English, are both derived from a name given to it by the natives of Brazil, *acaju, acajaiba,* quoted by early travellers.[2] The species is certainly wild in the forests of tropical America, and indeed occupies a wide area in that region ; it is found, for example, in Brazil, Guiana, the Isthmus of Panama, and the West Indies.[3] Dr. Ernst[4] believes it is only indigenous in the basin of the Amazon River, although he had seen it also in Cuba, Panama, Ecuador, and New Granada. His opinion is founded upon the absence of all mention of the plant in Spanish authors of the time of the Conquest—a negative proof, which establishes a mere probability.

Rheede and Rumphius had also indicated this plant in the south of Asia. The former says it is common on the Malabar coast.[5] The existence of the same tropical arborescent species in Asia and America was so little probable, that it was at first suspected that there was a difference of species, or at least of variety ; but this was not confirmed. Different historical and philological proofs have convinced me that its origin is not Asiatic.[6] Moreover, Rumphius, who is always accurate, spoke of an ancient introduction by the Portuguese into the Malay Archipelago from America. The Malay name he gives,

[1] *Géogr. Bot. Rais.,* p. 873.

[2] Piso and Marcgraf, *Hist. rer. Natur. Brasil,* 1648, p. 57.

[3] Vide Piso and Marcgraf; Aublet, *Guyane,* p. 392 ; Seemann, *Bot. of the Herald,* p. 106 ; Jacquin, *Amér.,* p. 124 ; Macfadyen, *Pl. Jamaic.,* p. 119 ; Greisbach, *Fl. of Brit. W. Ind.,* p. 176.

[4] Ernst in Seemann, *Journ. of Bot.,* 1867, p. 273.

[5] Rheede, *Malabar,* iii. pl. 54.

[6] Rumphius, *Herb. Amboin.,* i. pp. 177, 178.

cadju, is American; that used at Amboyna means Portugal
fruit, that of Macassar was taken from the resemblance of
the fruit to that of the *jambosa*. Rumphius says that the
species was not widely diffused in the islands. Garcia ab
Orto did not find it at Goa in 1550, but Acosta afterwards saw it at Couchin, and the Portuguese propagated
it in India and the Malay Archipelago. According to
Blume and Miquel, the species is only cultivated in Java.
Rheede, it is true, says it is abundant (*provenit ubique*)
on the coast of Malabar, but he only quotes one name
which seems to be Indian, *kapa mava ;* all the others
are derived from the American name. Piddington gives
no Sanskrit name. Lastly, Anglo-Indian colonists, after
some hesitation as to its origin, now admit the importation
of the species from America at an early period. They
add that it has become naturalized in the forests of
British India.[1]

It is yet more doubtful that the tree is indigenous
in Africa, indeed it is easy to disprove the assertion.
Loureiro [2] had seen the species on the east coast of this
continent, but he supposed it to have been of American
origin. Thonning had not seen it in Guinea, nor Brown
in Congo.[3] It is true that specimens from the last-named
country and from the islands in the Gulf of Guinea were
sent to the herbarium at Kew, but Oliver says it is cultivated there.[4] A tree which occupies such a large area
in America, and which has become naturalized in several
districts of India within the last two centuries, would
exist over a great extent of tropical Africa if it were indigenous in that quarter of the globe.

Mango—*Mangifera indica*, Linnæus.

Belonging to the same order as the *Cashew*, this tree
nevertheless produces a true fruit, something the colour
of the apricot.[5]

It is impossible to doubt that it is a native of the
south of Asia or of the Malay Archipelago, when we see

---

[1] Beddone, *Flora Sylvatica*, t. 163; Hooker, *Fl. Brit. Ind.*, ii. p. 20.
[2] Loureiro, *Fl. Cochin.*, p. 304.   [3] Brown, *Congo*, pp. 12, 49.
[4] Oliver, *Fl. of Trop. Afr.*, i. p. 443.
[5] See plate 4510 of the *Botanical Magazine.*

the multitude of varieties cultivated in these countries, the number of ancient common names, in particular a Sanskrit name,[1] its abundance in the gardens of Bengal, of the Dekkan Peninsula, and of Ceylon, even in Rheede's time.   Its cultivation was less diffused in the direction of China, for Loureiro only mentions its existence in Cochin-China.   According to Rumphius,[2] it had been introduced into certain islands of the Asiatic Archipelago within the memory of living men. Forster does not mention it in his work on the fruits of the Pacific Islands at the time of Cook's expedition. The name common in the Philippine Isles, *manga*,[3] shows a foreign origin, for it is the Malay and Spanish name.   The common name in Ceylon is *ambe*, akin to the Sanskrit *amra*, whence the Persian and Arab *amb*,[4] the modern Indian names, and perhaps the Malay, *mangka, manga, manpelaan*, indicated by Rumphius. There are, however, other names used in the Sunda Islands, in the Moluccas, and in Cochin-China.   The variety of these names argues an ancient introduction into the East Indian Archipelago, in spite of the opinion of Rumphius.

The *Mangifera* which this author had seen wild in Java, and *Mangifera sylvatica* which Roxburgh had discovered at Silhet, are other species; but the true mango is indicated by modern authors as wild in the forests of Ceylon, the regions at the base of the Himalayas, especially towards the east, in Arracan, Pegu, and the Andaman Isles.[5]   Miquel does not mention it as wild in any of the islands of the Malay Archipelago.   In spite of its growing in Ceylon, and the indications, less positive certainly, of Sir Joseph Hooker in the *Flora of British India*, the species is probably rare or only naturalized in the Indian Peninsula.   The size of the stone is too great to allow of its being transported by

[1] Roxburgh, *Flora Indica*, edit. 2, vol. ii. p. 435 ; Piddington, *Index*.
[2] Rumphius, *Herb. Amboin.*, i. p. 95.
[3] Blanco, *Fl. Filip.*, p. 181.   [4] Rumphius ; Forskal, p. cvii.
[5] Thwaites, *Enum. Plant. Ceyl.*, p. 75 ; Brandis, *Forest Flora*, p. 126
Hooker, *Fl. Brit. Ind.*, ii. p. 13 ; Kurz, *Forest Flora Brit. Burmah*, i. p. 304.

birds, but the frequency of its cultivation causes a dispersion by man's agency. If the mango is only naturalized in the west of British India, this must have occurred at a remote epoch, as the existence of a Sanskrit name shows. On the other hand, the peoples of Western Asia must have known it late, since they did not transport the species into Egypt or elsewhere towards the west.

It is cultivated at the present day in tropical Africa, and even in Mauritius and the Seychelles, where it has become to some extent naturalized in the woods.[1]

In the new world it was first introduced into Brazil, for the seeds were brought thence to Barbados in the middle of the last century.[2] A French vessel was carrying some young trees from Bourbon to Saint Domingo in 1782, when it was taken by the English, who took them to Jamaica, where they succeeded wonderfully. When the coffee plantations were abandoned, at the time of the emancipation of the slaves, the mango, whose stones the negroes scattered everywhere, formed forests in every part of the islands, and these are now valued both for their shade and as a form of food.[3] It was not cultivated in Cayenne in the time of Aublet, at the end of the eighteenth century, but now there are mangoes of the finest kind in this colony. They are grafted, and it is observed that their stones produce better fruit than that of the original stock.[4]

Tahiti Apple—*Spondias dulcis*, Forster.

This tree belongs to the family of the *Anacardiaceœ* and is indigenous in the Society, Friendly, and Fiji Islands.[5] The natives consumed quantities of the fruit at the time of Cook's voyage. It is like a large plum, of

---

[1] Oliver, *Flora of Trop. Afr.*, i. p. 442; Baker, *Fl. of Maur. and Seych.*, p. 63.

[2] Hughes, *Barbados*, p. 177.

[3] Macfadyen, *Fl. of Jam.*, p. 221; Sir J. Hooker, *Speech at the Royal Institute*.

[4] Sagot, *Jour. de la Soc. Centr. d'Agric. de France*, 1872.

[5] Forster, *De Plantis Esculentis Insularum Oceani Australis*, p. 33; Seemann, *Flora Vitiensis*, p. 51; Nadaud, *Enum. des Plantes de Taïti*, p. 75.

the colour of an apple, and contains a stone covered with long hooked bristles.[1]   The flavour, according to travellers, is excellent.   It is not among the fruits most widely diffused in tropical colonies.   It is, however, cultivated in Mauritius and Bourbon, under the primitive Polynesian name *evi* or *hevi*,[2] and in the West Indies.   It was introduced into Jamaica in 1782, and thence into Saint Domingo.   Its absence in many of the hot countries of Asia and Africa is probably owing to the fact that the species was discovered, only a century ago, in small islands which have no communications with other countries.

Strawberry—*Fragaria vesca*, Linnæus.

Our common strawberry is one of the most widely diffused plants, partly owing to the small size of its seeds, which birds, attracted by the fleshy part on which they are found, carry to great distances.

It grows wild in Europe, from Lapland and the Shetland Isles [3] to the mountain ranges in the south; in Madeira, Spain, Sicily, and in Greece.[4]   It is also found in Asia, from Armenia and the north of Syria [5] to Dahuria.   The strawberries of the Himalayas and of Japan,[6] which several authors have attributed to this species, do not perhaps belong to it,[7] and this makes me doubt the assertion of a missionary [8] that it is found in China.   It is wild in Iceland,[9] in the north-east of the United States,[10] round Fort Cumberland, and on the north-west coast,[11] perhaps even in the Sierra-Nevada of

[1] There is a good coloured illustration in Tussac's *Fl. des Antilles,* iii. pl. 28.

[2] Boyer, *Hortus Mauritianus,* p. 81.

[3] H. C. Watson, *Compendium Cybele Brit.,* i. p. 160 ; Fries, *Summa Veg. Scand.,* p. 44.

[4] Lowe, *Man. Fl. of Madeira,* p. 246; Willkomm and Lange, *Prodr. Fl. Hisp.,* iii. p. 224; Moris, *Fl. Sardoa,* ii. p. 17.

[5] Boissier, *Fl. Orient.*   [6] Ledebour, *Fl. Ross.,* ii. p. 64.

[7] Gay; Hooker, *Fl. Brit. Ind.,* ii. p. 344; Franchet and Savatier, *Enum. Pl. Japon.,* i. p. 129.

[8] Perny, *Propag. de la Foi,* quoted in Decaisne's *Jardin Fruitier du Mus.,* p. 27.   Gay does not give China.

[9] Babington, *Journ. of Linnæan Society,* ii. p. 303 ; J. Gay.

[10] Asa Gray, *Botany of the Northern States,* edit. 1868, p. 156.

[11] Sir W. Hooker, *Fl. Bor. Amer.,* i. p. 184.

California.[1] Thus its area extends round the north pole, except in Eastern Siberia and the basin of the river Amur, since the species is not mentioned by Maximowicz in his *Primitiæ Floræ Amurensis.* In America its area is extended along the highlands of Mexico ; for *Fragaria mexicana,* cultivated in the *Jardin des Plantes,* and examined by Gay, is *F. vesca.* It also grows round Quito, according to the same botanist, who is an authority on this question.[2]

The Greeks and Romans did not cultivate the strawberry. Its cultivation was probably introduced in the fifteenth or sixteenth century. Champier, in the sixteenth century, speaks of it as a novelty in the north of France,[3] but it already existed in the south, and in England.[4]

Transported into gardens in the colonies, the strawberry has become naturalized in a few cool localities far from dwellings. This is the case in Jamaica,[5] in Mauritius,[6] and in Bourbon, where some plants had been placed by Commerson on the table-land known as the Kaffirs' Plain. Bory Saint-Vincent relates that in 1801 he found districts quite red with strawberries, and that it was impossible to cross them without staining the feet red with the juice, mixed with volcanic dust.[7] It is probable that similar cases of naturalization may be seen in Tasmania and New Zealand.

The genus Fragaria has been studied with more care than many others, by Duchesne (*fils*), the Comte de Lambertye, Jacques Gay, and especially by Madame Eliza Vilmorin, whose faculty of observation was worthy of the name she bore. A summary of their works, with excellent coloured plates, is published in the *Jardin*

[1] A. Gray, *Bot. Calif.,* i. p. 176.

[2] J. Gay, in Decaisne, *Jardin Fruitier du Muséum,* Fraisier, p. 30.

[3] Le Grand d'Aussy, *Hist. de la Vie Privée des Français,* i. pp. 233 and 3.

[4] Olivier de Serres, *Théâtre d'Agric.,* p. 511 ; Gerard, from Phillips, *Pomarium Britannicum,* p. 334.

[5] Purdie, in Hooker's *London Journal of Botany,* 1844, p. 515.

[6] Bojer, *Hortus Mauritianus,* p. 121.

[7] Bory Saint-Vincent, *Comptes Rendus de l'Acad. des. Sc. Nat.,* 1836, *sem.* ii. p. 109.

Fruitier du Muséum by Decaisne. These authors have overcome great difficulties in distinguishing the varieties and hybrids which are multiplied in gardens from the true species, and in defining these by well-marked characters. Some strawberries whose fruit is poor have been abandoned, and the finest are the result of the crossing of the species of Virginia and Chili, of which I am about to speak.

Virginian Strawberry—*Fragaria virginiana*, Ehrarht.

The scarlet strawberry of French gardens. This species, indigenous in Canada and in the eastern States of America, and of which one variety extends west as far as the Rocky Mountains, perhaps even to Oregon,[1] was introduced into English gardens in 1629.[2] It was much cultivated in France in the last century, but its hybrids with other species are now more esteemed.

Chili Strawberry—*Fragaria Chiloensis*, Duchesne.

A species common in Southern Chili, at Conception, Valdivia, and Chiloe,[3] and often cultivated in that country. It was brought to France by Frezier in the year 1715. Cultivated in the Museum of Natural History in France, it spread to England and elsewhere. The large size of the berry and its excellent flavour have produced by different crossings, especially with *F. virginiana*, the highly prized varieties *Ananas, Victoria, Trollope, Rubis*, etc.

Bird-Cherry—*Prunus avium*, Linnæus; *Süsskirschbaum* in German.

I use the word cherry because it is customary, and has no inconvenience when speaking of cultivated species or varieties, but the study of allied wild species confirms the opinion of Linnæus, that the cherries do not form a separate genus from the plums.

All the varieties of the cultivated cherry belong to two species, which are found wild: 1. *Prunus avium*, Linnæus, tall, with no suckers from the roots, leaves

---

[1] Asa Gray, *Manual of Botany of the Northern States*, edit. 1868, p. 155 ; *Botany of California*, i. p. 177.

[2] Phillips, *Pomar. Brit.*, p. 335.

[3] Cl. Gay, *Hist. Chili, Botanica*, ii. p. 305.

downy on the under side, the fruit sweet; 2. *Prunus cerasus*, Linnæus, shorter, with suckers from the roots, leaves glabrous, and fruit more or less sour or bitter.

The first of these species, from which the white and black cherries are developed, is wild in Asia; in the forest of Ghilan (north of Persia), in the Russian provinces to the south of the Caucasus and in Armenia;[1] in Europe in the south of Russia proper, and generally from the south of Sweden to the mountainous parts of Greece, Italy, and Spain.[2] It even exists in Algeria.[3]

As we leave the district to the south of the Caspian and Black Seas, the bird-cherry becomes less common, less natural, and determined more perhaps by the birds which seek its fruit and carry the seeds from place to place.[4] It cannot be doubted that it was thus naturalized, from cultivation, in the north of India,[5] in many of the plains of the south of Europe, in Madeira,[6] and here and there in the United States;[7] but it is probable that in the greater part of Europe this took place in prehistoric times, seeing that the agency of birds was employed before the first migrations of nations, perhaps before there were men in Europe. Its area must have extended in this region as the glaciers diminished.

The common names in ancient languages have been the subject of a learned article by Adolphe Pictet,[8] but nothing relative to the origin of the species can be deduced from them ; and besides, the different species and varieties have often been confused in popular nomenclature. It is far more important to know whether archæology can tell us anything about the presence of the bird-cherry in Europe in prehistoric times.

[1] Ledebour, *Fl. Ross.*, ii. p. 6; Boissier, *Fl. Orient.*, ii. p. 649.
[2] Ledebour, *ibid.* ; Fries, *Summa Scand.*, p. 46 ; Nyman, *Conspec. Fl. Eur.*, p. 213 ; Boissier, *ibid.* ; Willkomm and Lange, *Prodr. Fl. Hisp.*, iii. p. 245.
[3] Munby, *Catal. Alger.*, edit. 2, p. 8.
[4] As the cherries ripen after the season when birds migrate, they disperse the stones chiefly in the neighbourhood of the plantations.
[5] Sir J. Hooker, *Fl. of Brit. India.*
[6] Lowe, *Manual of Madeira*, p. 235.
[7] Darlington, *Fl. Cestrica*, edit. 3, p. 73.
[8] Ad. Pictet, *Origines Indo-Europ.*, edit. 2, vol. i. p. 281.

Heer gives an illustration of the stones of *Prunus avium*, in his paper on the lake-dwellings of Western Switzerland.[1] From what he was kind enough to write to me, April 14, 1881, these stones were found in the peat formed above the ancient deposits of the age of stone. De Mortillet[2] found similar cherry-stones in the lake-dwellings of Bourget belonging to an epoch not very remote, more recent than the stone age. Dr. Gross sent me some from the locality, also comparatively recent, of Corcelette on Lake Neuchâtel, and Strobel and Pigorini discovered some in the "terramare" of Parma.[3] All these are settlements posterior to the stone age, and perhaps belonging to historic time. If no more ancient stones of this species are found in Europe, it will seem probable that naturalization took place after the Aryan migrations.

Sour Cherry—*Prunus cerasus*, Linnæus ; *Cerasus vulgaris*, Miller ; *Baumweischel, Sauerkirschen*, in German.

The *Montmorency* and *griotte* cherries, and several other kinds known to horticulturists, are derived from this species.[4]

Hohenacker[5] saw *Prunus cerasus* at Lenkoran, near the Caspian Sea, and Koch[6] in the forests of Asia Minor, that is to say, in the north-east of that country, as that was the region in which he travelled. Ancient authors found it at Elisabethpol and Erivan, according to Ledebour.[7] Grisebach[8] indicates it on Mount Olympus of Bithynia, and adds that it is nearly wild on the plains of Macedonia. The true and really ancient habitation seems to extend from the Caspian Sea to the environs of Constantinople ; but in this very region *Prunus avium* is more common. Indeed, Boissier and Tchihatcheff do not appear to have seen *P. cerasus* even in the

---

[1] Heer, *Pflanzen der Pfahlbauten*, p. 24, figs. 17, 18, and p. 26.

[2] In Perrin, *Études Préhist. sur la Savoie*, p. 22.

[3] *Atte Soc. Ital. Sc. Nat.*, vol. vi.

[4] For the numerous varieties which have common names in France, varying with the different provinces, see *Duhamel, Traité des Arbres*, edit. 2, vol. v., in which are good coloured illustrations.

[5] Hohenacker, *Plantæ Talysch.*, p. 128.

[6] Koch, *Dendrologie*, i. p. 110.   [7] Ledebour, *Fl. Ross.*, ii. p. 6.

[8] Grisebach, *Spicil. Fl. Rumel.*, p. 86.

Pontus, though they received or brought back several specimens of *P. avium*.[1]

In the north of India, *P. cerasus* exists only as a cultivated plant.[2] The Chinese do not appear to have been acquainted with our two kinds of cherry. Hence it may be assumed that it was not very early introduced into India, and the absence of a Sanskrit name confirms this. We have seen that, according to Grisebach, *P. cerasus* is nearly wild in Macedonia. It was said to be wild in the Crimea, but Steven [3] only saw it cultivated; and Rehmann [4] gives only the allied species, *P. chamœcerasus*, Jacquin, as wild in the south of Russia. I very much doubt its wild character in any locality north of the Caucasus. Even in Greece, where Fraas said he saw this tree wild, Heldreich only knows it as a cultivated species.[5] In Dalmatia,[6] a particular variety or allied species, *P. Marasca*, is found really wild; it is used in making Maraschino wine. *P. cerasus* is wild in mountainous parts of Italy [7] and in the centre of France,[8] but farther to the west and north, and in Spain, the species is only found cultivated, and naturalized here and there as a bush. *P. cerasus*, more than the bird-cherry, evidently presents itself in Europe, as a foreign tree not completely naturalized.

None of the often-quoted passages [9] in Theophrastus, Pliny, and other ancient authors appear to apply to *P. cerasus*.[10] The most important, that of Theophrastus, belongs to *Prunus avium*, because of the height of the tree, a character which distinguishes it from *P. cerasus*. *Kerasos* being the name for the bird-cherry

[1] Boissier, *Fl. Orient.*, ii. p. 649; Tchihatcheff, *Asie Mineure, Bot.*, p. 198.

[2] Sir J. Hooker, *Fl. of Brit. India*, ii. p. 313.

[3] Steven, *Verzeichniss Halbinselm*, etc., p. 147.

[4] Rehmann, *Verhandl. Nat. Ver. Brunn*, x. 1871.

[5] Heldreich, *Nutzpfl. Griech.*, p. 69; *Pflanzen d'Attisch. Ebene.*, p. 477.

[6] Viviani, *Fl. Dalmat.*, iii. p. 258.   [7] Bertoloni, *Fl. Ital.*, v. p. 131.

[8] Lecoc and Lamotte, *Catal. du Plat. Centr. de la France*, p. 148.

[9] Theophrastes, *Hist. Pl.*, lib. 3, c. 13; Pliny, lib. 15, c. 25, and others quoted in Lenz, *Bot. der Alten Gr. and Röm.*, p. 710.

[10] Part of the description of Theophrastus shows a confusion with other trees. He says, for instance, that the nut is soft.

in Theophrastus, as now *kerasaia* among the modern Greeks, I notice a linguistic proof of the antiquity of *P. cerasus.* The Albanians, descendants of the Pelasgians, call the latter *vyssine*, an ancient name which reappears in the German *Wechsel*, and the Italian *visciolo*.[1] As the Albanians have also the name *kerasie* for *P. avium*, it is probable that their ancestors very clearly distinguished the two species by different names, perhaps before the arrival of the Hellenes in Greece.

Another indication of antiquity may be seen in Virgil (*Geor.* ii. 17)—

> " Pullulat ab radice aliis densissima silva
> Ut cerasis ulmisque "—

which applies to *P. cerasus,* not to *P. avium.*

Two paintings of the cherry tree were found at Pompeii, but it seems that it cannot be discovered to which of the two species they should be attributed.[2] Comes calls them *Prunus cerasus.*

Any archæological discovery would be more convincing. The stones of the two species present a difference in the furrow or groove, which has not escaped the observation of Heer and Sordelli. Unfortunately, only one stone of *P. cerasus* has been found in the prehistoric settlements of Italy and Switzerland, and what is more, it is not quite certain from what stratum it was taken. It appears that it was a non-archæological stratum.[3]

From all these data, somewhat contradictory and sufficiently vague, I am inclined to admit that *Prunus cerasus* was known and already becoming naturalized at the beginning of Greek civilization, and a little later in Italy before the epoch when Lucullus brought a cherry tree from Asia Minor. Pages might be transcribed from authors, even modern ones, who attribute, after Pliny, the introduction of the cherry into Italy to

---

[1] Ad. Pictet quotes forms of the same name in Persian, Turkish, and Russian, and derives from the same source the French word *guigne*, now used for certain varieties of the cherry.

[2] Schouw, *Die Erde*, p. 44; Comes, *Ill. delle Piante*, etc., in 4to, p. 56.

[3] Sordelli, *Piante della torbiera di Lagozza*, p. 40.

this rich Roman, in the year 65 B.C.   Since this error is
perpetuated by its incessant repetition in classical schools,
it must once more be said that cherry trees (at least the
bird-cherry) existed in Italy before Lucullus, and that
the famous *gourmet* did not need to go far to seek the
species with sour or bitter fruit.   I have no doubt that
he pleased the Romans with a good variety cultivated
in the Pontus, and that cultivators hastened to propagate
it by grafting, but Lucullus' share in the matter was
confined to this.

From what is now known of Kerasunt and the
ancient names of the cherry tree, I venture to maintain,
contrary to the received opinion, that it was a variety
of the bird-cherry of which the fleshy fruit is of a sweet
flavour.   I am inclined to think so because *Kerasos* in
Theophrastus is the name of *Prunus avium*, which is
far the commoner of the two in Asia Minor.   The town
of Kerasunt took its name from the tree, and it is
probable that the abundance of *Prunus avium* in the
neighbouring woods had induced the inhabitants to seek
the trees which yielded the best fruits in order to plant
them in their gardens.   Certainly, if Lucullus brought
fine white-heart cherries to Rome, his countrymen who
only knew the little wild cherry may well have said,
" It is a fruit which we have not." Pliny affirms nothing
more.

I must not conclude without suggesting a hypothesis
about the two kinds of cherry.   They differ but little in
character, and, what is very rare, their two ancient
habitations, which are most clearly proved, are similar
(from the Caspian Sea to Western Anatolia).   The two
species have spread towards the West, but unequally.
That which is commonest in its original home and the
stronger of the two (*P. avium*) has extended further and
at an earlier epoch, and has become better naturalized.
P. cerasus is, therefore, perhaps derived from the
other in prehistoric times.   I come thus, by a different
road, to an idea suggested by Caruel;[1]   only, instead
of saying that it would perhaps be better to unite them

---

[1] Caruel, *Flora Toscana*, p. 48.

now in one species, I consider them actually distinct, and content myself with supposing a descent, which for the rest it would not be easy to prove.

Cultivated Plums.

Pliny[1] speaks of the immense quantity of plums known in his time: *ingens turba prunorum.* Horticulturists now number more than three hundred. Some botanists have tried to attribute these to distinct wild species, but they have not always agreed, and judging from the specific names especially they seem to have had very different ideas. This diversity is on two heads; first as to the descent of a given cultivated variety, and secondly as to the distinction of the wild forms into species or varieties.

I do not pretend to classify the innumerable cultivated forms, and I think that labour useless when dealing with the question of geographical origin, for the differences lie principally in the shape, size, colour, and taste of the fruit, in characters, that is to say, which it has been the interest of horticulturists to cultivate when they occur, and even to create as far as it was in their power to do so. It is better to insist upon the distinction of the forms observed in a wild state, especially upon those from which man derives no advantage, and which have probably remained as they were before the existence of gardens.

It is probably only for about thirty years that botanists have given really comparative characters for the three species or varieties which exist in nature.[2] They may be summed up as follows:—

Prunus domestica, Linnæus. Tree or tall shrub, without thorns; young branches glabrous; flowers appearing with the leaves, their peduncles usually downy; fruit pendulous, ovoid and of a sweet flavour.

Prunus insititia, Linnæus. Tree or tall shrub, without thorns; young shoots covered with a velvet down; flowers appearing with the leaves, with peduncles covered

---

[1] *Hist.*, lib. 15, c. 13.
[2] Koch, *Syn. Fl. Germ.*, edit. 2, p. 228; Cosson and Germain, *Flore des Environs de Paris,* i. p. 165.

with a fine down, or glabrous; fruit pendulous, round or slightly elliptical, of a sweet flavour.

Prunus spinosa, Linnæus. A thorny shrub, with branches spreading out at right angles; young shoots downy; flowers appearing before the leaves; pedicles glabrous; fruit upright, round, and very sour.

This third form, so common in our hedges (sloe or blackthorn), is very different from the other two. Therefore, unless we interpret by hypothesis what may have happened before all observation, it seems to me impossible to consider the three forms as constituting one and the same species, unless we can show transitions from one to the other in those organs which have not been modified by cultivation, and hitherto this has not been done. At most the fusion of the two first categories can be admitted. The two forms with naturally sweet fruit occur in few countries. These must have tempted cultivators more than *Prunus spinosa*, whose fruit is so sour. It is, therefore, in these that we must seek to find the originals of cultivated plums. For greater clearness I shall speak of them as two species.[1]

Common Plum—*Prunus domestica*, Linnæus; *Zwetchen* in German.

Several botanists[2] have found this variety wild throughout Anatolia, the region to the south of the Caucasus and Northern Persia, in the neighbourhood of Mount Elbruz, for example.

I know of no proof for the localities of Kashmir, the country of the Kirghis and of China, which are mentioned in some floras. The species is often doubtful, and it is probably rather *Prunus insititia;* in other cases it is its true and ancient wild character which is uncertain, for the stones have evidently been dispersed from cultivation. Its area does not appear to extend as far as Lebanon, although the plums cultivated at Damascus (damascenes, or damsons) have a reputation which dates

---

[1] Hudson, *Fl. Anglic.*, 1778, p. 212, unites them under the name *Prunus communis*.

[2] Ledebour, *Fl. Ross.*, ii. p. 5; Boissier, *Fl. Orient.*, ii. p. 652; K. Koch, *Dendrologie*, i. p. 94; Boissier and Bühse, *Aufzähl Transcaucasien*, p. 80.

from the days of Pliny. It is supposed that this was the
species referred to by Dioscorides [1] under the name of
Syrian coccumelea, growing at Damascus.   Karl Koch
relates that the merchants trading on the borders of
China told him that the species was common in the
forests of the western part of the empire.   It is true that
the Chinese have cultivated different kinds of plums
from time immemorial, but we do not know them well
enough to judge of them, and we cannot be sure that
they are indigenous.   As none of our kinds of plum has
been found wild in Japan or in the basin of the river
Amur, it is very probable that the species seen in China
are different to ours.   This appears also to be the result
of Bretschneider's statements. [2]

It is very doubtful if *Prunus domestica* is indigenous in Europe.   In the south, where it is given, it
grows chiefly in hedges, near dwellings, with all the
appearance of a tree scarcely naturalized, and maintained
here and there by the constant bringing of stones from
plantations.   Authors who have seen the species in the
East do not hesitate to say that it is "subspontaneous."
Fraas [3] affirms that it is not wild in Greece, and this is
confirmed as far as Attica is concerned by Heldreich. [4]
Steven [5] says the same for the Crimea.   If this is the
case near Asia Minor, it must be the more readily
admitted for the rest of Europe.

In spite of the abundance of plums cultivated formerly
by the Romans, no kind is found represented in the
frescoes at Pompeii. [6]   Neither has *Prunus domestica*
been found among the remains of the lake-dwellings of
Italy, Switzerland, and Savoy, where, however, stones
of *Prunus insititia* and *spinosa* have been discovered.
From these facts, and the small number of words at-
tributable to this species in Greek authors, it may be

---

[1] Dioscorides, p. 174.
[2] Bretschneider, *On the Study,* etc., p. 10.
[3] Fraas, *Syn. Fl. Class.,* p. 69.
[4] Heldreich, *Pflanzen Attischen Ebene.*
[5] Steven, *Verzeichniss Halbinseln,* i. p. 172.
[6] Comes, *Ill. Piante Pompeiane.*

inferred that its half-wild or half-naturalized state dates in Europe from two thousand years at most.

Prunes and damsons are ranked with this species.

Bullace—*Prunus insititia,* Linnæus;[1] *Pflauenbaum* and *Haferschlehen* in German.

This kind of plum grows wild in the south of Europe.[2] It has also been found in Cilicia, Armenia, to the south of the Caucasus, and in the province of Talysch near the Caspian Sea.[3] It is especially in Turkey in Europe and to the south of the Caucasus that it appears to be truly wild. In Italy and in Spain it is perhaps less so, although trustworthy authors who have seen the plant growing have no doubt about it. In the localities named north of the Alps, even as far as Denmark, it is probably naturalized from cultivation. The species is commonly found in hedges not far from dwellings, and apparently not truly wild.

All this agrees with archæological and historical data. The ancient Greeks distinguished the *Coccumelea* of their country from those of Syria,[4] whence it is inferred that the former were *Prunus insititia.* This seems the more likely that the modern Greeks call it *coromeleia.*[5] The Albanians say *corombile,*[6] which has led some people to suppose an ancient Pelasgian origin. For the rest, we must not insist upon the common names of the plum which each nation may have given to one or another species, perhaps also to some cultivated variety, without any rule. The names which have been much commented upon in learned works generally, appear to me to apply to any plum or plum tree without having any very defined meaning.

No stones of *P. insititia* have yet been found in

---

[1] *Insititia* = foreign. A curious name, since every plant is foreign to all countries but its own.

[2] Willkomm and Lange, *Prodr. Fl. Hisp.,* iii. p. 244; Bertoloni, *Fl. Ital.,* v. p. 135; Grisebach, *Spicel. Fl. Rumel.,* p. 85; Heldreich, *Nutzpfl. Griech.,* p. 68.

[3] Boissier, *Fl. Orient.,* ii. p. 651; Ledebour, *Fl. Ross.,* ii. p. 5; Hohenacker, *Pl. Talysch,* p. 128.

[4] Dioscorides, p. 173; Fraas, *Fl. Class.,* p. 69.

[5] Heldreich, *Nutzpflanzen Griechenlands,* p. 68.   [6] *Ibid.*

the *terra-mare* of Italy, but Heer has described and given illustrations of some which were found in the lake-dwellings of Robenhausen.[1] The species does not seem to be now indigenous in this part of Switzerland, but we must not forget that, as we saw in the history of flax, the lake-dwellers of the canton of Zurich, in the age of stone, had communications with Italy. These ancient Swiss were not hard to please in the matter of food, for they also gathered the berries of the blackthorn, which are, as we think, uneatable. It is probable that they ate them cooked.

Apricot—*Prunus armeniaca*, Linnæus; *Armenica vulgaris*, Lamarck.

The Greeks and Romans received the apricot about the beginning of the Christian era. Unknown in the time of Theophrastus, Dioscorides [2] mentions it under the name of *mailon armeniacon*. He says that the Latins called it *praikokion*. It is, in fact, one of the fruits mentioned briefly by Pliny,[3] under the name of *præcocium*, so called from the precocity of the species.[4] Its Armenian origin is indicated by the Greek name, but this name might mean only that the species was cultivated in Armenia. Modern botanists have long had good reason to believe that the species is wild in that country. Pallas, Güldenstädt, and Hohenacker say they found it in the neighbourhood of the Caucasus Mountains, on the north, on the banks of the Terek, and to the south between the Caspian and Black Seas.[5] Boissier [6] admits all these localities, but without saying anything about the wild character of the species. He saw a specimen gathered by Hohenacker, near Elisabethpol. On the

---

[1] Heer, *Pflanzen der Pfahlbauten*, p. 27, fig. 16, c.

[2] Dioscorides, lib. 1, c. 165.   [3] Pliny, lib. 2, cap. 12.

[4] The Latin name has passed into modern Greek (*prikokkia*). The Spanish and French names, etc. (*albaricoque, abricot*), seem to be derived from *arbor præcox*, or *præcocium*, while the old French word *armegne*, and the Italian *armenilli*, etc., come from *mailon armeniacon*. See further details about the names of the species in my *Géographie Botanique Raisonnée*, p. 880.

[5] Ledebour, *Fl. Ross.*, ii. p. 3.

[6] Boissier, *Fl. Orient.*, ii. p. 652.

other hand, Tchihatcheff[1] who has crossed Anatolia and Armenia several times, does not seem to have seen the wild apricot; and what is still more significant, Karl Koch, who travelled through the region to the south of the Caucasus, in order to observe facts of this nature, expresses himself as follows:[2] "Native country unknown. At least, during my long sojourn in Armenia, I nowhere found the apricot wild, and I have rarely seen it even cultivated."

A traveller, W. J. Hamilton,[3] said he found it wild near Orgou and Outch Hisar in Anatolia: but this assertion has not been verified by a botanist. The supposed wild apricot of the ruins of Baalbek, described by Eusèbe de Salle[4] is, from what he says of the leaf and fruit, totally different to the common apricot. Boissier, and the different collectors who sent him plants from Syria and Lebanon, do not appear to have seen the species. Spach[5] asserts that it is indigenous in Persia, but he gives no proof. Boissier and Buhse[6] do not mention it in their list of the plants of Transcaucasia and Persia. It is useless to seek its origin in Africa. The apricots which Reynier[7] says he saw, "almost wild," in Upper Egypt must have sprung from stones grown in cultivated ground, as is seen in Algeria.[8] Schweinfurth and Ascherson,[9] in their catalogue of the plants of Egypt and Abyssinia, only mention the species as cultivated. Besides, if it had existed formerly in the north of Africa it would have been early known to the Hebrews and the Romans. Now there is no Hebrew name, and Pliny says its introduction at Rome took place thirty years before he wrote.

Carrying our researches eastward, we find that Anglo-

---

[1] Tchihatcheff, *Asie Mineure, Botanique*, vol. i.
[2] K. Koch, *Dendrologie*, i. p. 87.
[3] *Nouv. Ann. des Voyages*, Feb., 1839, p. 176.
[4] E. de Salle, *Voyage*, i. p. 140.
[5] Spach, *Hist. des Végét. Phanér.*, i. p. 389.
[6] Boissier and Buhse, *Aufzählung*, etc., in 4to, 1860.
[7] Reynier, *Économie des Égyptiens*, p. 371.
[8] Munby, *Catal. Fl. d'Algér.*, edit. 2, p. 49.
[9] Schweinfurth and Ascherson, *Beitrage z. Fl. Æthiop.*, in 4to., 1867, p. 259.

Indian botanists [1] are agreed in considering that the apricot, which is generally cultivated in the north of India and in Thibet, is not wild in those regions; but they add that it has a tendency to become naturalized, and that it is found upon the site of ruined villages. Messrs. Schlagintweit brought specimens from the north-west provinces of India, and from Thibet, which West-mael verified,[2] but he was kind enough to write to me that he cannot affirm that it was wild, since the collector's label gives no information on that head.

Roxburgh,[3] who did not neglect the question of origin, says, speaking of the apricot, "native of China as well as the west of Asia." I read in Dr. Bretschneider's curious little work,[4] drawn up at Pekin, the following passage, which seems to me to decide the question in favour of a Chinese origin:—"*Sing*, as is well known, is the apricot (*Prunus armeniaca*). The character (a Chinese sign printed on p. 10) does not exist as indicating a fruit, either in the *Shu-king*, or in the *Shi-king*, *Cihouli*, etc., but the *Shan-hai-king* says that several *sings* grow upon the hills (here a Chinese character). Besides, the name of the apricot is represented by a particular sign which may show that it is indigenous in China." The *Shan-hai-king* is attributed to the Emperor Yü, who lived in 2205–2198 B.C. Decaisne,[5] who was the first to suspect the Chinese origin of the apricot, has recently received from Dr. Bretschneider some specimens accompanied by the following note:—"No. 24, apricot wild in the mountains of Pekin, where it grows in abundance; the fruit is small (an inch and a quarter in diameter), the skin red and yellow; the flesh salmon colour, sour, but eatable. No. 25, the stone of the apricot cultivated round Pekin. The fruit is twice as large as

---

[1] Royle, *Ill. of Himalaya*, p. 205; Aitchison, *Catal. of Punjab and Sindh*, p. 56; Sir Joseph Hooker, *Fl. of Brit. Ind.*, ii. p. 313; Brandis, *Forest Flora of N. W. and Central India*, 191.
[2] Westmael, in *Bull. Soc. Bot. Belgiq.*, viii., p. 219.
[3] Roxburgh, *Fl. Ind.*, edit. 2, v. ii. p. 501.
[4] Bretschneider, *On the Study and Value*, etc., pp. 10, 49.
[5] Decaisne, *Jardin Fruitier du Muséum*, vol. viii., art. *Abricotier.*

that of the wild tree."[1]　Decaisne adds, in the letter
he was good enough to write to me, "In shape and
surface the stones are exactly like those of our small
apricots; they are smooth and not pitted." The leaves
he sent me are certainly those of the apricot.

The apricot is not mentioned in Japan, or in the basin
of the river Amoor.[2] Perhaps the cold of the winter is
too great. If we recollect the absence of communication
in ancient times between China and India, and the
assertions that the plant is indigenous in both countries,
we are at first tempted to believe that the ancient area
extended from the north-west of India to China. However, if we wish to adopt this hypothesis, we must also
admit that the culture of the apricot spread very late
towards the West.[3] For no Sanskrit or Hebrew name is
known, but only a Hindu name, *zard alu*, and a Persian
name, *mischmisch*, which has passed into Arabic.[4] How
is it to be supposed that so excellent a fruit, and one
which grows in abundance in Western Asia, spread so
slowly from the north-west of India towards the Græco-
Roman world? The Chinese knew it two or three
thousand years before the Christian era. Changkien
went as far as Bactriana, a century before our era, and
he was the first to make the West known to his fellow-
countrymen.[5] It was then, perhaps, that the apricot was
introduced in Western Asia, and that it was cultivated
and became naturalized here and there in the north-west
of India, and at the foot of the Caucasus, by the scatter-
ing of the stones beyond the limits of the plantations.

Almond—*Amygdalus communis*, Linnæus; *Pruni
species*, Baillon; *Prunus Amygdalus*, Hooker.

---

[1] Dr. Bretschneider confirms this in a recent work, *Notes on Botanical
Questions*, p. 3.

[2] *Prunus armeniaca* of Thunberg is *P. mume* of Siebold and Zuccha-
rini. The apricot is not mentioned in the *Enumeratio*, etc., of Franchet
and Savatier.

[3] Capus (*Ann. Sc. Nat.*, sixth series, vol. xv. p. 206) found it wild in
Turkestan at the height of four thousand to seven thousand feet, which
weakens the hypothesis of a solely Chinese origin.

[4] Piddington, *Index*; Roxburgh, *Fl. Ind.*; Forskal, *Fl. Ægyp.*; Delile,
Ill. Egypt.

[5] Bretschneider, *On the Study and Value*, etc.

The almond grows apparently wild or half wild in the warm, dry regions of the Mediterranean basin and of western temperate Asia. As the nuts from cultivated trees naturalize the species very easily, we must have recourse to various indications to discern its ancient home.

We may first discard the notion of its origin in Eastern Asia. Japanese floras make no mention of the almond. That which M. de Bonge saw cultivated in the north of China was the *Persica Davidiana*.[1] Dr. Bretschneider,[2] in his classical work, tells us that he has never seen the almond cultivated in China, and that the compilation entitled *Pent-sao,* published in the tenth or eleventh century of our era, describes it as a tree of the country of the Mahometans, which signifies the north-west of India, or Persia.

Anglo-Indian botanists[3] say that the almond is cultivated in the cool parts of India, but some add that it does not thrive, and that many almonds are brought from Persia.[4] No Sanskrit name is known, nor even any in the languages derived from Sanskrit. Evidently the north-west of India is not the original home of the species.

On the other hand, there are many localities in the region extending from Mesopotamia and Turkestan to Algeria, where excellent botanists have found the almond tree quite wild. Boissier[5] has seen specimens gathered in rocky ground in Mesopotamia, Aderbijan, Turkestan, Kurdistan, and in the forests of the Anti-Lebanon. Karl Koch[6] has not found it wild to the south of the Caucasus, nor Tchihatcheff in Asia Minor. Cosson[7] found natural woods of almond trees near Saida in Algeria. It

[1] Bretschneider, *Early European Researches,* p. 149.
[2] Bretschneider, *Study and Value,* etc., p. 10; and *Early Europ. Resear.,* p. 149.
[3] Brandis, *Forest Flora*; Sir J. Hooker, *Fl. of Brit. Ind.,* iii. p. 313.
[4] Roxburgh, *Fl. Ind.,* edit. 2, vol. ii. p. 500; Royle, *Ill. Himal.,* p. 204.
[5] Boissier, *Fl. Orien.,* iii. p. 641.
[6] K. Koch, *Dendrologie,* i. p. 80; Tchihatcheff, *Asie Mineure Botanique,* i. p. 108.
[7] *Ann. des Sc. Nat.,* 3rd series, vol. xix. p. 108.

is also regarded as wild on the coasts of Sicily and of
Greece;[1] but there, and still more in the localities in
which it occurs in Italy, Spain, and France, it is probable,
and almost certain, that it springs from the casual dispersal of the nuts from cultivation.

The antiquity of its existence in Western Asia is
proved by Hebrew names for the almond tree—*schaked,
luz* or *lus* (which recurs in the Arabic *louz*), and *sche-
kedim* for the nut.[2]   The Persians have another name,
badam, but I do not know how old this is.   Theophrastus and Dioscorides [3] mention the almond by an entirely
different name, *amugdalai*, translated by the Latins into
amygdalus. It may be inferred from this that the Greeks
did not receive the species from the interior of Asia, but
found it in their own country, or at least in Asia Minor.
The almond tree is represented in several frescoes found
at Pompeii.[4]   Pliny [5] doubts whether the species was
known in Italy in Cato's time, because it was called the
Greek nut.   It is very possible that the almond was introduced into Italy from the Greek islands.   Almonds
have not been found in the *terra-mare* of the neighbourhood of Parma, even in the upper layers.

The late introduction of the species into Italy, and the
absence of naturalization in Sardinia and Spain,[6] incline
me to doubt whether it is really indigenous in the north
of Africa and Sicily.   In the latter countries it was more
probably naturalized some centuries ago.   In confirmation of this hypothesis, I note that the Berber name of
the almond, *talouzet*,[7] is evidently connected with the
Arabic *louz*, that is to say with the language of the
conquerors who came after the Romans.   In Western
Asia, on the contrary, and even in some parts of Greece,

---

[1] Gussone, *Synopsis Floræ Siculæ*, i. p. 552 ; Heldreich, *Nutzpflanzen
Griechenlands*, p. 67.

[2] Hiller, *Hierophyton*, i. p. 215 ; Rosenmüller, *Handb. Bibl. Alterth.*,
iv. p. 263.

[3] Theophrastus, *Hist.*, lib. 1, c. 11, 18, etc.; Dioscorides, lib. 1, c. 176.

[4] Schouw, *Die Erde*, etc. ; Comes, *Ill. Piante nei dipinti Pomp.*, p. 13.

[5] Pliny, *Hist.*, lib. 16, c. 22.

[6] Moris, *Flora Sardoa*, ii. p. 5 ; Willkomm and Lange, *Prodr. Fl. Hisp.*,
ii. p. 243.

[7] *Dictionnaire Français Berbère*, 1844.

it may be regarded as indigenous from prehistoric time. I do not say primitive, for everything was preceded by something else. I remark finally that the difference between bitter and sweet almonds was known to the Greeks and even to the Hebrews.

Peach—*Amygdalus persica*, Linnæus; *Persica vulgaris*, Miller; *Prunus persica*, Bentham and Hooker.

I will quote the article in which I formerly[1] attributed a Chinese origin to the peach, a contrary opinion to that which prevailed at the time, and which people who are not on a par with modern science continue to reproduce. I will afterwards give the facts discovered since 1855.

" The Greeks and Romans received the peach shortly after the beginning of the Christian era. The names *persica, malum persicum*, indicate whence they had it. I need not dwell upon those well-known facts.[2] Several kinds of peach are now cultivated in the north of India,[3] but, what is remarkable, no Sanskrit name is known;[4] whence we may infer that its existence and its cultivation are of no great antiquity in these regions. Roxburgh, who is usually careful to give the modern Indian names, only mentions Arab and Chinese names. Piddington gives no Indian name, and Royle only Persian names. The peach does not succeed, or requires the greatest care to ensure success, in the north-east of India.[5] In China, on the contrary, its cultivation dates from the remotest antiquity. A number of superstitious ideas and of legends about the properties of its different varieties exist in that country.[6] These varieties are very

---

[1] Alph. de Candolle, *Géogr. Bot. Rais.*, p. 881.

[2] Theophrastus, *Hist.*, iv. c. 4; Dioscorides, lib. 1, c. 164; Pliny, Geneva edit., bk. 15, c. 13.

[3] Royle, *Ill. Him.*, p. 204.

[4] Roxburgh, *Fl. Ind.*, 2nd. edit., ii. p. 500; Piddington, *Index;* Royle, *ibid.*

[5] Sir Joseph Hooker, *Journ. of Bot.*, 1850, p. 54.

[6] Rose, the head of the French trade at Canton, collected these from Chinese manuscripts, and Noisette (*Jard. Fruit.*, i. p. 76) has transcribed a part of his article. The facts are of the following nature. The Chinese believe the oval peaches, which are very red on one side, to be a symbol of a long life. In consequence of this ancient belief, peaches are used in all ornaments in painting and sculpture, and in congratulatory pre-

numerous;[1] and in particular the singular variety with compressed or flattened fruit,[2] which appears to be further removed than any other from the natural state of the peach; lastly, a simple name, *to*, is given to the common peach.[3]

" From all these facts, I am inclined to believe that the peach is of Chinese rather than of western Asiatic origin. If it had existed in Persia or Armenia from all time, the knowledge and cultivation of so pleasant a fruit would have spread earlier into Asia Minor and Greece. The expedition of Alexander probably was the means of making it known to Theophrastus (332 B.C.), who speaks of it as a Persian fruit. Perhaps this vague idea of the Greeks dates from the retreat of the ten thousand (401 B.C.); but Xenophon does not mention the peach. Nor do the Hebrew writings speak of it. The peach has no Sanskrit name, yet the peoples who spoke this language came into India from the north-west; that is to say, from the generally received home of the species. On this hypothesis, how are we to account for the fact that neither the Greeks of the early times of Greece, nor the Hebrews, nor the Sanskrit-speaking peoples, who all radiated from the upper part of the Euphrates valley or communicated with it, did not cultivate the peach ? On the other hand, it is very possible that the stones of a fruit tree cultivated in China from the remotest times, should have been carried over the mountains from the centre of Asia into Kashmir, Bokhara, and Persia. The Chinese had very early discovered this route. The importation would have taken place between the epoch of the Sanskrit emigrations and the relations of the Persians with the Greeks. The cultivation of the peach, once

sents, etc. According to the work of Chin-noug-king, the peach *Yu* prevents death. If it is not eaten in time, it at least preserves the body from decay until the end of the world. The peach is always mentioned among the fruits of immortality, with which were entertained the hopes of Tsinchi-Hoang, Vouty, of the Hans and other emperors who pretended to immortality, etc.

[1] Lindley, *Trans. Hort. Soc.*, v. p. 121.
[2] *Trans. Hort. Soc. Lond.*, iv. p. 512, tab. 19.
[3] Roxburgh, *Fl. Ind.*

established in Persia, would have easily spread on the one side towards the west; on the other, through Cabul towards the north of India, where it is not so very ancient.

" In confirmation of the hypothesis of a Chinese origin, it may be added that the peach was introduced into Cochin-China from China,[1] and that the Japanese give the Chinese name *Tao* [2] to the peach. M. Stanislas Julien was kind enough to read to me in French some passages of the Japanese encyclopædia (bk. lxxxvi. p. 7), in which the peach tree *tao* is said to be a tree of Western countries, which should be understood to mean the interior of China as compared to the eastern coast, since the passage is taken from a Chinese author. The *tao* occurs in the writings of Confucius in the fifth century before the Christian era, and even in the *Ritual* in the tenth century before Christ. Its wild nature is not specified in the encyclopædia of which I have just spoken; but Chinese authors pay little attention to this point."

After a few details about the common names of the peach in different languages, I went on to say, " The absence of Sanskrit and Hebrew names remains the most important fact, whence we may infer an introduction into Western Asia from a more distant land, that is to say, from China.

" The peach has been found wild in different parts of Asia; but it is always a question whether it is indigenous there, or whether it sprang from the dispersion of stones produced by cultivated trees. The question is the more necessary since the stones germinate easily, and several of the modifications of the peach are hereditary.[3] Apparently wild peach trees have often been found in the neighbourhood of the Caucasus. Pallas [4] saw several on the banks of the Terek, where the inhabitants give

[1] Loureiro, *Fl. Cochin.*, p. 386.

[2] Kæmpfer, *Amœn.*, p. 798; Thunberg, *Fl. Jap.*, p. 199. Kæmpfer and Thunberg also give the name *momu*, but Siebold (*Fl. Jap.*, i. p. 29) attributes a somewhat similar name, *mume*, to a plum tree, *Prunus mume*, Sieb. and Z.

[3] Noisette, *Jard. Fr.*, p. 77; *Trans. Soc. Hort. Lond.*, iv. p. 513.

[4] Pallas, *Fl. Rossica*, p. 13.

it a name which he calls Persian, *scheptata*.[1] It fruit is velvety, sour, not very fleshy, and hardly larger than a walnut; the tree small. Pallas suspects that this tree has degenerated from cultivated peaches. He adds that it is found in the Crimea, to the south of the Caucasus, and in Persia; but Marshall, Bicberstein, Meyer, and Hohenacker do not give the wild peach in the neighbourhood of the Caucasus. Early travellers, Gmelin, Guldenstadt, and Georgi, quoted by Ledebour, mentioned it. C. Koch [2] is the only modern botanist who said he found the peach tree in abundance in the Caucasian provinces. Ledebour, however, prudently adds, Is it wild? The stones which Brugnière and Olivier brought from Ispahan, which were sown in Paris and yielded a good velvety peach, were not, as Bosc [3] asserted, taken from a peach tree wild in Persia, but from one growing in a garden at Ispahan.[4] I do not know of any proof of a peach tree found wild in Persia, and if travellers mention any it is always to be feared that these are only sown trees. Dr. Royle [5] says that the peach grows wild in several places south of the Himalayas, notably near Mussouri, but we have seen that its culture is not ancient in these regions, and neither Roxburgh nor Don's *Flora Nepalensis* mention the peach. Bunge [6] only found cultivated trees in the north of China. This country has hardly been explored, and Chinese legends seem sometimes to indicate wild peaches. Thus the *Chou-y-ki*, according to the author previously quoted, says, 'Whosoever eats of the peaches of Mount Kouoliou shall obtain eternal life.' For Japan, Thunberg [7] says, *Crescit ubique vulgaris, præcipue juxta Nagasaki. In omni horto colitur ob elegantiam florum.* It seems from this passage that the species grows both in and out of gardens, but perhaps in the first case he only alludes to peaches growing in the open air and without shelter.

---

[1] *Shuft aloo* is, according to Royle (*Ill. Him.* p. 204), the Persian name for the nectarine.

[2] Ledebour, *Fl. Ross.*, i. p. 3. See p. 228, the subsequent opinion of Koch.

[3] Bosc, *Dict. d'Agric.*, ix. p. 481.   [4] Thouin, *Ann. Mus.*, viii. p. 433.

[5] Royle, *Ill. Him.*, p. 204.   [6] Bunge, *Enum. Pl. Chin.*, p. 23.

[7] Thunberg, *Fl. Jap.* 199.

"I have said nothing hitherto of the distinction to be established between the different varieties or species of the peach, since most of them are cultivated in all countries—at least the clearly defined kinds, which may be considered as botanical species. Thus the great distinction between the downy and smooth-skinned fruits (peaches proper and nectarines), on which it is proposed to found two species (*Persica vulgaris*, Mill, and *P. levis*, D. C.), exists in Japan [1] and in Europe, as in most of the intermediate countries.[2] Less importance is attached to distinctions founded on the adherence or non-adherence of the skin, on the white, yellow, or red colour of the flesh, and on the general form of the fruit. The great division into peaches and nectarines presents most of these modifications in Europe, in Western Asia, and probably in China. It is certain that in the latter country the form of the fruit varies more than elsewhere; for there are as in Europe oval peaches, and also the peaches of which I spoke just now, which are quite flattened, in which the top of the stone is not even covered with flesh.[3] The colour also varies greatly.[4] In Europe the most distinct varieties, nectarines and peaches, freestones and clingstones, existed three centuries ago, for J. Bauhin enumerates them very clearly;[5] and before him Dalechamp, in 1587, also gave the principal ones.[6] At that time nectarines were called *Nucipersica*, because of their resemblance in shape, size, and colour to the walnut. It is in the same sense that the Italians call them *pescanoce*.

"I have sought in vain for a proof that the nectarine existed in Italy in the time of ancient Rome. Pliny,[7] who confounds in his compilation peaches, plums, the *Laurus Persea*,[8] and perhaps other trees, says nothing

[1] Thunberg, *Fl. Jap.*, 199.
[2] The accounts about China which I have consulted do not mention the nectarine; but as it exists in Japan, it is extremely probable that it does also in China.
[3] Noisette, *Jard. Fr.*, p. 77; *Trans. Hort. Soc.*, iv. p. 512, tab. 19.
[4] Lindley, *Trans. Hort. Soc.*, v. p. 122.   [5] J. Bauhin,*Hist.*, i. pp. 162,163.
[6] Dalechamp, *Hist.*, i. p. 295.   [7] Pliny, lib. xv. cap. 12 and 13.
[8] Pliny, *De Div. Gen. Malorum*, lib. ii. cap. 14.

which can apply to such a fruit. Sometimes people have thought they recognized it in the *tuberes* of which he speaks. It was a tree imported from Syria in the time of Augustus. There were both red and white *tuberes.* Others (*tuberes?* or *mala?*) of the neighbourhood of Verona were downy. Some graceful verses of Petronus,[1] quoted by Dalechamp,[1] clearly prove that the *tuberes* of the Romans in Nero's time were a smooth-skinned fruit; but this might be the jujube (*Zizyphus*), *Diospyros,* or some *Cratægus,* just as well as the smooth-skinned peach. Each author in the time of the Renaissance had his opinion on this point, or criticized that of the others.[2] Perhaps there were two or three species of *tuberes,* as Pliny says, and one of them which was grafted on plum trees was the nectarine (?)[3] but I doubt whether this question can ever be cleared up.[4]

"Even admitting that the *Nucipersica* was only introduced into Europe in the Middle Ages, we cannot help remarking that in European gardens for centuries, and in Japan from time unknown, there was an intermixture of all the principal kinds of peach. It seems that its different qualities were produced everywhere from a primitive species, which was probably the downy peach. If the two kinds had existed from the beginning, either they would have been in different countries, and their cultivation would have been established separately, or they would have been in the same country, and in this case it is probable that one kind would have been anciently introduced into this country and the other into that."

I laid stress, in 1855, on other considerations in support of the theory that the nectarine is derived from the common peach; but Darwin has given such a large number of cases in which a branch of nectarine has

---

[1] Dalechamp, *Hist.*, i. p. 358.

[2] Dalechamp, *ibid.;* Matthioli, p. 122; Cæsalpinus, p. 107; J. Bauhin, p. 163, etc.

[3] Pliny, lib. xvii. cap. 10.

[4] I have not been able to discover an Italian name for a glabrous or other fruit derived from *tuber,* or *tuberes,* which is singular, as the ancient names of fruits are usually preserved under some form or other.

unexpectedly appeared upon a peach tree, that it is useless to insist longer upon this point, and I will only add that the nectarine has every appearance of an artificial tree. Not only is it not found wild, but it never becomes naturalized, and each tree lives for a shorter time than the common peach. It is, in fact, a weakened form.

" The facility," I said, " with which our peach trees are multiplied from seed in America, and have produced fleshy fruits, sometimes very fine ones, without the resource of grafting, inclines me to think that the species is in a natural state, little changed by a long cultivation or by hybrid fertilization. In Virginia and the neighbouring states there are peaches grown on trees raised from seed and not grafted, and their abundance is so great that brandy is made from them.[1] On some trees the fruit is magnificent.[2] At Juan Fernandez, says Bertero,[3] the peach tree is so abundant that it is impossible to form an idea of the quantity of fruit which is gathered; it is usually very good, although the trees have reverted to a wild condition. From these instances it would not be surprising if the wild peaches with indifferent fruit found in Western Asia were simply naturalized trees in a climate not wholly favourable, and that the species was of Chinese origin, where its cultivation seems most ancient."

Dr. Bretschneider,[4] who at Pekin has access to all the resources of Chinese literature, merely says, after reading the above passages, " *Tao* is the peach tree. De Candolle thinks that China is the native country of the peach. He may be right."

The antiquity of the existence of the species and its wild nature in Western Asia have become more doubtful since 1855. Anglo-Indian botanists speak of the peach solely as a cultivated tree,[5] or as cultivated and becoming naturalized and apparently wild in the north-west of India.[6] Boissier[7] mentions specimens gathered in Ghilan

[1] Braddick, *Trans. Hort. Soc. Lond.*, ii. p. 205.   [2] *Ibid.*, pl. **13.**
[3] Bertero, *Annales Sc. Nat.*, xxi. p. 350.
[4] Bretschneider, *On the Study and Value*, etc., p. **10.**
[5] Sir J. Hooker, *Flora of Brit. Ind.*, ii. p. 313.
[6] Brandis, *Forest Flora*, etc., p. 191.   [7] Boissier, *Fl. Orient.*, ii. p. 640.

and to the south of the Caucasus, but he says nothing as
to their wild nature ; and Karl Koch,[1] after travelling
through this district, says, speaking of the peach,
" Country unknown, perhaps Persia.   Boissier saw trees
growing in the gorges on Mount Hymettus, near Athens."

The peach spreads easily in the countries in which it
is cultivated, so that it is hard to say whether a given
tree is of natural origin and anterior to cultivation, or
whether it is naturalized. But it certainly was first cultivated in China ; it was spoken of there two thousand
years before its introduction into the Greco-Roman world,
a thousand years perhaps before its introduction into the
lands of the Sanskrit-speaking race.

The group of peaches (genus or subgenus) is composed
of five forms, which Decaisne[2] regards as species, but
which other botanists are inclined to call varieties.   The
one is the common peach ; the second the nectarine, which
we know to be derived ; the third is the flattened peach
(*P. platycarpa*, Decaisne) cultivated in China; and the
two last are indigenous in China (*P. simonii*, Decaisne,
and *P. Davidii*, Carrière).   It is, therefore, essentially a
Chinese group.

It is difficult, from all these facts, not to admit the
Chinese origin of the common peach, as I had formerly
inferred from more scanty data.   Its arrival in Italy at
the beginning of the Christian era is now confirmed by
the absence of peach stones in the *terra-mare* or lake-
dwellings of Parma and Lombardy, and by the represen-
tations of the peach tree in the paintings on the walls of
the richer houses in Pompeii.[3]

I have yet to deal with an opinion formerly expressed
by Knight, and supported by several horticulturists, that
the peach is a modification of the almond.   Darwin[4]
collected facts in support of this idea, not omitting to
mention one which seems opposed to it.   They may be
concisely put as follows :—(1) Crossed fertilization, which

[1] K. Koch, *Dendrologie*, i. p. 83.
[2] Decaisne, *Jard. Fr. du Mus.*, *Pêchers*, p. 42.
[3] Comes, *Illus. Piante nei Dipinti Pompeiani*, p. 14.
[4] Darwin, *Variation of Plants and Animals*, etc., i. p. 338.

presented Knight with somewhat doubtful results; (2)
intermediate forms, as to the fleshiness of the fruit and
the size of the nut or stone, obtained by sowing peach
stones, or by chance in plantations, forms of which the
almond-peach is an example which has long been known.
Decaisne [1] pointed out differences between the almond
and peach in the size and length of the leaves independently of the fruit. He calls Knight's theory a "strange
hypothesis."

Geographical botany opposes his hypothesis, for the
almond tree has its origin in Western Asia; it was not
indigenous in the centre of the Asiatic continent, and its
introduction into China as a cultivated species was not
anterior to the Christian era. The Chinese, however, had
already possessed for thousands of years different varieties
of the common peach besides the two wild forms I have
just mentioned. The almond and the peach, starting
from two such widely separated regions, can hardly be
considered as the same species. The one was established
in China, the other in Syria and in Anatolia. The peach,
after being transported from China into Central Asia,
and a little before the Christian era into Western Asia,
cannot, therefore, have produced the almond, since the
latter existed already in Syria. And if the almond of
Western Asia had produced the peach, how could the
latter have existed in China at a very remote period
while it was not known to the Greeks and Latins?

Pear—*Pyrus communis*, Linnæus.

The pear grows wild over the whole of temperate
Europe and Western Asia, particularly in Anatolia, to the
south of the Caucasus and in the north of Persia,[2] perhaps even in Kashmir,[3] but this is very doubtful. Some
authors hold that its area extends as far as China. This
opinion is due to the fact that they regard *Pyrus
sinensis*, Lindley, as belonging to the same species. An
examination of the leaves alone, of which the teeth are

---

[1] Decaisne, *ubi supra*, p. 2.
[2] Ledebour, *Fl. Ross.*, ii. p. 94; Boissier, *Fl. Orient.*, ii. p. 653. He
has verified several specimens.
[3] Sir J. Hooker, *Fl. Brit. Ind.*, ii. p. 374.

covered with a fine silky down, convinced me of the specific difference of the two trees.[1]

Our wild pear does not differ much from some of the cultivated varieties. Its fruit is sour, spotted, and narrowing towards the stalk, or nearly spherical on the same tree.[2] With many other cultivated species, it is hard to distinguish the individuals of wild origin from those which the chance transport of seeds has produced at a distance from dwellings. In the present case it is not difficult. Pear trees are often found in woods, and they attain to a considerable height, with all the conditions of fertility of an indigenous plant.[3] Let us examine, however, whether in the wide area they occupy a less ancient existence may be suspected in some countries than in others.

No Sanskrit name for the pear is known, whence it may be concluded that its cultivation is of no long standing in the north-west of India, and that the indication, which is moreover very vague, of wild trees in Kashmir is of no importance. Neither are there any Hebrew or Aramaic names,[4] but this is explained by the fact that the pear does not flourish in the hot countries in which these tongues were spoken.

Homer, Theophrastus, and Dioscorides mention the pear tree under the names *ochnai, apios,* or *achras.* The Latins called it *pyrus* or *pirus,*[5] and cultivated a great

---

[1] *P. sinensis* described by Lindley is badly drawn with regard to the indentation of the leaves in the plate in the *Botanical Register,* and very well in that of Decaisne's *Jardin Fruitier du Muséum.* It is the same species as *P. ussuriensis,* Maximowicz, of Eastern Asia.

[2] Well drawn in Duhamel, *Traité des Arbres,* edit. 2, vi. pl. 59; and in Decaisne, *Jard. Frui. du Mus.,* pl. 1, figs. B and C. *P. balansœ,* pl. 6 of the same work, appears to be identical, as Boissier observes.

[3] This is the case in the forests of Lorraine, for instance, according to the observations of Godron, *De l'Origine Probable des Poiriers Cultivés,* 8vo pamphlet, 1873, p. 6.

[4] Rosenmüller, *Bibl. Alterth.*; Löw, *Aramaeische Pflanzennamen,* 1881.

[5] The spelling *Pyrus,* adopted by Linnæus, occurs in Pliny, *Historia,* edit. 1631, p. 301. Some botanists, purists in spelling, write *pirus,* so that in referring to a modern work it is necessary to look in the index for both forms, or run the risk of believing that the pears are not in the work. In any case the ancient name was a common name; but the true botanical name is that of Linnæus, founder of the received nomenclature, and Linnæus wrote *Pyrus.*

number of varieties, at least in Pliny's time.   The mural paintings at Pompeii frequently represent the tree with its fruit.[1]

The lake-dwellers of Switzerland and Italy gathered wild apples in great quantities, and among their stores pears are sometimes, but rarely, found.   Heer has given an illustration of one which cannot be mistaken, found at Wangen or Robenhausen.   It is a fruit narrowing towards the stalk, 28 mm. (about an inch and a half) long by 19 mm. (an inch) wide, cut longitudinally so as to show the small quantity of pulp as compared to the cartilaginous central part.[2]   None have been found in the lake-dwellings of Bourget in Savoy.   In those of Lombardy, Professor Raggazzoni [3] found a pear cut length-ways, 25 mm. by 16.   This was at Bardello, Lago di Varese. The wild pears figured in Duhamel, *Traite des Arbrés*, edit. 2, are 30 to 33 by 30 to 32 mm.; and those of Laristan, figured in the *Jardin Fruitier du Muséum* under the name *P. balansæ*, which seem to me to be of the same species, and undoubtedly wild, are 26 to 27 mm. by 24 to 25.   In modern wild pears the fleshy part is a little thicker, but the ancient lake-dwellers dried their fruits after cutting them lengthways, which must have caused them to shrink a little.   No knowledge of metals or of hemp is shown in the settlements where these were found; but, con-sidering their distance from the more civilized centres of antiquity, especially in the case of Switzerland, it is possible that these remains are not more ancient than the Trojan war, or than the foundation of Rome.

I have mentioned three Greek and one Roman name, but there are many others; for instance, *pauta* in Armenian and Georgian; *vatzkor* in Hungarian; in Slav languages *gruscha* (Russian), *hrusska* (Bohemian), *kruska* (Illyrian).   Names similar to the Latin *pyrus* recur in the Keltic languages; *peir* in Erse, *per* in Kymric and Armorican.[4]   I leave philologists to conjecture the Aryan

[1] Comes, *Ill. Piante nei Dipinti Pompeiani*, p. 59.
[2] Heer, *Pfahlbauten*, pp. 24, 26, fig. 7.
[3] Sordelli, *Notizie Stat. Lacustre di Lagozza*.
[4] Nemnich, *Polyglott. Lex. Naturgesch.*; Ad. Pictet, *Origines Indo-Europ.*, i. p. 277; and my manuscript dictionary of common names.

origin of some of these names, and of the German *Birn;*
I merely note their number and diversity as an indication of the very ancient existence of the species from the
Caspian Sea to the Atlantic. The Aryans certainly did
not carry pears nor pear pips with them in their wander-
ings westward; but if they found in Europe a fruit they
knew, they would have given it the name or names they
were accustomed to use, while other earlier names may
have survived in some countries. As an example of the
latter case, I may mention two Basque names, *udarea* and
madaria,[1] which have no analogy with any known
European or Asiatic name. The Basques being probably
the descendants of the conquered Iberians who were
driven back to the Pyrenees by the Kelts, the antiquity
of their language is very great, and it is clear that their
names for the species in question were not derived from
Keltic or Latin.

The modern area of the pear extending from the
north of Persia to the western coast of temperate Europe,
principally in mountainous regions, may therefore be considered as prehistoric, and anterior to all cultivation. It
must be added, however, that in the north of Europe and
in the British Isles an extensive cultivation must have
extended and multiplied naturalizations in comparatively
modern times which can scarcely be now distinguished.

I cannot accept Godron's hypothesis that the
numerous cultivated varieties come from an unknown
Asiatic species.[2] It seems that they may be ranked, as
Decaisne says, either with *P. communis* or *P. nivalis* of
which I am about to speak, taking into account the
effect of accidental crossing, of cultivation, and of long-
continued selection. Besides, Western Asia has been
explored so thoroughly that it is probable it contains
no other species than those already described.

Snow Pear—*Pyrus nivalis,* Jacquin.

This variety of pear is cultivated in Austria, in the
north of Italy, and in several departments of the east and

---

[1] From a list of plant-names sent by M. d'Abadie to Professor Clos,
of Toulouse.
[2] Godron, *ubi supra,* p. 28.

centre of France. It was named *Pyrus nivalis* by Jacquin [1] from the German name *Schneebirn*, given to it because the Austrian peasants eat the fruit when the snow is on the ground. It is called in France *Poirier sauger*, because the under side of the leaves is covered with a white down which makes them like the sage (Fr. *sauge*). Decaisne [2] considered all the varieties of *P. nivalis* to be derived from *P. kotschyana*, Boissier,[3] which grows wild in Asia Minor. The latter in this case should take the name of *nivalis*, which is the older.

The snowy pears cultivated in France to make the drink called perry have become wild in the woods here and there.[4] They constitute the greater number of the so-called "cider pears," which are distinguished by the sour taste of the fruit independent of the character of the leaf. The descriptions of the Greeks and Romans are too imperfect for us to be certain if they possessed this species. It may be presumed that they did, however, since they made cider.[5]

Sandy Pear, Chinese Pear—*Pyrus sinensis*, Lindley.[6]

I have already mentioned this species, which is nearly allied to the common pear. It is wild in Mongolia and Mantchuria,[7] and cultivated in China and Japan. Its fruit, large rather than good, is used for preserving. It has also been recently introduced into European gardens for experiments in crossing it with our species. This will very likely take place naturally.

Apple—*Pyrus Malus*, Linnæus.

The apple tree grows wild throughout Europe

[1] Jacquin, *Flora Austriaca*, ii. pp. 4, 107.

[2] Decaisne, *Jardin Fruitier du Muséum, Poiriers*, pl. 21.

[3] Decaisne, *ibid.*, p. 18, and Introduction, p. 30. Several varieties of this species, of which a few bear a large fruit, are figured in the same work.

[4] Boreau, *Fl. du Centre de la France*, edit. 3, vol. ii. p. 236.

[5] Palladius, *De re Rustica*, lib. 3, c. 25. For this purpose "*pira sylvestria vel asperi generis*" were used.

[6] The Chinese quince had been called by Thonin *Pyrus sinensis*. Lindley has unfortunately given the same name to a true *pyrus*.

[7] Decaisne (*Jardin Fruitier du Muséum, Poiriers*, pl. 5) saw specimens from both countries. Franchet and Savatier give it as only cultivated in Japan.

(excepting in the extreme north), in Anatolia, the south of the Caucasus, and the Persian province of Ghilan.[1] Near Trebizond, the botanist Bourgeau saw quite a small forest of them.[2]   In the mountains of the north-west of India it is "apparently wild," as Sir Joseph Hooker writes in his *Flora of British India*.   No author mentions it as growing in Siberia, in Mongolia, or in Japan.[3]

There are two varieties wild in Germany, the one with glabrous leaves and ovaries, the other with leaves downy on the under side, and Koch adds that this down varies considerably.[4]   In France accurate authors also give two wild varieties, but with characters which do not tally exactly with those of the German flora.[5]   It would be easy to account for this difference if the wild trees in certain districts spring from cultivated varieties whose seeds have been accidentally dispersed.   The question is, therefore, to discover to what degree the species is probably ancient and indigenous in different countries, and, if it is not more ancient in one country than another, how it was gradually extended by the accidental sowing of forms changed by the crossing of varieties and by cultivation.

The country in which the apple appears to be most indigenous is the region lying between Trebizond and Ghilan.   The variety which there grows wild has leaves downy on the under side, short peduncles, and sweet fruit,[6] like *Malus communis* of France, described by Boreau.   This indicates that its prehistoric area extended from the Caspian Sea nearly to Europe.

Piddington gives in his *Index* a Sanskrit name for the apple, but Adolphe Pictet[7] informs us that this

---

[1] Nyman, *Conspectus Floræ Europeæ*, p. 240; Ledebour, *Flora Rossica*, ii. p. 96; Boissier, *Flora Orientalis*, ii. p. 656; Decaisne, *Nouv. Arch. Mus.*, x. p. 153.

[2] Boissier, *ibid.*

[3] Maximowicz, *Prim. Ussur.*; Regel, *Opit. Flori*, etc., on the plants of the Ussuri collected by Maak; Schmidt, *Reisen Amur*.   Franchet and Savatier do not mention it in their *Enum. Jap.*   Bretschneider quotes a Chinese name which, he says, applies also to other species.

[4] Koch, *Syn. Fl. Germ.*, i. p. 261.

[5] Boreau, *Fl. du Centre de la France,* edit. 3, vol. ii. p. 236.

[6] Boissier, *ubi supra*.   [7] *Orig. Indo-Eur.*, i. p. 276.

name *seba* is Hindustani, and comes from the Persian *sêb, sêf*. The absence of an earlier name in India argues that the now common cultivation of the apple in Kashmir and Thibet, and especially that in the north-west and central provinces of India, is not very ancient. The tree was probably known only to the western Aryans.

This people had in all probability a name of which the root was *ab, af, av, ob*, as this root recurs in several European names of Aryan origin. Pictet gives *aball, ubhall,* in Erse; *afal* in Kymric; *aval* in Armorican; *aphal* in old High German; *appel* in old English; *apli* in Scandinavian; *obolys* in Lithuanian; *iabluko* in ancient Slav; *iabloko* in Russian. It would appear from this that the western Aryans, finding the apple wild or already naturalized in the north of Europe, kept the name under which they had known it. The Greeks had *mailea* or *maila,* the Latins *malus, malum,* words whose origin, according to Pictet, is very uncertain. The Albanians, descendants of the Pelasgians, have *molé*.[1] Theophrastus[2] mentions wild and cultivated *maila*. Lastly, the Basques (ancient Iberians) have an entirely different name, *sagara,* which implies an existence in Europe prior to the Aryan invasions.

The inhabitants of the *terra-mare* of Parma, and of the palafittes of the lakes of Lombardy, Savoy, and Switzerland, made great use of apples. They always cut them lengthways, and preserved them dried as a provision for the winter. The specimens are often carbonized by fire, but the internal structure of the fruit is only the more clearly to be distinguished. Heer,[3] who has shown great penetration in observing these details, distinguishes two varieties of the apple known to the inhabitants of the lake-dwellings before they possessed metals. The smaller kind are 15 to 24 mm. in their longitudinal diameter, and about 3 mm. more across (in their dried and carbonized state); the larger, 29 to 32 mm. lengthways by 36 wide (dried, but not carbonized). The latter

---

[1] Heldreich, *Nutzpflanzen Griechenlands,* i. p. 64.
[2] Theophrastus, *De Causis,* lib. 6, cap. 24.
[3] Heer, *Pfahlbauten,* p. 24, figs. 1–7.

corresponds to an apple of German-Swiss orchards, now called *campaner*. The English wild apple, figured in *English Botany*, pl. 179, is 17 mm. long by 22 wide. It is possible that the little apples of the lake-dwellings were wild; however, their abundance in the stores makes it doubtful. Dr. Gross sent me two apples from the more recent palafittes of Lake Neuchâtel; the one is 17 the other 22 mm. in longitudinal diameter. At Lagozza, in Lombardy, Sordelli[1] mentions two apples, the one 17 mm. by 19, the other 19 mm. by 27. In a prehistoric deposit of Lago Varese, at Bardello, Ragazzoni found an apple in the stores a little larger than the others.

From all these facts, I consider the apple to have existed in Europe, both wild and cultivated, from pre-historic times. The lack of communication with Asia before the Aryan invasion makes it probable that the tree was indigenous in Europe as in Anatolia, the south of the Caucasus, and Northern Russia, and that its culti-vation began early everywhere.

Quince—*Cydonia vulgaris*, Persoon.

The quince grows wild in the woods in the north of Persia, near the Caspian Sea, in the region to the south of the Caucasus, and in Anatolia.[2] A few botanists have also found it apparently wild in the Crimea, and in the north of Greece;[3] but naturalization may be suspected even in the east of Europe, and the further we advance towards Italy, especially towards the south-west of Europe and Algeria, the more it becomes probable that the species was naturalized at an early period round villages, in hedges, etc.

No Sanskrit name is known for the quince, whence it may be inferred that its area did not extend towards the centre of Asia. Neither is there any Hebrew name, though the species is wild upon Mount Taurus.[4] The Persian name is *haivah*,[5] but I do not know whether

---

[1] Sordelli, *Sulle Piante della Stazione di Lagozza*, p. 35.
[2] Boissier, *Fl. Orient.*, ii. p. 656; Ledebour, *Fl. Ross.*, ii. p. 55.
[3] Steven, *Verzeichniss Taurien*, p. 150; Sibthorp, *Prodr. Fl. Græcæ*, i. p. 344.
[4] Boissier, *ibid.*
[5] Nemnich, *Polyglott Lexicon.*

it is as old as Zend. The same name, *aiva*, exists in Russian for the cultivated quince, while the name of the wild plant is *armud*, from the Armenian *armuda*.[1] The Greeks grafted upon a common variety, *strution*, a superior kind, which came from Cydon, in Crete, whence κυδώνιον, translated by the Latin *malum cotoneum*, by *cydonia*, and all the European names, such as *codogno* in Italian, *coudougner*, and later *coing* in French, *quitte* in German, etc. There are Polish, *pigwa*, Slav, *tunja*,[2] and Albanian (Pelasgian ?), *ftua*,[3] names which differ entirely from the others. This variety of names points to an ancient knowledge of the species to the west of its original country, and the Albanian name may even indicate an existence prior to the Hellenes.

Its antiquity in Greece may also be gathered from the superstition, mentioned by Pliny and Plutarch, that the fruit of the quince was a preservation from evil influences, and from its entrance into the marriage rites prescribed by Solon. Some authors go so far as to maintain that the apple disputed by Hera, Aphrodite, and Athene was a quince. Those who are interested in such questions will find details in Comes's paper on the plants represented in the frescoes at Pompeii.[4] The quince tree is figured twice in these, which is not surprising, as the tree was known in Cato's time.[5]

It seems to me probable that it was naturalized in the east of Europe before the epoch of the Trojan war. The quince is a fruit which has been little modified by cultivation; it is as harsh and acid when fresh as in the time of the ancient Greeks.

Pomegranate—*Punica granatum*, Linnæus.

The pomegranate grows wild in stony ground in Persia, Kurdistan, Afghanistan, and Beluchistan.[6] Burnes saw groves of it in Mazanderan, to the south of the Caspian Sea.[7] It appears equally wild to the south

---

[1] Nemnich, *Poly. Lex.*   [2] *Ibid.*   [3] Heldreich, *Nutz. Griech.*, p. 64.
[4] In 4to, Napoli, 1879.   [5] *De re Rustica*, lib. 7, cap. 2.
[6] Boissier, *Fl. Orient.*, ii. p. 737 ; Sir J. Hooker, *Fl. of Brit. Ind.*, ii. p. 581.
[7] Quoted from Royle, *Illus. Himal.*, p. 208.

of the Caucasus.[1] Westwards, that is to say, in Asia Minor, in Greece, and in the Mediterranean basin generally, in the north of Africa and in Madeira, the species appears rather to have become naturalized from cultivation, and by the dispersal of the seeds by birds. Many floras of the south of Europe speak of it as a "subspontaneous" or naturalized species. Desfontaines, in his *Atlantic Flora*, gives it as wild in Algeria, but subsequent authors think [2] rather it is naturalized.[3] I doubt its being wild in Beluchistan, where the traveller Stocks found it, for Anglo-Indian botanists do not allow it to be indigenous east of the Indus, and I note the absence of the species in the collections from Lebanon and Syria which Boissier is always careful to quote.

In China the pomegranate exists only as a cultivated plant. It was introduced from Samarkhand by Chang-Kien, a century and a half before the Christian era.[4]

The naturalization in the Mediterranean basin is so general that it may be termed an extension of the original area. It probably dates from a very remote period, for the cultivation of the species dates from a very early epoch in Western Asia.

Let us see whether historical and philological data can give us any information on this head.

I note the existence of a Sanskrit name, *darimba*, whence several modern Indian names are derived.[5] Hence we may conclude that the species had long been known in the regions traversed by the Aryans in their route towards India. The pomegranate is mentioned several times in the Old Testament, under the name of *rimmon*,[6] whence the Arabic *rumman* or *rûman*. It was one of the fruit trees of the promised land, and the Hebrews had learnt to appreciate it in Egyptian gardens. Many localities in Palestine took their name from this

---

[1] Ledebour, *Fl. Ross.*, ii. p. 104.

[2] Munby, *Fl. Alger.*, p. 49 ; *Spicilegium Flora Maroccanœ*, p. 458.

[3] Boissier, *ibid.*

[4] Bretschneider, *On Study and Value*, etc., p. 16.

[5] Piddington, *Index*.

[6] Rosenmüller, *Bibl. Naturge.*, i. p. 273 ; Hamilton, *La Bot. de la Bible*, Nice, 1871, p. 48.

shrub, but the Scriptures only mention it as a cultivated species. The flower and the fruit figured in the religious rites of the Phœnicians, and the goddess Aphrodite had herself planted it in the isle of Cyprus,[1] which implies that it was not indigenous there. The Greeks were acquainted with the species in the time of Homer. It is twice mentioned in the *Odyssey* as a tree in the gardens of Phæacia and Phrygia. They called it *roia* or *roa*, which philologists believe to be derived from the Syrian and Hebrew name,[2] and also *sidai*,[3] which seems to be Pelasgic, for the modern Albanian name is *sige*.[4] There is nothing to show that the species was wild in Greece, where Fraas and Heldreich affirm that it is now only naturalized.[5]

The pomegranate enters into the myths and religious ceremonies of the ancient Romans.[6] Cato speaks of its properties as a vermifuge. According to Pliny,[7] the best pomegranates came from Carthage, hence the name *Malum punicum ;* but it should not be supposed, as it has been assumed, that the species came originally from Northern Africa. Very probably the Phœnicians had introduced it at Carthage long before the Romans had anything to do with this town, and it was doubtless cultivated as in Egypt.

If the pomegranate had formerly been wild in Northern Africa and the south of Europe, the Latins would have had more original names for it than *granatum* (from *granum ?*) and *Malum punicum.* We should have perhaps found local names derived from ancient Western tongues ; whereas the Semitic name *rimmon* has prevailed in Greek and in Arabic, and even occurs, through Arab influence, among the Berbers.[8] It must be admitted that the African origin is one of the errors caused by the erroneous popular nomenclature of the Romans.

Leaves and flowers of a pomegranate, described by

[1] Hehn, *Cultur und Hansthiere aus Asien,* edit. 3, p. 106.
[2] Hehn, *ibid.*   [3] Lenz, *Bot. der Alten Grie. und Röm.,* p. 681.
[4] Heldreich, *Die Nutzpflanzen Griechenlands,* p. 64.
[5] Fraas, *Fl. Class.,* p. 79 ; Heldreich, *ibid.*
[6] Hehn, *ibid.*   [7] Pliny, lib. **13,** c. **19.**
[8] *Dictionnaire Français-Berbère,* published by the French Government.

Saporta[1] as a variety of the modern *Punica granatum*, have been discovered in the pliocene strata of the environs of Meximieux. The species, therefore, existed under this form, before our epoch, along with several species, some extinct, others still existing in the south of Europe, and others in the Canaries, but the continuity of existence down to our own day is not thereby proved.

To conclude, botanical, historical, and philological data agree in showing that the modern species is a native of Persia and some adjacent countries. Its cultivation began in prehistoric time, and its early extension, first towards the west and afterwards into China, has caused its naturalization in cases which may give rise to errors as to its true origin, for they are frequent, ancient, and enduring. I arrived at these conclusions in 1869,[2] which has not prevented the repetition of the erroneous African origin in several works.

Rose Apple—*Eugenia Jambos*, Linnæus; *Jambosa vulgaris*, de Candolle.

This small tree belongs to the family of Myrtaceæ. It is cultivated in tropical regions of the old and new worlds, as much perhaps for the beauty of its foliage as for its fruit, of which the rose-scented pulp is too scanty. There is an excellent illustration and a good description of it in the *Botanical Magazine*, pl. 3356. The seed is poisonous.[3]

As the cultivation of this species is of ancient date in Asia, there was no doubt of its Asiatic origin; but the locality in which it grew wild was formerly unknown. Loureiro's assertion that it grew in Cochin-China and some parts of India required confirmation, which has been afforded by some modern writers.[4] The *jambos* is wild in Sumatra, and elsewhere in the islands of the Malay Archipelago. Kurz did not meet with it in the forests of British Burmah, but when Rheede saw this tree in gardens in Malabar he noticed that it was called *Malacca-schambu*, which shows that it came origi-

[1] De Saporta, *Bull. Soc. Géol. de France*, April 5, 1869, pp. 767–769.

[2] *Géogr. Bot. Rais.*, p. 191.

[3] Descourtilz, *Flore Médicale des Antilles*, v. pl. 315.

[4] Miquel, *Sumatra*, p. 118; *Flora Indiæ-Batavæ*, i. p. 425   Blume, *Museum Lugd.-Bat.*, i. p. 93.

nally from the Malay Peninsula. Lastly, Brandis says it is wild in Sikkim, to the north of Bengal. Its natural area probably extends from the islands of the Malay Archipelago to Cochin-China, and even to the north-east of India, where, however, it is probably naturalized from cultivation and by the agency of birds. Naturalization has also taken place elsewhere—at Hong-kong, for instance, in the Seychelles, Mauritius, and Rodriguez, and in several of the West India Islands.[1]

Malay Apple—*Eugenia malaccensis,* Linnæus; *Jambosa malaccensis,* de Candolle

A species allied to *Eugenia jambos,* but differing from it in the arrangement of its flowers, and in its fruit, of an obovoid instead of ovoid form; that is to say, the smaller end is attached to the stalk. The fruit is more fleshy and is also rose-scentèd, but it is much[2] or little[3] esteemed according to the country and varieties. These are numerous, differing in the red or pink colour of the flowers, and in the size, shape, and colour of the fruit.

The numerous varieties show an ancient cultivation in the Malay Archipelago, where the species is indigenous. In confirmation, it must be noted that Forster found it established in the Pacific Islands, from Otahiti to the Sandwich Isles, at the time of Cook's voyages.[4] The Malay apple grows wild in the forests of the Malay Archipelago, and in the peninsula of Malacca.[5]

Tussac says that it was brought to Jamaica from Otahiti in 1793. It has spread and become naturalized in several of the West India Islands, also in Mauritius and the Seychelles.[6]

Guava—*Psidium guayava,* Raddi.

Ancient authors, Linnæus, and some later botanists,

---

[1] Hooker, *Fl. Brit. Ind.,* ii. p. 474; Baker, *Fl. of Maurit.,* etc., p. 115; Grisebach, *Fl. of Brit. W. Ind. Isles,* p. 235.

[2] Rumphius, *Amboin,* i. p. 121, t. 37.

[3] Tussac, *Flore des Antilles,* iii. p. 89, pl. 25.

[4] Forster, *Plantis Esculentis,* p. 36.

[5] Blume, *Museum Lugd.-Bat.,* i. p. 91; Miquel, *Fl. Indiæ-Batav.,* i. p. 411; Hooker, *Flora of British India,* ii. p. 472.

[6] Grisebach, *Fl. Brit. W. Indies,* p. 235; Baker, *Fl. of Mauritius,* p. 115.

admitted two species of this fruit tree of the family
of Myrtaceæ, the one with elliptical or spherical fruit,
with red flesh, *Psidium pomiferum;* the other with a
pyriform fruit and white or pink flesh, more agreeable
to the taste. Such diversity is also observed in pears,
apples, or peaches; so it was decided to consider all the
Psidii as forming a single species. Raddi saw a proof
that there was no essential difference, for he observed
pyriform and round fruits growing on the same tree in
Brazil.[1] The majority of botanists, especially those who
have observed the guava in the colonies, follow the
opinion of Raddi,[2] to which I was inclined, even in 1855,
from reasons drawn from the geographical distribution.[3]

Lowe,[4] in his *Flora of Madeira,* maintains with some
hesitation the distinction into two species, and asserts
that each can be raised from seed. They are, therefore,
races like those of our domestic animals, and of many
cultivated plants. Each of these races comprehends
several varieties.[5]

The study of the origin of the guava presents in the
highest degree the difficulty which exists in the case of
many fruit trees of this nature: their fleshy and somewhat aromatic fruits attract omnivorous animals which
cast their seeds in places far from cultivation. Those of
the guava germinate rapidly, and fructify in the third
or fourth year. Its area has thus spread, and is still
spreading by naturalization, principally in those tropical
countries which are neither very hot nor very damp.

In order to simplify the search after the origin of the
species, I may begin by eliminating the old world, for it
is sufficiently evident that the guava came from America.

---

[1] Raddi, *Di Alcune Specie di Pero Indiano,* in 4to, Bologna, 1821, p. 1.

[2] Martius, *Syst. Nat. Medicæ Bras.,* p. 32; Blume, *Museum Lugd.-
Bat.,* i. p. 71; Hasskarl, in *Flora,* 1844, p. 589; Sir J. Hooker, *Fl. of Brit.
Ind.,* ii. p. 468.

[3] *Géogr. Bot. Rais.,* p. 893.

[4] Lowe, *Flora of Madeira,* p. 266.

[5] See Blume, *ibid.;* Descourtilz, *Flore Médicale des Antilles,* ii. p. 20,
in which there is a good illustration of the pyriform guava. Tussac,
Flore des Antilles, gives a good plate of the round form. These two
latter works furnish interesting details on the use of the guava, on the
vegetation of the species, etc.

Out of sixty species of the genus Psidium, all those which have been carefully studied are American. It is true that botanists from the sixteenth century have found plants of *Psidium guayava* (varieties *pomiferum* and *pyriferum*) more or less wild in the Malay Archipelago and the south of Asia,[1] but everything tends to show that these were the result of recent naturalization. In each locality a foreign origin was admitted; the only doubt was whether this origin was Asiatic or American. Other considerations justify this idea. The common names in Malay are derived from the American word *guiava*. Ancient Chinese authors do not mention the guava, though Loureiro said a century and a half ago that they were growing wild in Cochin-China. Forster does not mention them among the cultivated plants of the Pacific Isles at the time of Cook's voyage, which is significant when we consider how easy this plant is to cultivate and its ready dispersion. In Mauritius and the Seychelles there is no doubt of their recent introduction and naturalization.[2]

It is more difficult to discover from what part of America the guava originally came. In the present century it is undoubtedly wild in the West Indies, in Mexico, in Central America, Venezuela, Peru, Guiana, and Brazil.[3] But whether this is only since Europeans extended its cultivation, or whether it was previously diffused by the agency of the natives and of birds, seems to be no more certain than when I spoke on the subject in 1855.[4] Now, however, with a little more experience in questions of this nature, and since the specific unity of the two varieties of guava is recognized, I shall endeavour to show what seems most probable.

J. Acosta,[5] one of the earliest authors on the natural history of the new world, expresses himself as follows, about the spherical variety of the guava: "There are

---

[1] Rumphius, *Amboin*, i. p. 141; Rheede, *Hortus Malabariensis*, iii. t. 34.

[2] Bojer, *Hortus Mauritianus;* Baker, *Flora of Mauritius*, p. 112.

[3] All the floras, and Berg in *Flora Brasiliensis*, vol. xiv. p. 196.

[4] *Géogr. Bot. Rais.*, p. 894.

[5] Acosta, *Hist. Nat. et Morale des Indes Orient. et Occid.*, French trans., 1598, p. 175.

mountains in San Domingo and the other islands entirely covered with guavas, and the natives say that there were no such trees in the islands before the arrival of the Spaniards, who brought them, I know not whence." The mainland seems, therefore, to have been the original home of the species. Acosta says that it grows in South America, adding that the Peruvian guavas have a white flesh superior to that of the red fruit. This argues an ancient cultivation on the mainland. Hernandez[1] saw both varieties wild in Mexico in the warm regions of the plains and mountains near Quauhnaci. He gives a description and a fair drawing of *P. pomiferum.* Piso and Marcgraf[2] also found the two guavas wild in the plains of Brazil; but they remark that it spreads readily. Marcgraf says that they were believed to be natives of Peru or of North America, by which he may mean the West Indies or Mexico. Evidently the species was wild in a great part of the continent at the time of the discovery of America. If the area was at one time more restricted, it must have been at a far more remote epoch.

Different common names were given by the different native races. In Mexico it was *xalxocotl;* in Brazil the tree was called *araca-iba,* the fruit *araca guacu;* lastly, the name *guajavos,* or *guajava,* is quoted by Acosta and Hernandez for the guavas of Peru and San Domingo without any precise indication of origin. This diversity of names confirms the hypothesis of a very ancient and extended area.

From what ancient travellers say of an origin foreign to San Domingo and Brazil (an assertion, however, which we may be permitted to doubt), I suspect that the most ancient habitation extended from Mexico to Columbia and Peru, possibly including Brazil before the discovery of America, and the West Indies after that event. In its earliest state, the species bore spherical, highly coloured fruit, harsh to the taste. The other form is perhaps the result of cultivation.

---

[1] Hernandez, *Novæ Hispaniæ Thesaurus,* p. 85.
[2] Piso, *Hist. Brasil,* p. 74; Marcgraf, *ibid.,* p. 105.

Gourd,[1] or **Calabash**—*Lagenaria vulgaris,* Seringe; *Cucurbita lagenaria,* Linnæus.

The fruit of this *Curcubitacea* has taken different forms in cultivation, but from a general observation of the other parts of the plant, botanists have ranked them in one species which comprises several varieties.[2] The most remarkable are the *pilgrim's gourd,* in the form of a bottle, the *long-necked gourd,* the *trumpet gourd,* and the *calabash,* generally large and without a neck. Other less common varieties have a flattened, very small fruit, like the *snuff-box gourd.* The species may always be recognized by its white flower, and by the hardness of the outer rind of the fruit, which allows of its use as a vessel for liquids, or a reservoir of air suitable as a buoy for novices in swimming. The flesh is sometimes sweet and eatable, sometimes bitter and even purgative.

Linnæus[3] pronounced the species to be American. De Candolle[4] thought it was probably of Indian origin, and this opinion has since been confirmed.

Lagenaria vulgaris has been found wild on the coast of Malabar and in the humid forests of Deyra Doon.[5] Roxburgh[6] considered it to be wild in India, although subsequent floras give it only as a cultivated species. Lastly, Rumphius[7] mentions wild plants of it on the sea-shore in one of the Moluccas. Authors generally note that the pulp is bitter in these wild plants, but this is sometimes the case in cultivated forms. The Sanskrit language already distinguished the common gourd, *ulavou,* and another, bitter, *kutou-toumbi,* to which Pictet also attributes the name *tiktaka* or *tiktika.*[8] Seemann[9] saw

---

[1] The word *gourd* is also used in English for *Cucurbita maxima.* This is one of the examples of the confusion in common names and the greater accuracy of scientific terms.

[2] Naudin, *Annales des Sc. Nat.,* 4th series, vol. xii. p. 91; Cogniaux, in our *Monog. Phanérog.,* iii. p. 417.

[3] Linnæus, *Species Plantarum,* p. 1434, under *Cucurbita.*

[4] A. P. de Candolle, *Flora Française* (1805), vol. iii. p. 692.

[5] Rheede, *Malabar,* iii. pls. 1, 5; Royle, *Ill. Himal.,* p. 218.

[6] Roxburgh, *Fl. Ind.,* edit. 1832, vol. iii. p. 719.

[7] Rumphius, *Amboin,* vol. v. p. 397, t. 144.

[8] Piddington, *Index,* at the word *Cucurbita lagenaria;* Ad. Pictet, *Origines Indo-Europ.,* edit. 3, vol. i. p. 386.

[9] Seemann, *Flora Vitiensis,* p. 106.

the species cultivated and naturalized in the Fiji Isles. Thozet gathered it on the coast of Queensland,[1] but it had perhaps spread from neighbouring cultivation. The localities in continental India seem more certain and more numerous than those of the islands to the south of Asia.

The species has also been found wild in Abyssinia, in the valley of Hieha by Dillon, and in the bush and stony ground of another district by Schimper.[2]

From these two regions of the old world it has been introduced into the gardens of all tropical countries and of those temperate ones where there is a sufficiently high temperature in summer. It has occasionally become naturalized from cultivation, as is seen in America.[3]

The earliest Chinese work which mentioned the gourd is that of Tchong-tchi-chou, of the first century before Christ, quoted in a work of the fifth or sixth century according to Bretschneider.[4] He is speaking here of cultivated plants. The modern varieties of the gardens at Pekin are the trumpet gourd, which is eatable, and the bottle gourd.

Greek authors do not mention the plant, but Romans speak of it from the beginning of the empire. It is clearly alluded to in the often-quoted lines[5] of the tenth book of Columella. After describing the different forms of the fruit, he says—

> " Dabit illa capacem,
> Nariciæ picis, aut Actæi mellis Hymetti,
> Aut habilem lymphis hamulam, Bacchove lagenam,
> Tum pueros eadem fluviis innare docebit."

Pliny[6] speaks of a *Cucurbitacea,* of which vessels and

---

[1] Bentham, *Flora Australiensis,* iii. p. 316.

[2] Described first under the name *Lagenaria idolatrica.* A. Richard, *Tentamen Fl. Abyss.,* i. p. 293, and later, Naudin and Cogniaux, recognized its identity with *L. vulgaris.*

[3] Torrey and Gray, *Fl. of N. Amer.,* i. p. 543 ; Grisebach, *Flora of Brit. W. Ind. Is.,* p. 288.

[4] Bretschneider, letter of the 23rd of August, 1881.

[5] Tragus, *Stirp.,* p. 285 ; Ruellius, *De Natura Stirpium,* p. 498 ; Naudin, *ibid.*

[6] Pliny, *Hist. Plant.,* l. 19, c. 5.

flasks for wine were made, which can only apply to this species.

It does not appear that the Arabs were early acquainted with it, for Ibn Alawâm and Ibn Baithar say nothing of it.[1] Commentators of Hebrew works attribute no name to this species with certainty, and yet the climate of Palestine is such as to popularize the use of gourds had they been known. From this it seems to me doubtful that the ancient Egyptians possessed this plant, in spite of a single figure of leaves observed on a tomb which has been sometimes identified with it.[2] Alexander Braun, Ascherson, and Magnus, in their learned paper on the Egyptian remains of plants in the Berlin Museum,[3] indicate several Cucurbitaceæ without mentioning this one. The earliest modern travellers, such as Rauwolf,[4] in 1574, saw it in the gardens of Syria, and the so-called pilgrim's gourd, figured in 1539 by Brunfels, was probably known in the Holy Land from the Middle Ages.

All the botanists of the sixteenth century give illustrations of this species, which was more generally cultivated in Europe at that time than it is now. The common name in these older writings is *Cameraria,* and three kinds of fruit are distinguished. From the white colour of the flower, which is always mentioned, there can be no doubt of the species. I also note an illustration, certainly a very indifferent one, in which the flower is wanting, but with an exact representation of the fruit of the pilgrim's gourd, which has the great interest of having appeared before the discovery of America. It is pl. 216 of *Herbarius Pataviæ Impressus,* in 4to, 1485—a rare work.

In spite of the use of similar names by some authors, I do not believe that the gourd existed in America before the arrival of the Europeans. The *Taquera* of Piso [5]

---

[1] Ibn Alawâm, in E. Meyer, *Geschichte der Botanik,* iii. p. 60; Ibn Baithar, Sondtheimer's translation.

[2] Unger, *Pflanzen des Alten Ægyptens,* p. 59; Pickering, *Chronol. Arrang.,* p. 137.

[3] In 8vo, 1877, p. 17.   [4] Rauwolf, *Fl. Orient.,* p. 125.

[5] Piso, *Indiæ Utriusque.,* etc., edit. 1658, p. 264.

and *Cucurbita lagenæforma* of Marcgraf[1] are perhaps *Lagenaria vulgaris* as monographs say,[2] and the
specimens from Brazil which they mention should be
certain, but that does not prove that the species was in
the country before the voyage of Amerigo Vespucci in
1504. From that time until the voyages of these two
botanists in 1637 and 1638, a much longer time elapsed
than is needed to account for the introduction and diffusion of an annual species of a curious form, easy of
cultivation, and of which the seeds long retain the faculty
of germination. It may have become naturalized from
cultivation, as has taken place elsewhere. It is still
more likely that *Cucurbita siceratia,* Molina, attributed
sometimes to the species under consideration, sometimes
to *Cucurbita maxima,*[3] may have been introduced into
Chili between 1538, the date of the discovery of that
country, and 1787, the date of the Italian edition of
Molina. Acosta[4] also speaks of calabashes which the
Peruvians used as cups and vases, but the Spanish
edition of his book appeared in 1591, more than a
hundred years after the Conquest. Among the first
naturalists to mention the species after the discovery of
America (1492) is Oviedo,[5] who had visited the mainland, and, after dwelling at Vera Paz, came back to
Europe in 1515, but returned to Nicaragua in 1539.[6]
According to Ramusio's compilation[7] he spoke of *zueche,*
freely cultivated in the West India Islands and Nicaragua
at the time of the discovery of America, and used as
bottles. The authors of the floras of Jamaica in the
seventeenth century say that the species was cultivated
in that island. P. Brown,[8] however, mentions a large
cultivated gourd, and a smaller one with a bitter and
purgative pulp, which was found wild.

[1] Marcgraf, *Hist. Nat. Brasiliæ,* 1648, p. 44.
[2] Naudin, *ibid.;* Cogniaux, *Flora Brasil.,* fasc. 78, p. 7; and de Candolle,
Monogr. Phanér., iii. p. 418.
[3] Cl. Gay, *Flora Chilena,* ii. p. 403.
[4] Jos. Acosta, French trans., p. 167.
[5] Pickering, *Chronol. Arrang.,* p. 861.   [6] Pickering, *ibid.*
[7] Ramusio, vol. iii. p. 112.
[8] P. Brown, *Jamaica,* edit. ii. p. 354.

Lastly, Elliott [1] writes as follows, in 1824, in a work on the Southern States of America: "*L. vulgaris* is rarely found in the woods, and is certainly not indigenous. It seems to have been brought by the early inhabitants of our country from a warmer climate. The species has now become wild near dwellings, especially in islands." The expression, "inhabitants of our country," seems to refer rather to the colonists than to the natives. Between the discovery of Virginia by Cabot in 1497, or the travels of Raleigh in 1584, and the floras of modern botanists, more than two centuries elapsed, and the natives would have had time to extend the cultivation of the species if they had received it from Europeans. But the fact of its cultivation by Indians at the time of the earliest dealings with them is doubtful. Torrey and Gray [2] mentioned it as certain in their flora published in 1830–40, and later the second of these able botanists, [3] in an article on the *Cucurbitaceæ* known to the natives, does not mention the calabash, or *Lagenaria*. I remark the same omission in another special article on the same subject, published more recently. [4]

[In the learned articles by Messrs. Asa Gray and Trumbull on the present volume (*American Journal of Science*, 1883, p. 370), they give reasons for supposing the species known and indigenous in America previous to the arrival of the Europeans. Early travellers are quoted more in detail than I had done. From their testimony it appears that the inhabitants of Peru, Brazil, and of Paria possessed gourds, in Spanish *calabazas*, but I do not see that this proves that this was the species called by botanists *Cucurbita lagmaria*. The only character independent of the exceedingly variable form of the fruit is the white colour of the flowers, and this character is not mentioned.—AUTHOR'S NOTE, 1884.]

Gourd—*Cucurbita maxima*, Duchesne.

In enumerating the species of the genus *Cucurbita*, I

[1] Elliott, *Sketch of the Botany of South Carolina and Georgia*, ii. p. 663.
[2] Torrey and Gray, *Flora of N. America*, i. p. 544.
[3] Asa Gray, in the *American Journal of Science*, 1857, vol. xxiv. p. 442.
[4] Trumbull, in *Bull. Torrey Bot. Club*, vol. vi. p. 69.

should explain that their distinction, formerly exceedingly difficult, has been established by M. Naudin [1] in a very scientific manner, by means of an assiduous cultivation of varieties and of experiments upon their crossed fertilization. Those groups of forms which cannot fertilize each other, or of which the product is not fertile and stable, are regarded by him as species, and the forms which can be crossed and yield a fertile and varied product, as races, breeds, or varieties. Later experiments [2] showed him that the establishment of species on this basis is not without exceptions, but in the genus *Cucurbita* physiological facts agree with exterior differences. M. Naudin has established the true distinctive characters of *C. maxima* and *C. Pepo*. The leaves of the first have rounded lobes, the peduncles are smooth and the lobes of the corolla are curved outwards; the second has leaves with pointed lobes, the peduncles marked with ridges and furrows, the corolla narrowed towards the base and with lobes nearly always upright.

The principal varieties of *Cucurbita maxima* are the great yellow gourd, which sometimes attains to an enormous size,[3] the Spanish gourd, the turban gourd, etc.

Since common names and those in ancient authors do not agree with botanical definitions, we must mistrust the assertions formerly put forth on the origin and early cultivation of such and such a gourd at a given epoch in a given country. For this reason, when I considered the subject in 1855, the home of these plants seemed to me either unknown or very doubtful. At the present day it is more easy to investigate the question.

According to Sir Joseph Hooker,[4] *Cucurbita maxima* was found by Barter on the banks of the Niger in Guinea, apparently indigenous, and by Welwitsch in Angola without any assertion of its wild character. In works on Abyssinia, Egypt, or other African countries in which the species is commonly cultivated, I find no

[1] Naudin, *Ann. Sc. Nat.*, 4th series, vol. vi. p. 5; vol. xii. p. 84.
[2] *Ibid.*, 4th series, vol. xviii. p. 160; vol. xix. p. 180.
[3] As much as 200 lbs., according to the *Bon Jardinier*, 1850, p. 180.
[4] Hooker, *Fl. of Trop. Afr.*, ii. p. 555.

indication that it is found wild.   The Abyssinians used the word *dubba,* which is applied in Arabic to gourds in general.

The plant was long supposed to be of Indian origin, because of such names as Indian gourd, given by sixteenth-century botanists, and in particular the *Pepo maximus indicus,* figured by Lobel,[1] which answers to the modern species; but this is a very insufficient proof, since popular indications of origin are very often erroneous.   The fact is that though pumpkins are cultivated in Southern Asia, as in other parts of the tropics, the plant has not been found wild.[2]   No similar species is indicated by ancient Chinese authors, and the modern names of gourds and pumpkins now grown in China are of foreign and southern origin.[3]   It is impossible to know to what species the Sanskrit name *kurkarou* belonged, although Roxburgh attributes it to *Cucurbita Pepo;* and there is no less uncertainty with respect to the gourds, pumpkins, and melons cultivated by the Greeks and Romans. It is not certain if the species was known to the ancient Egyptians, but perhaps it was cultivated in that country and in the Græco-Roman world.   The *Pepones,* of which Charlemagne commanded the cultivation in his farms,[4] were perhaps some kind of pumpkin or marrow, but no figure or description of these plants which may be clearly recognized exists earlier than the sixteenth century.

This tends to show its American origin.   Its existence in Africa in a wild state is certainly an argument to the contrary, for the species of the family of *Cucurbitaceæ* are very local; but there are arguments in favour of America, and I must examine them with the more care since I have been reproached in the United States for not having given them sufficient weight.

In the first place, out of the ten known species of the genus *Cucurbita,* six are certainly wild in America

---

[1] Lobel, *Icones,* t. 641.   The illustration is reproduced in Dalechamp's *Hist.,* i. p. 626.

[2] Clarke, Hooker's *Fl. Brit. Ind.,* ii. p. 622.

[3] Bretschneider, letter of Aug. 23, 1881.

[4] The list is given by E. Meyer, *Geschichte du Botanik,* iii. p. 401. The Cucurbita of which he speaks must have been the gourd, *Lagenaria.*

(Mexico and California); but these are perennial species, while the cultivated pumpkins are annuals.

The plant called *jurumu* by the Brazilians, figured by Piso and Marcgraf[1] is attributed by modern writers to *Cucurbita maxima*. The drawing and the short account by the two authors agree pretty well with this theory, but it seems to have been a cultivated plant. It may have been brought from Europe or from Africa by Europeans, between the discovery of Brazil in 1504, and the travels of the above-named authors in 1637 and 1638. No one has found the species wild in North or South America. I cannot find in works on Brazil, Guiana, or the West Indies any sign of an ancient cultivation or of wild growth, either from names, or from traditions or more or less distinct belief. In the United States those men of science who best know the languages and customs of the natives, Dr. Harris for instance, and more recently Trumbull,[2] maintain that the *Cucurbitaceœ* called *squash* by the Anglo-Americans, and *macock*, or *cashaw*, *cushaw*, by early travellers in Virginia, are *pumpkins*. Trumbull says that *squash* is an Indian word. I have no reason to doubt the assertion, but neither the ablest linguists, nor the travellers of the seventeenth century, who saw the natives provided with fruits which they called *gourds and pumpkins*, have been able to prove that they were such and such species recognized as distinct by modern botanists. All that we learn from this is that the natives a century after the discovery of Virginia, and twenty to forty years after its colonization by Sir Walter Raleigh, made use of some fruits of the *Cucurbitaceœ*. The common names are still so confused in the United States, that Dr. Asa Gray, in 1868, gives *pumpkin* and *squash* as answering to different species of *Cucurbita*,[3] while Darlington[4] attributes the name *pumpkin* to the common *Cucurbita Pepo*, and that of *squash* to the varieties of the

[1] Piso, *Brazil*, edit. 1658, p. 264; Marcgraf, edit. 1648, p. 44.

[2] Harris, *American Journal*, 1857, vol. xxiv. p. 441; Trumbull, *Bull. of Torrey Bot. Club*, 1876, vol. vi. p. 69.

[3] Asa Gray, *Botany of the Northern States*, edit. 1868, p. 186.

[4] Darlington, *Flora Cestrica*, 1853, p. 94.

latter which correspond to the forms of *Melopepo* of early botanists. They attribute no distinct common name to *Cucurbita maxima*.

Finally, without placing implicit faith in the indigenous character of the plant on the banks of the Niger, based upon the assertion of a single traveller, I still believe that the species is a native of the old world, and introduced into America by Europeans.

[The testimony of early travellers touching the existence of *Cucurbita maxima* in America before the arrival of Europeans has been collected and supplemented by Messrs. Asa Gray and Trumbull (*American Journal of Science*, 1883, p. 372). They confirm the fact already known, that the natives cultivated species of *Cucurbita* under American names, of which some remain in the modern idiom of the United States. None of these early travellers has noted the botanical characters by which Naudin established the distinction between *C. maxima* and *C. Pepo*, and consequently it is still doubtful to which species they referred. For various reasons I had already admitted that *C. Pepo* was of American origin, but I retain my doubts about *C. maxima*. After a more attentive perusal of Tragus and Matthiolo than I had bestowed upon them, Asa Gray and Trumbull notice that they call *Indian* whatever came from America. But if these two botanists did not confound the East and West Indies, several others, and the public in general, did make this confusion, which occasioned errors touching the origin of species which botanists were liable to repeat. A further indication in favour of the American origin of *C. maxima* is communicated by M. Wittmack, who informs me that seeds, certified by M. Naudin to belong to this species, have been found in the tombs of Ancon. This would be conclusive if the date of the latest burials at Ancon were certain. See on this head the article on *Phaseolus vulgaris.*—AUTHOR'S NOTE, 1884.]

Pumpkin—*Cucurbita Pepo* and *C. Melopepo*, Linnæus. Modern authors include under the head of *Cucurbita Pepo* most of the varieties which Linnæus designated by this name, and also those which he called *C. Melopepo*.

These varieties are very different as to the shape of the
fruit, which shows a very ancient cultivation. There is
the Patagonian pumpkin, with enormous cylindrical fruit;
the *sugared pumpkin*, called Brazilian; the vegetable
marrow, with smaller long-shaped fruit; the *Barberine*,
with knobby fruit; the *Elector's hat*, with a curiously
shaped conical fruit, etc. No value should be attached
to the local names in this designation of varieties, for we
have often seen that they express as many errors as
varieties. The botanical names attributed to the species
by Naudin and Cogniaux are numerous, on account of the
bad habit which existed not long ago of describing as
species purely garden varieties, without taking into
account the wonderful effects of cultivation and selection
upon the organ for the sake of which the plant is
cultivated.

Most of these varieties exist in the gardens of the
warm and temperate regions of both hemispheres. The
origin of the species is considered to be doubtful. I
hesitated in 1855 [1] between Southern Asia and the
Mediterranean basin. Naudin and Cogniaux [2] admit
Southern Asia as probable, and the botanists of the
United States on their side have given reasons for their
belief in an American origin. The question requires
careful investigation.

I shall first seek for those forms now attributed to
the species which have been found growing anywhere in
a wild state.

The variety *Cucurbita ovifera*, Linnæus, was
formerly gathered by Lerche, near Astrakhan, but no
modern botanist has confirmed this fact, and it is
probable it was a cultivated plant. Moreover, Linnæus
does not assert it was wild. I have consulted all the
Asiatic and African floras without finding the slightest
mention of a wild variety. From Arabia, or even from
the coast of Guinea to Japan, the species, or the varieties
attributed to it, are always said to be cultivated. In

---

[1] *Géogr. Bot. Raisonnée*, p. 902.
[2] Naudin, *Ann. Sc. Nat.*, 3rd series, vol. vi. p. 9; Cogniaux, in de
Candolle, *Monogr. Phanér.*, iii. p. 546.

India, Roxburgh remarked this, and certainly Clarke, in his recent flora of British India, has good reasons for indicating no locality for it outside cultivation.

It is otherwise in America. A variety, *C. texana*,[1] very near to the variety *ovata*, according to Asa Gray, and which is now unhesitatingly attributed to *C. Pepo*, was found by Lindheimer "on the edges of thickets, in damp woods, on the banks of the upper Guadaloupe, apparently an indigenous plant." Asa Gray adds, however, that it is perhaps the result of naturalization. However, as several species of the genus *Cucurbita* grow wild in Mexico and in the south-west of the United States, we are naturally led to consider the collector's opinion sound. It does not appear that other botanists found this plant in Mexico, or in the United States. It is not mentioned in Hemsley's *Biologia Centrali-Americana*, nor in Asa Gray's recent flora of California.

Some synonyms or specimens from South America, attributed to *C. Pepo*, appear to me very doubtful. It is impossible to say what Molina[2] meant by the names *C. Siceratia* and *C. mammeata*, which appear, moreover, to have been cultivated plants. Two species briefly described in the account of the journey of Spix and Martius (ii. p. 536), and also attributed to *C. Pepo*,[3] are mentioned among cultivated plants on the banks of the Rio Francisco. Lastly, the specimen of Spruce, 2716, from the river Uaupes, a tributary of the Rio Negro, which Cogniaux[4] does not mention having seen, and which he first attributed to the *C. Pepo*, and afterwards to the *C. moschata*, was perhaps cultivated or naturalized from cultivation, or by transport, in spite of the paucity of inhabitants in this country.

Botanical indications are, therefore, in favour of a Mexican or Texan origin. It remains to be seen if

[1] Asia Gray, *Plantæ Lindheimerianæ*, part ii. p. 193.
[2] Molina, *Hist. Nat. du Chili*, p. 377.
[3] Cogniaux, in *Monogr. Phanér.* and *Flora Brasil*, fasc. 78, p. 21.
[4] Cogniaux, *Fl. Bras.* and *Monogr. Phanér.*, iii., p. 547.

historical records are in agreement with or contrary to
this idea.

It is impossible to discover whether a given Sanskrit,
Greek, or Latin name for the pumpkin belongs to one
species rather than to another. The form of the fruit is
often the same, and the distinctive characters are never
mentioned by authors.

There is no figure of the pumpkin in the *Herbarius
Pataviæ Impressus* of 1485, before the discovery of
America, but sixteenth-century authors have published
plates which may be attributed to it. There are three
forms of *Pepones* figured on page 406 of Dodoens,
edition 1557. A fourth, *Pepo rotundus major*, added
in the edition of 1616, appears to me to be *C. maxima*.
In the drawing of *Pepo oblongus* of Lobel, *Icones*, 641,
the character of the peduncle is clearly defined. The
names given to these plants imply a foreign origin ; but
the authors could make no assertions on this head, all
the more that the name of " the Indies " applied both to
Southern Asia and America.

Thus historical data do not gainsay the opinion of an
American origin, but neither do they adduce anything
in support of it.

If the belief that it grows wild in America is confirmed, it may be confidently asserted that the pumpkins
cultivated by the Romans and in the Middle Ages were
Cucurbita maxima, and those of the natives of North
America, seen by different travellers in the seventeenth
century, were *Cucurbita Pepo.*

Musk, or **Melon Pumpkin** — *Cucurbita moschata,*
Duchesne.

The *Bon Jardinier* quotes as the principal varieties
of this species pumpkin *muscade de Provence, pleine
de Naples,* and *de Barbarie.* It is needless to say that
these names show nothing as to origin. The species is
easily recognized by its fine soft down, the pentagonal
peduncle which supports the fruit broadening at the
summit ; the fruit is more or less covered with a glaucous
efflorescence, and the flesh is somewhat musk-scented.
The lobes of the calyx are often terminated by a leafy

border.[1] Cultivated in all tropical countries, it is less successful than other pumpkins in temperate regions.

Cogniaux[2] suspects that it comes from the south of Asia, but he gives no proof of this. I have searched through the floras of the old and new worlds, and I have nowhere been able to discover the mention of the species in a truly wild state. The indications which approach most nearly to it are: (1) In Asia, in the island of Bangka, a specimen verified by Cogniaux, and which Miquel[3] says is not cultivated; (2) in Africa, in Angola, specimens which Welwitsch says are quite wild, but "probably due to an introduction;" (3) in America, five specimens from Brazil, Guiana, or Nicaragua, mentioned by Cogniaux, without knowing whether they were cultivated, naturalized, or indigenous. These indications are very slight Rumphius, Blume, Clarke (*Flora of British India*) in Asia, Schweinfurth (Oliver's *Flora of Trop. Africa*) in Africa, only know it as a cultivated plant. Its cultivation is recent in China,[4] and American floras rarely mention the species.

No Sanskrit name is known, and the Indian, Malay, and Chinese names are neither very numerous nor very original, although the cultivation of the plant seems to be more diffused in Southern Asia than in other parts of the tropics. It was already grown in the seventeenth century according to the *Hortus Malabaricus*, in which there is a good plate (vol. viii. pl. 2). It does not appear that this species was known in the sixteenth century, for Dalechamp's illustration (*Hist.*, i. p. 616) which Seringe attributed to it has not its true characters, and I can find no other figure which resembles it.

Fig-leaved Pumpkin — *Cucurbita ficifolia*, Bouché; *Cucurbita melanosperma*, Braun.

About thirty years ago this pumpkin with black or brown seeds was introduced into gardens. It differs

[1] See the excellent plate in Wight's *Icones*, t. 507, under the erroneous name of *Cucurbita maxima*.
[2] Cogniaux, in *Monogr. Phanér.*, iii. p. 547.
[3] Miquel, *Sumatra*, under the name *Gymnopetalum*, p. 332.
[4] Cogniaux, *in Monogr. Phanér.*

from other cultivated species in being perennial. It is
sometimes called the *Siamese melon.* The *Bon Jardinier*
says that it comes from China. Dr. Bretschneider does
not mention it in his letter of 1881, in which he enumerates the pumpkins grown by the Chinese.

Hitherto no botanist has found it wild. I very much
doubt its Asiatic origin as all the known perennial species
of *Cucurbita* are from Mexico or California.

Melon—*Cucumis Melo,* Linnæus.

The aspect of the question as to the origin of the
melon has completely changed since the experiments of
Naudin. The paper which he published in 1859, in the
Annales des Sciences Naturelles, 4th series, vol. ii., on
the genus *Cucumis,* is as remarkable as that on the genus
Cucurbita. He gives an account of the observations and
experiments of several years on the variability of forms
and the crossed fecundation of a multitude of species,
breeds, or varieties coming from all parts of the world. I
have already spoken (p. 250) of the physiological principle
on which he believes it possible to distinguish those groups
of forms which he terms species, although certain exceptions have occurred which render the criterion of fertilization less absolute. In spite of these exceptional cases,
it is evident that if nearly allied forms can be easily
crossed and produce fertile individuals, as we see, for
example, in the human species, they must be considered
as constituting a single species.

In this sense *Cucumis Melo,* according to the experiments and observations made by Naudin upon about
two thousand living plants, constitutes a species which
comprehends an extraordinary number of varieties and
even of breeds; that is to say, forms which are preserved by heredity. These varieties or races can be fertilized by each other, and yield varied and variable products.
They are classed by the author into ten groups, which he
calls *canteloups, melons brodés, sucrins, melons d'hiver,
serpents, forme de concombre, Chito, Dudaim, rouges de
Perse,* and *sauvages,* each containing varieties or nearly
allied races. These have been named in twenty-five or
thirty different ways by botanists, who, without noticing

transitions of form, the faculty of crossing or of change
under cultivation, have distinguished as species all the
varieties which occur in a given time or place.

Hence it results that several forms found wild, and
which have been described as species, must be the types
and sources of the cultivated forms ; and Naudin makes
the very just observation that these wild forms, which
differ more or less the one from the other, may have produced different cultivated varieties.   This is the more
probable that they sometimes inhabit countries remote
from each other as Southern Asia and tropical Africa,
so that differences in climate and isolation may have
created and consolidated varieties.

The following are the forms which Naudin enumerates as wild : 1. Those of India, which are named by
Wildenow *Cucumis pubescens*, and by Roxburgh *C. turbinatus* or *C. maderas-patanus*.   The whole of British
India and Beluchistan is their natural area.   Its natural
wildness is evident even to non-botanical travellers.[1]
The fruit varies from the size of a plum to that of a
lemon.   It is either striped or barred, or all one colour,
scented or odourless.   The flesh is sweet, insipid, or
slightly acid, differences which it has in common with
the cultivated Cantelopes.   According to Roxburgh the
Indians gather and have a taste for the fruits of *C. turbinatus* and of *C. maderas-patanus*, though they do not
cultivate it.

Referring to the most recent flora of British India,
in which Clarke has described the *Cucurbitaceæ* (ii. p.
619), it seems that this author does not agree with M.
Naudin about the Indian wild forms, although both have
examined the numerous specimens in the herbarium at
Kew.   The difference of opinion, more apparent than real,
arises from the fact that the English author attributes
to a nearly and certainly wild allied species, *C. trigonus*,
Roxburgh, the varieties which Naudin classes under
C. Melo.   Cogniaux,[2] who afterwards saw the same speci-

Gardener's Chronicle, articles signed " I. H. H.," 1857, p. 153 ; 1858,
p. 130.
    [2] Cogniaux, *Monogr. Phanér.*, iii. p. 485.

mens, attributes only *C. turbinatus* to *trigonus*. The specific difference between *C. Melo* and *C. trigonus* is unfortunately obscure, from the characters given by these three authors. The principal difference is that *C. Melo* is an annual, the other perennial, but this duration does not appear to be very constant. Mr. Clarke says himself that *C. Melo* is perhaps derived by cultivation from *C. trigonus ;* that is to say, according to him, from the forms which Naudin attributes to *C. Melo.*

The experiments made during three consecutive years by Naudin [1] upon the products of *Cucumis trigonus*, fertilized by *C. Melo,* seem in favour of the opinion which admits a specific diversity ; for if fertilization took place the products were of different forms, and often reverted to one or other of the original parents.

2. The African forms. Naudin had no specimens in sufficiently good condition, or of which the wild state was sufficiently certain to assert positively the habitation of the species in Africa. He admits it with hesitation. He includes in the species cultivated forms, or other wild ones, of which he had not seen the fruit. Sir Joseph Hooker [2] subsequently obtained specimens which prove more. I am not speaking of those from the Nile Valley,[3] which are probably cultivated, but of plants gathered by Barter in Guinea in the sands on the banks of the Niger. Thonning [4] had previously found, in sandy soil in Guinea, a *Cucumis* to which he had given the name *arenarius ;* and Cogniaux,[5] after having seen a specimen brought home by this traveller, had classed it with *C. Melo,* as Sir J. Hooker thought. The negroes eat the fruit of the plant found by Barter. The smell is that of a fresh green melon. In Thonning's plant the fruit is ovoid, the size of a plum. Thus in Africa as in India the species bears small fruit in a wild state, as we might expect. The *Dudaim* among cultivated varieties is allied to it.

[1] Naudin, *Ann. Sc. Nat.*, 4th series, vol. xviii. p. 171.
[2] Hooker, in Oliver, *Fl. of Trop. Afr.*, ii. p. 546.
[3] Schweinfurth and Ascherson, *Aufzählung*, p. 267.
[4] Schumacher and Thonning, *Guineiske Planten.*, p.
[5] Cogniaux, in de Candolle, *Monogr. Phanér.*, p. 483.

The majority of the species of the genus *Cucumis* are found in Africa; a small minority in Asia or in America. Other species of *Cucurbitaceæ* are divided between Asia and America, although as a rule, in this family, the areas of species are continuous and restricted. *Cucumis Melo* was once perhaps, like *Citrullus Colocynthis* of the same family, wild from the west coast of Africa as far as India without any break.

I formerly hesitated to admit that the melon was indigenous in the north of the Caucasus, as it is asserted by ancient authors—an assertion which has not been confirmed by subsequent botanists. Hohenacker, who was said to have found the species near Elisabethpolis, makes no mention of it in his paper upon the province of Talysch. M. Boissier does not include *Cucumis Melo* in his Oriental flora. He merely says that it is easily naturalized on rubbish-heaps and waste ground. The same thing has been observed elsewhere, for instance in the sands of Ussuri, in Eastern Asia. This would be a reason for mistrusting the locality of the sands of the Niger, if the small size of the fruit in this case did not recall the wild forms of India.

The culture of the melon, or of different varieties of the melon, may have begun separately in India and Africa.

Its introduction into China appears to date only from the eighth century of our era, judging from the epoch of the first work which mentions it.[1] As the relations of the Chinese with Bactriana, and the north-west of India by the embassy of Chang-kien, date from the second century, it is possible that the culture of the species was not then widely diffused in Asia. The small size of the wild fruit offered little inducement. No Sanskrit name is known, but there is a Tamul name, probably less ancient, *molam*,[2] which is like the Latin *Melo*.

It is not proved that the ancient Egyptians cultivated the melon. The fruit figured by Lepsius[3] is not recognizable. If the cultivation had been customary and

---

[1] Bretschneider, letter of Aug. 26, 1881.   [2] Piddington, *Index*.
[3] See the copy in Unger's *Pflanzen des Alten Ægyptens*, fig. 25.

ancient in that country, the Greeks and Romans would
have early known it. Now, it is doubtful whether the
Sikua of Hippocrates and Theophrastus, or the *Pepon* of
Dioscorides, or the *Melopepo* of Pliny, was the melon.
The passages referring to it are brief and insignificant;
Galen[1] is less obscure, when he says that the inside of
the *Melopepones* is eaten, but not of the *Pepones*. There
has been much discussion about those names,[2] but we
want facts more than words. The best proof which I
have been able to discover of the existence of the melon
among the Romans is a very accurate representation of
a fruit in the beautiful mosaic of fruits in the Vatican.
Moreover, Dr. Comes certifies that the half of a melon
is represented in a painting at Herculaneum.[3] The
species was probably introduced into the Græco-Roman
world at the time of the Empire, in the beginning of the
Christian era. It was probably of indifferent quality, to
judge from the silence or the faint praise of writers in
a country where *gourmets* were not wanting. Since
the Renaissance, an improved cultivation and relations
with the East have introduced better varieties into our
gardens. We know, however, that they often degenerate
either from cold or bad conditions of soil, or by crossing
with inferior varieties of the species.

Water-Melon—*Citrullus vulgaris*, Schrader; *Cucurbita Citrullus*, Linnæus.

The origin of the water-melon was long mistaken
or unknown. According to Linnæus, it was a native
of Southern Italy.[4] This assertion was taken from
Matthiole, without observing that this author says it was
a cultivated species. Seringe,[5] in 1828, supposed it
came from India and Africa, but he gives no proof.
I believed it came from Southern Asia, because of its

---

[1] Galen, *De Alimentis*, 1. 2, c. 5.
[2] See all the Vergilian floras, and Naudin, *Ann. Sc. Nat.*, 4th series,
vol. xii. p. 111.
[3] Comes, *Ill. Piante nei Dipinti Pompeiani*, in 4to, p. 20, in the *Museo
Nation.*, vol. iii. pl. 4.
[4] Habitat in Apulia, Calabria, Sicilia (Linnæus, *Species*, edit. 1763,
p. 1435).
[5] Seringe, in *Prodromus*, iii. p. 301.

very general cultivation in this region. It was not known in a wild state. At length it was found indigenous in tropical Africa, on both sides of the equator, which settles the question.[1] Livingstone[2] saw districts literally covered with it, and the savages and several kinds of wild animals eagerly devoured the wild fruit. They are sometimes, but not always, bitter, and this cannot be detected from the appearance of the fruit. The negroes strike it with an axe, and taste the juice to see whether it is good or bad. This diversity in the wild plant, growing in the same climate and in the same soil, is calculated to show the small value of such a character in cultivated *Cucurbitaceæ*. For the rest, the frequent bitterness of the water-melon is not at all extraordinary, as the most nearly allied species is *Citrullus Colocynthis*. Naudin obtained fertile hybrids from crossing the bitter water-melon, wild at the Cape, with a cultivated species which confirms the specific unity suggested by the outward appearance.

The species has not been found wild in Asia.

The ancient Egyptians cultivated the water-melon, which is represented in their paintings.[3] This is one reason for believing that the Israelites knew the species, and called it *abbatitchim*, as is said; but besides the Arabic name, *battich, batteca,* evidently derived from the Hebrew, is the modern name for the water-melon. The French name, *pastèque,* comes through the Arabic from the Hebrew. A proof of the antiquity of the plant in the north of Africa is found in the Berber name, *tadelaât*,[4] which differs too widely from the Arabic name not to have existed before the Conquest. The Spanish names *zandria, cindria,* and the Sardinian *sindria*,[5] which I cannot connect with any others, show also an ancient culture in the eastern part of the Mediterranean basin. Its

[1] Naudin, *Ann. sc. Nat.,* 4th series, vol. xii. p. 101 ; Sir J. Hooker, in Oliver, *Flora of Trop. Afr.,* ii. p. 549.

[2] French trans., p. 56.

[3] Unger has copied the figures from Lepsius' work in his memoir, *Die Pflanzen des Alten Ægyptens,* figs. 30, 31, 32.

[4] *Dictionnaire Français-Berber,* at the word *pastèque.*

[5] Moris, *Flora Sardoa.*

cultivation early spread into Asia, for there is a Sanskrit
name, *chayapula*,[1] but the Chinese only received the
plant in the tenth century of the Christian era. They
call it *si-kua*, that is melon of the West.[2]

As the water-melon is an annual, it ripens out of the
tropics wherever the summer is sufficiently hot. The
modern Greeks cultivate it largely, and call it *carpousia*
or *carpousea*,[3] but this name does not occur in ancient
authors, nor even in the Greek of the decadence and of
the Middle Ages.[4]   It is the same as the *karpus* of the
Turks of Constantinople,[5] which we find again in the
Russian *arbus*,[6] and in Bengali and Hindustani as *tarbuj,
turbouz*.[7]   Another Constantinople name, mentioned by
Forskal, *chimonico*, recurs in Albanian *chimico*.[8]   The
absence of an ancient Greek name which can with
certainty be attributed to this species, seems to show
that it was introduced into the Græco-Roman world
about the beginning of the Christian era. The poem
Copa, attributed to Virgil and Pliny, perhaps mentions
it (lib. 19, cap. 5), as Naudin thinks, but it is doubtful.

Europeans have introduced the water-melon into
America, where it is now cultivated from Chili to the
United States. The *jacé* of the Brazilians, of which
Piso and Marcgraf have a drawing, is evidently in-
troduced, for the first-named author says it is cultivated
and partly naturalized.[9]

Cucumber—*Cucumis sativus*, Linnæus.

In spite of the very evident difference between the
melon and cucumber, which both belong to the genus
Cucumis, cultivators suppose that the species may be
crossed, and that the quality of the melon is thus some-

[1] Piddingtòn, *Index.*
[2] Bretschneider, *Study and Value*, etc., p. 17.
[3] Heldreich, *Pflanz. d. Attisch. Ebene.*, p. 591; *Nutzpfl. Griechenl.*,
p. 50.
[4] Langkavel, *Bot. der Spät. Griechen.*
[5] Forskal, *Flora Ægypto-Arabica.*, part i. p. 34.
[6] Nemnich, *Polyg. Lexic.*, i. p. 1309.
[7] Piddington, *Index;* Pickering, *Chronol. Arrang.*, p. 72.
[8] Heldreich, *Nutzpfl.*, etc., p. 50.
[9] "*Sativa planta et tractu temporis quasi nativa facta*" (Piso,
edit. 1658, p. 233).

times spoilt. Naudin[1] ascertained by experiments that this fertilization is not possible, and has also shown that the distinction of the two species is well founded.

The original country of *Cucumis sativus* was unknown to Linnæus and Lamarck. In 1805, Wildenow[2] asserted it was indigenous in Tartary and India, but without furnishing any proof. Later botanists have not confirmed the assertion. When I went into the question in 1855, the species had not been anywhere found wild. For various reasons deduced from its ancient culture in Asia and in Europe, and especially from the existence of a Sanskrit name, *soukasa*,[3] I said, "Its original habitat is probably the north-west of India, for instance Cabul, or some adjacent country. Everything seems to show that it will one day be discovered in these regions which are as yet but little known."

This conjecture has been realized if we admit, with the best-informed modern authors, that *Cucumis Hardwickii*, Royle, possesses the characteristics of *Cucumis sativus*. A coloured illustration of this cucumber found at the foot of the Himalayas may be seen in Royle's *Illustrations of Himalayan Plants*, p. 220, pl. 47. The stems, leaves, and flowers are exactly those of *C. sativus*. The fruit, smooth and elliptical, has a bitter taste; but there are similar forms of the cultivated cucumber, and we know that in other species of the same family, the water-melon, for instance, the pulp is sweet or bitter. Sir Joseph Hooker, after describing the remarkable variety which he calls the *Sikkim* cucumber,[4] adds that the variety *Hardwickii*, wild from Kumaon to Sikkim, and of which he has gathered specimens, does not differ more from the cultivated plant than certain varieties of the latter differ from others; and Cogniaux, after seeing the plants in the herbarium at Kew, adopts this opinion.[5]

The cucumber, cultivated in India for at least three

---

[1] Naudin, in *Ann. Sc. Nat.*, 4th series, vol. xi. p. 31.
[2] Wildenow, *Species*, iv. p. 615.   [3] Piddington, *Index*.
[4] *Bot. Mag.*, pl. 6206.
[5] Cogniaux, in de Candolle, *Monogr. Phanér.*, iii. p. 499.

thousand years, was only introduced into China in the second century before Christ, when the ambassador Chang-kien returned from Bactriana.[1] The species spread more rapidly towards the West. The ancient Greeks cultivated the cucumber under the name of *sikuos*,[2] which remains as *sikua* in the modern language. The modern Greeks have also the name *aggouria*, from an ancient Aryan root which is sometimes applied to the water-melon, and which recurs for the cucumber in the Bohemian *agurka*, the German *Gurke*, etc. The Albanians (Pelasgians?) have quite a different name, *kratsavets*,[3] which we recognize in the Slav *Krastavak*. The Latins called the cucumber *cucumis*. These different names show the antiquity of the species in Europe. There is even an Esthonian name, *uggurits, ukkurits, urits.*[4] It does not seem to be Finnish, but to belong to the same Aryan root as *aggouria*. If the cucumber came into Europe before the Aryans, there would perhaps be some name peculiar to the Basque language, or seeds would have been found in the lake-dwellings of Switzerland and Savoy; but this is not the case. The peoples in the neighbourhood of the Caucasus have names quite different to the Greek; in Tartar *kiar*, in Kalmuck *chaja*, in Armenian *karan*.[5] The name *chiar* exists also in Arabic for a variety of the cucumber.[6] This is, therefore, a Turanian name anterior to the Sanskrit, whereby its culture in Western Asia would be more than three thousand years old.

It is often said that the cucumber is the *kischschuim*, one of the fruits of Egypt regretted by the Israelites in the desert.[7] However, I do not find any Arabic name among the three given by Forskal which can be connected with this, and hitherto no trace has been found of the presence of the cucumber in ancient Egypt.

---

[1] Bretschneider, letters of Aug. 23 and 26, 1881.
[2] Theophrastus, *Hist.*, lib. 7, cap. 4; Lenz, *Bot. der Alten*, p. 492.
[3] Heldreich, *Nutzpfl. Griechen.*, p. 50.
[4] Nemnich, *Polygl. Lex.*, i. p. 1306.
[5] Nemnich, *ibid.*   [6] Forskal, *Fl. Ægypt.*, p. 76.
[7] Rosenmüller, *Biblische Alterth.*, i. p. 97; Hamilton, *Bot. de la Bible*, p. 34.

West Indian Gherkin—*Cucumis Anguria*, Linnæus.

This small species of cucumber is designated in the *Bon Jardinier* under the name of the cucumber *Arada*. The fruit, of the size of an egg, is very prickly. It is eaten cooked or pickled. As the plant is very productive, it is largely cultivated in the American colonies. Descourtilz and Sir Joseph Hooker have published good coloured illustrations of it, and M. Cogniaux a plate with a detailed analysis of the flower.[1]

Several botanists affirm that it is wild in the West Indies. P. Browne,[2] in the last century, spoke of the plant as the "little wild cucumber" (in Jamaica). Descourtilz said, "The cucumber grows wild everywhere, and principally in the dry savannahs and near rivers, whose banks afford a rich vegetation." The inhabitants call it the "maroon cucumber." Grisebach[3] saw specimens in several other West India Isles, and appears to admit their wild character. M. E. André found the species growing in the sand of the sea-shore at Porto-Cabello, and Burchell in a similar locality in Brazil, and Riedel near Rio di Janeiro.[4] In the case of a number of other specimens gathered in the east of America from Brazil to Florida, it is unknown whether they were wild or cultivated. A wild Brazilian plant, badly drawn by Piso,[5] is mentioned as belonging to the species, but I am very doubtful of this.

Botanists from Tournefort down to our own day have considered the Anguria to be of American origin, a native of Jamaica in particular. M. Naudin[6] was the first to point out that all the other species of *Cucumis* are of the old world, and principally African. He wondered whether this one had not been introduced into America by the negroes, like many other plants which have become

---

[1] Descourtilz, *Fl. Méd. des Antilles*, v. pl. 329; Hooker, *Bot. Mag.*, t. 5817; Cogniaux, in *Fl. Brasil*, fasc. 78, pl. 2.

[2] Browne, *Jamaica*, edit. 2, p. 353.

[3] Grisebach, *Fl. of Brit. W. India Is.*, p. 288.

[4] Cogniaux, *ubi supra*.

[5] *Guanerva-oba*, in Piso, *Brasil*, edit. 1658, p. 264; Marcgraf, edit. 1648, p. 44, without illustration, calls it *Cucumis sylvestris Brasiliæ*.

[6] Naudin, *Ann. Sc. Nat.*, 4th series, vol. ii. p. 12.

naturalized. However, unable to find any similar African plant, he adopted the general opinion. Sir Joseph Hooker, on the contrary, is inclined to believe that *C. Anguria* is a cultivated and modified form of some African species nearly allied to *C. prophetarum* and *C. Figarei*, although these are perennial. In favour of this hypothesis, I may add: (1) The name *maroon* cucumber, given in the French West India Islands, indicates a plant which has become wild, for this is the meaning of the word *maroon* as applied to the negroes; (2) its extended area in America from Brazil to the West Indies, always along the coast where the slave trade was most brisk, seems to be a proof of foreign origin. If the species grew in America previous to its discovery, it would, with such an extensive habitat, have been also found upon the west coast of America, and inland, which is not the case.

The question can only be solved by a more complete knowledge of the African species of *Cucumis*, and by experiments upon fertilization, if any have the patience and ability necessary to do for the genus *Cucumis* what Naudin has done for the genus *Cucurbita*.

Lastly, I would point out the absurdity of a common name for the Anguria in the United States—*Jerusalem Cucumber*.[1] After this, is it possible to take popular names as a guide in our search for origins?

White Gourd-melon, or **Benincasa**—*Benincasa hispida,* Thunberg; *Benincasa cerifera,* Savi.

This species, which is the only one of the genus Benincasa, is so like the pumpkins that early botanists took it for one,[2] in spite of the waxy efflorescence on the surface of the fruit. It is very generally cultivated in tropical countries. It was, perhaps, a mistake to abandon its cultivation in Europe after having tried it, for Naudin and the *Bon Jardinier* both recommend it.

It is the *cumbalam* of Rheede, the *camolenga* of Rumphius, who had seen it cultivated in Malabar and the Sunda Islands, and give illustrations of it.

---

[1] Darlington, *Agric. Bot.*, p. 58.
[2] *Cucurbita Pepo* of Loureiro and Roxburgh.

From several works, even recent ones,[1] it might be supposed that it had never been found in a wild state, but if we notice the different names under which it has been described we shall find that this is not the case. Thus *Cucurbita hispida*, Thunberg, and *Lagenaria dasystemon*, Miquel, from authentic specimens seen by Cogniaux,[2] are synonyms of the species, and these plants are wild in Japan.[3] *Cucurbita littoralis*, Hass-karl,[4] found among shrubs on the sea-shore in Java, and *Gymnopetalum septemlobum*, Miquel, also in Java, are the *Benincasa* according to Cogniaux. As are also *Cucurbita vacua*, Mueller,[5] and *Cucurbita pruriens*, Forster, of which he has seen authentic specimens found at Rockingham, in Australia, and in the Society Islands. Nadeaud[6] does not mention the latter. Temporary naturalization may be suspected in the Pacific Isles and in Queensland, but the localities of Java and Japan seem quite certain. I am the more inclined to believe in the latter, that the cultivation of the Benincasa in China dates from the remotest antiquity.[7]

Towel Gourd—*Momordica cylindrica*, Linnæus; *Luffa cylindrica*, Rœmer.

Naudin[8] says, "*Luffa cylindrica*, which in some of our colonies has retained the Indian name *pétole*, is probably a native of Southern Asia, and perhaps also of Africa, Australia, and Polynesia. It is cultivated by the peoples of most hot countries, and it appears to be naturalized in many places where it doubtless did not exist originally." Cogniaux[9] is more positive. "An indigenous species," he says, "in all the tropical regions

---

[1] Clarke, in *Fl. of Brit. Ind.*, ii. p. 616.
[2] Cogniaux, in de Candolle, *Monogr. Phanér.*, iii. p. 513.
[3] Thunberg, *Fl. Jap.*, p. 322 ; Franchet and Savatier, *Enum. Pl. Jap.*, i. p. 173.
[4] Hasskarl, *Catal. Horti. Bogor. Alter.*, p. 190 ; Miquel, *Flora Indo-Batav.*
[5] Mueller, *Fragm.*, vi. p. 186; Forster, *Prodr.* (no description); Seemann, *Jour. of Bot.*, ii. p. 50.
[6] Nadeaud, *Plan. Usu. des Taitiens*, *Enum. des Pl. Indig. à Taiti.*
[7] Bretschneider, letter of Aug. 26, 1881.
[8] Naudin, *Ann. Sc. Nat.*, 4th series, vol. xii. p. 121.
[9] Cogniaux, *Monogr. Phanér.*, iii. p. 458.

of the old world; often cultivated and half wild in America between the tropics." In consulting the works quoted in these two monographs, and herbaria, its character as a wild plant will be found sometimes conclusively certified.

With regard to Asia,[1] Rheede saw it in sandy places, in woods and other localities in Malabar; Roxburgh says it is wild in Hindustan; Kurz, in the forests of Burmah; Thwaites, in Ceylon. I have specimens from Ceylon and Khasia. There is no Sanskrit name known, and Dr. Bretschneider, in his work *On the Study and Value of Chinese Botanical Works*, and in his letters mentions no luffa either wild or cultivated in China. I suppose, therefore, that its cultivation is not ancient even in India.

The species is wild in Australia, on the banks of rivers in Queensland,[2] and hence it is probable it will be found wild in the Asiatic Archipelago, where Rumphius, Miquel, etc., only mention it as a cultivated plant.

Herbaria contain a great number of specimens from tropical Africa, from Mozambique to the coast of Guinea, and even as far as Angola, but collectors do not appear to have indicated whether they were cultivated or wild plants. In the Delessert herbarium, Heudelot indicates it as growing in fertile ground in the environs of Galam. Sir Joseph Hooker [3] quotes this without affirming anything. Schweinfurth and Ascheron,[4] who are always careful in this matter, say the species is only a cultivated one in the Nile Valley. This is curious, because the plant was seen in the seventeenth century in Egyptian gardens under the Arabian name of *luff*,[5] whence the genus was called *Luffa*, and the species *Luffa ægyptica*. The ancient Egyptian monuments show no trace of it. The

---

[1] Rheede, *Hort. Malab.*, viii. p. 15, t. 8; Roxburgh. *Fl. Ind.*, iii. p. 714, as *L. clavata*; Kurz, *Contrib.*, ii. p. 100; Thwaites, *Enum.*

[2] Mueller, *Fragmenta*, iii. p. 107; Bentham, *Fl. Austr.*, iii. p. 317, under names which Naudin and Cogniaux regard as synonyms of *L. cylindrica*.

[3] Hooker, in Oliver, *Fl. of Trop. Afr.*, ii. p. 530.

[4] Schweinfurth and Ascheron, *Aufzählung*, p. 268.

[5] Forskal, *Fl. Ægypt.*, p. 75.

absence of a Hebrew name is another reason for believing that its cultivation was introduced into Egypt in the Middle Ages. It is now grown in the Delta, not only for the fruit but also for the export of the seed, from which a preparation is made for softening the skin.

The species is cultivated in Brazil, Guiana, Mexico, etc., but I find no indication that it is indigenous in America. It appears to have been here and there naturalized, in Nicaragua for instance, from a specimen of Levy's.

In brief, the Asiatic origin is certain, the African very doubtful, that of America imaginary, or rather the effect of naturalization.

Angular Luffa—*Luffa acutangula*, Roxburgh.

The origin of this species, cultivated like the preceding one in all tropical countries, is not very clear, according to Naudin and Cogniaux.[1] The first gives Senegal, the second Asia, and, doubtfully, Africa. It is hardly necessary to say that Linnæus [2] was mistaken in indicating Tartary and China. Clarke, in Sir Joseph Hooker's flora, says without hesitation that it is indigenous in British India. Rheede [3] formerly saw the plant in sandy soil in Malabar. Its natural area seems to be limited, for Thwaites in Ceylon, Kurz in British Burmah, and Loureiro in China and Cochin-China,[4] only give the species as cultivated, or growing on rubbish-heaps near gardens. Rumphius [5] calls it a Bengal plant. No luffa has been long cultivated in China, according to a letter of Dr. Bretschneider. No Sanskrit name is known. All these are indications of a comparatively recent culture in Asia.

A variety with bitter fruit is common in British India [6] in a wild state, since there is no inducement to

---

[1] Naudin, *Ann. Sc. Nat.*, 4th series, vol. xii. p. 122 ; Cogniaux, in de Candolle, *Monogr. Phanér.*, iii. p. 459.

[2] Linnæus, *Species*, p. 1436, as *Cucumis acutangulus*.

[3] Rheede, *Hort. Malab.*, viii. p. 13, t. 7.

[4] Thwaites, *Enum. Ceylan*, p. 126 ; Kurz, *Contrib.*, ii. p. 101 ; Loureiro, *Fl. Cochin.*, p. 727.

[5] Rumphius, *Amboin*, v. p. 408, t. 149.

[6] Clarke, in *Fl. Brit. Ind.*, ii. p. 614.

cultivate it. It exists also in the Sunda Islands. It is *Luffa amara*, Roxburgh, and *L. sylvestris*, Miquel. *L. subangulata*, Miquel, is another variety which grows in Java, which M. Cogniaux also unites with the others from authentic specimens which he saw.

M. Naudin does not say what traveller gives the plant as wild in Senegambia; but he says the negroes call it *papengaye*, and as this is the name of the Mauritius planters,[1] it is probable that the plant is cultivated in Senegal, and perhaps naturalized near dwellings. Sir Joseph Hooker, in the *Flora of Tropical Africa*, gives the species, but without proof that it is wild in Africa, and Cogniaux is still more brief. Schweinfurth and Ascheron[2] do not mention it either as wild or cultivated in Egypt, Nubia, and Abyssinia. There is no trace of its ancient cultivation in Egypt.

The species has often been sent from the West Indies, New Granada, Brazil, and other parts of America, but there is no indication that it has been long in these places, nor even that it occurs at a distance from gardens in a really wild state.

The conditions or probabilities of origin, and of date of culture, are, it will be seen, identical for the two cultivated species of luffa. In support of the hypothesis that the latter is not of African origin, I may say that the four other species of the genus are Asiatic or American; and as a sign that the cultivation of the luffa is not very ancient, I will add that the form of the fruit varies much less than in the other cultivated cucurbitacea.

Snake Gourd—*Trichosanthes anguina*, Linnæus.

An annual creeping *Cucurbitacea*, remarkable for its fringed corolla. It is called *petole* in Mauritius, from a Java name. The fruit, which is something like a long fleshy pod of some leguminous plants, is eaten cooked like a cucumber in tropical Asia.

As the botanists of the seventeenth century received the plant from China, they imagined that the plant was

---

[1] Bojer, *Hort. Maurit.*
[2] Schweinfurth and Ascherson, *Aufzählung*, p. 268.

indigenous there, but it was probably cultivated. **Dr.** Bretschneider [1] tells us that the Chinese name, *mankua,* means "cucumber of the southern barbarians." Its home must be India, or the Indian Archipelago. No author, however, asserts that it has been found in a distinctly wild state. Thus Clarke, in Hooker's *Flora of British India,* ii. p. 610, says only, "India, cultivated." Naudin,[2] before him, said, "Inhabits the East Indies, where it is much cultivated for its fruits. It is rarely found wild." Rumphius [3] is not more positive for Amboyna. Loureiro and Kurz in Cochin-China and Burmah, Blume and Miquel in the islands to the south of Asia, have only seen the plant cultivated. The thirty-nine other species of the genus are all of the old world, found between China or Japan, the west of India and Australia. They belong especially to India and the Malay Archipelago. I consider the Indian origin as the most probable one.

The species has been introduced into Mauritius, where it sows itself round cultivated places. Elsewhere it is little diffused. No Sanskrit name is known.

Chayote, or **Choco**—*Sechium edule,* Swartz.

This plant, of the order *Cucurbitaceæ,* is cultivated in tropical America for its fruits, shaped like a pear, and tasting like a cucumber. They contain only one seed, so that the flesh is abundant.

The species alone constitutes the genus Sechium. There are specimens in every herbarium, but generally collectors do not indicate whether they are naturalized, or really wild, and apparently indigenous in the country. Without speaking of works in which this plant is said to come from the East Indies, which is entirely a mistake, several of the best give Jamaica [4] as the original home. However, P. Browne,[5] in the middle of the last century, said positively that it was cultivated there, and Sloane does not mention it. Jacquin [6] says that it "inhabits

---

[1] Bretschneider, *Study and Value,* etc., p. 17.
[2] Naudin, *Ann. Sc. Nat.,* 4th series, vol. xviii. p. 190.
[3] Rumphius, *Amboin,* v. pl. 148.
[4] Grisebach, *Flora of Brit. W. India Isl.,* p. 286.
[5] Browne, *Jamaica,* p. 355.
[6] Jacquin, *Stirp. Amer. Hist.,* p. 259.

Cuba, and is cultivated there," and Richard copies this phrase in the flora of R. de La Sagra without adding any proof. Naudin says,[1] "a Mexican plant," but he does not give his reasons for asserting this. Cogniaux,[2] in his recent monograph, mentions a great number of specimens gathered from Brazil to the West Indies without saying if he had seen any one of these given as wild. Seemann[3] saw the plant cultivated at Panama, and he adds a remark, important if correct, namely, that the name *chayote*, common in the isthmus, is the corruption of an Aztec word, *chayotl*. This is an indication of an ancient existence in Mexico, but I do not find the word in Hernandez, the classic author on the Mexican plants anterior to the Spanish conquest. The *chayote* was not cultivated in Cayenne ten years ago.[4] Nothing indicates an ancient cultivation in Brazil. The species is not mentioned by early writers, such as Piso and Marcgraf, and the name *chuchu*, given as Brazilian,[5] seems to me to come from *chocho*, the Jamaica name, which is perhaps a corruption of the Mexican word.

The plant is probably a native of the south of Mexico and of Central America, and was transported into the West India Islands and to Brazil in the eighteenth century. The species was afterwards introduced into Mauritius and Algeria, where it is very successful.[6]

Indian Fig, or **Prickly Pear**—*Opuntia ficus indica,* Miller.

This fleshy plant of the *Cactus* family, which produces the fruit known in the south of Europe as the Indian fig, has no connection with the fig tree, nor has the fruit with the fig. Its origin is not Indian but American. Everything is erroneous and absurd in this common name. However, since Linnæus took his botanical name from it, *Cactus ficus indica*, afterwards connected with the genus *Opuntia*, it was necessary to retain the specific

---

[1] Naudin, *Ann. Sc. Nat.*, 4th series, vol. xviii. p. 205.
[2] In *Monogr. Phanér.*, iii. p. 902.
[3] Seemann, *Bot. of Herald*, p. 128.
[4] Sagot, *Journal de la Soc. d'Hortic. de France*, 1872.
[5] Cogniaux, *Fl. Brasil*, fasc. 78.   [6] Sagot, *ibid.*

name to avoid changes which are a source of confusion, and to recall the popular denomination. The prickly forms, and those more or less free from spines, have been considered by some authors as distinct species, but an attentive examination leads us to regard them as one.[1]

The species existed both wild and cultivated in Mexico before the arrival of the Spaniards. Hernandez[2] describes nine varieties of it, which shows the antiquity of its cultivation. The cochineal insect appears to feed on one of these, almost without thorns, more than on the others, and it has been transported with the plant to the Canary Isles and elsewhere. It is not known how far its habitat extended in America before man transported pieces of the plant, shaped like a racket, and the fruits, which are two easy ways of propagating it. Perhaps the wild plants in Jamaica, and the other West India Islands mentioned by Sloane,[3] in 1725, were the result of its introduction by the Spaniards. Certainly the species has become naturalized in this direction as far as the climate permits; for instance, as far as Southern Florida.[4]

It was one of the first plants which the Spaniards introduced to the old world, both in Europe and Asia. Its singular appearance was the more striking that no other species belonging to the family had before been seen.[5] All sixteenth-century botanists mention it, and the plant became naturalized in the south of Europe and in Africa as its cultivation was introduced. It was in Spain that the prickly pear was first known under the American name *tuna*, and it was probably the Moors who took it into Barbary when they were expelled from the peninsula. They called it fig of the Christians.[6] The custom of using the plant for fences, and the nourishing property of the fruits, which contain a large proportion of sugar, have determined its extension round the Mediterranean, and in general in all countries near the tropics.

---

[1] Webb and Berthelot, *Phytog. Canar.*, sect. 1, p. 208.
[2] Hernandez, *Theo. Novæ Hisp.*, p. 78.   [3] Sloane, *Jamaica*, ii. p. 150.
[4] Chapman, *Flora of Southern States*, p. 144.
[5] The *cactos* of the Greeks was quite a different plant.
[6] Steinheil, in Boissier, *Voyage Bot. en Espagne*, i. p. 25.

The cultivation of the cochineal, which was unfavourable to the production of the fruit,[1] is dying out since the
manufacture of colouring matters by chemical processes.

Gooseberry — *Ribes grossularia* and *R. Vacrispa,*
Linnæus.

The fruit of the cultivated varieties is generally
smooth, or provided with a few stiff hairs, while that of
the wild varieties has soft and shorter hairs ; but intermediate forms exist, and it has been shown by experiment that by sowing the seeds of the cultivated fruit,
plants with either smooth or hairy fruit are obtained.[2]
There is, therefore, but one species, which has produced
under cultivation one principal variety and several sub-
varieties as to the size, colour, or taste of the fruit.

The gooseberry grows wild throughout temperate
Europe, from Southern Sweden to the mountainous
regions of Central Spain, of Italy, and of Greece.[3]   It is
also mentioned in Northern Africa, but the last published
catalogue of Algerian plants[4] indicates it only in the
mountains of Aures, and Ball has found a variety in
the Atlas of Marocco.[5]   It grows in the Caucasus,[6] and
under more or less different forms in the western
Himalayas.[7]

The Greeks and Romans do not mention the species,.
which is rare in the South, and which is hardly worth
planting where grapes will ripen.   It is especially in
Germany, Holland, and England that it has been cultivated from the sixteenth century,[8] principally as a
seasoning, whence the English name, and the French
groseille à maquereaux (mackerel currant).   A wine
is also made from it.

The frequency of its cultivation in the British Isles
and in other places where it is found wild, which are

---

[1] Webb and Berthelot, *Phytog. Canar.*, vol. iii. sect. 1, p. 208
[2] Robson, quoted in *English Botany*, pl. 2057.
[3] Nyman, *Conspectus Fl. Europeæ*, p. 266 ; Boissier, *Fl. Or.*, ii. p. 815..
[4] Munby, *Catal.*, edit. 2, p. 15.
[5] Ball, *Spicilegium Fl. Maroc.*, p. 449.
[6] Ledebour, *Fl. Ross.*, ii. p. 194 ; Boissier, *ubi supra,*
[7] Clarke, in Hooker's *Fl. Brit. Ind.*, ii. p. 410.
[8] Phillips, *Account of Fruits*, p. 174.

often near gardens, has suggested to some English
botanists the idea of an accidental naturalization. This
is likely enough in Ireland;[1] but as it is an essentially
European species, I do not see why it should not have
existed in England, where the wild plant is more common,
since the establishment of most of the species of the
British flora; that is to say, since the end of the glacial
period, before the separation of the island from the
continent. Phillips quotes an old English name, *feaberry*
or *feabes*, which supports the theory of an ancient existence, and two Welsh names,[2] of which I cannot, however,
certify the originality.

Red Currant—*Ribes rubrum*, Linnæus.

The common red currant is wild throughout Northern
and Temperate Europe, and in Siberia[3] as far as Kamtschatka, and in America, from Canada and Vermont to
the mouth of the river Mackenzie.[4]

Like the preceding species, it was unknown to the
Greeks and Romans, and its cultivation was only introduced in the Middle Ages. The cultivated plant hardly
differs from the wild one. That the plant was foreign
to the south of Europe is shown by the name of *groseillier
d'outremer* (currant from beyond the sea), given in France[5]
in the sixteenth century. In Geneva the currant is still
commonly called *raisin de mare*, and in the canton of
Soleure *meertrübli*. I do not know why the species was
supposed, three centuries ago, to have come from beyond seas. Perhaps this should be understood to mean
that it was brought by the Danes and the Northmen,
and that these peoples from beyond the northern seas
introduced its cultivation. I doubt it, however, for the
Ribes rubrum is wild in almost the whole of Great
Britain[6] and in Normandy;[7] the English, who were in
constant communication with the Danes, did not cultivate
it as late as 1557, from a list of the fruits of that epoch

[1] Moore and More, *Contrib. to the Cybele Hybernica*, p. 113.
[2] Davies, *Welsh Botanology*, p. 24.
[3] Ledebour, *Fl. Ross.*, ii. p. 199.
[4] Torrey and Gray, *Fl. N. Amer.*, i. p. 150.   [5] Dodoneus, p. 748.
[6] Watson, *Cybele Brit.*
[7] Brebisson, *Flore de Normandie*, p. 99.

drawn up by Th. Tusser, and published by Phillips;[1] and even in the time of Gerard, in 1597,[2] its cultivation was rare, and the plant had no particular name.[3] Lastly, there are French and Breton names which indicate a cultivation anterior to the Normans in the west of France.

The old names in France are given in the dictionary by Ménage. According to him, red currants are called at Rouen *gardes,* at Caen *grades,* in Lower Normandy *gradilles,* and in Anjou *castilles.* Ménage derives all these names from *rubius, rubicus,* etc., by a series of imaginary transformations, from the word *ruber,* red. Legonidec[4] tells us that red currants are also called *Kastilez* (l. liquid) in Brittany, and he derives this name from Castille, as if a fruit scarcely known in Spain and abundant in the north could come from Spain. These words, found both in Brittany and beyond its limits, appear to me to be of Celtic origin; and I may mention, in support of this theory, that in Legonidec's dictionary *gardis* means *rough, harsh, pungent, sour,* etc., which gives a hint as to the etymology. The generic name *Ribes* has caused other errors. It was thought the plant might be one which was so called by the Arabs; but the word comes rather from a name for the currant very common in the north, *ribs* in Danish,[5] *risp* and *resp* in Swedish.[6] The Slav names are quite different and in considerable number.

Black Currant—*Cassis; Ribes nigrum,* Linnæus.

The black currant grows wild in the north of Europe, from Scotland and Lapland as far as the north of France and Italy; in Bosnia,[7] Armenia,[8] throughout Siberia, in the basin of the river Amur, and in the western Hima-

[1] Phillips, *Account of Fruits,* p. 136.
[2] Gerard, *Herbal,* p. 1143.
[3] That of *currant* is a later introduction, given from the resemblance to the grapes of Corinth (Phillips, *ibid.*).
[4] Legonidec, *Diction. Celto-Breton.*
[5] Moritzi, *Dict. Inédit des Noms Vulgaires.*
[6] Linnæus, *Flora Suecica,* n. 197.
[7] Watson, *Compend. Cybele,* i. p. 177; Fries, *Summa Veg. Scand.,* p. 39; Nyman, *Conspect. Fl. Europ.,* p. 266.
[8] Boissier, *Fl. Or.,* ii. p. 815.

layas ;[1] it often becomes naturalized, as for instance, in the centre of France.[2]

This shrub was unknown in Greece and Italy, for it is proper to colder countries. From the variety of the names in all the languages, even in those anterior to the Aryans, of the north of Europe, it is clear that this fruit was very early sought after, and its cultivation was probably begun before the Middle Ages. J. Bauhin[3] says it was planted in gardens in France and Italy, but most sixteenth-century authors do not mention it. In the *Histoire de la Vie Privée des Français*, by Le Grand d'Aussy, published in 1872, vol. i. p. 232, the following curious passage occurs: "The black currant has been cultivated hardly forty years, and it owes its reputation to a pamphlet entitled *Culture du Cassis*, in which the author attributed to this shrub all the virtues it is possible to imagine." Further on (vol. iii. p. 80), the author mentions the frequent use, since the publication of the pamphlet in question, of a liqueur made from the black currant. Bosc, who is always accurate in his articles in the *Dictionnaire d'Agriculture*, mentions this fashion under the head *Currant*, but he is careful to add, "It has been very long in cultivation for its fruit, which has a peculiar odour agreeable to some, disagreeable to others, and which is held to be stomachic and diuretic." It is also used in the manufacture of the liqueurs known as ratafia de Cassis.[4]

Olive—*Olea Europea*, Linnæus.

The wild olive, called in botanical books the variety

[1] Ledebour, *Fl. Ross.*, p. 200; Maximowicz, *Primitiæ Fl. Amur.*, p. 119; Clarke, in Hooker, *Fl. Brit. Ind.*, ii. p. 411.
[2] Boreau, *Flore du Centre de la France*, edit. 3, p. 262.
[3] Bauhin, *Hist. Plant.*, ii. p. 99.
[4] This name *Cassis* is curious. Littré says that it seems to have been introduced late into the language, and that he does not know its origin. I have not met with it in botanical works earlier than the middle of the seventeenth century. My manuscript collection of common names, among more than forty names for this species in different languages or dialects has not one which resembles it. Buchoz, in his *Dictionnaire des Plantes*, 1770, i. p. 289, calls the plant the *Cassis* or *Cassetier des Poitevins*. The old French name was *Poivrier* or *groseillier noir*. Larousse's dictionary says that good liqueurs were made at Cassis in Provence. Can this be the origin of the name?

sylvestris or *oleaster*, is distinguished from the cultivated
olive tree by a smaller fruit, of which the flesh is not so
abundant.   The best fruits are obtained by selecting the
seeds, buds, or grafts from good varieties.

The oleaster now exists over a wide area east and
west of Syria, from the Punjab and Beluchistan [1] as far
as Portugal and even Madeira, the Canaries and even
Marocco,[2] and from the Atlas northwards as far as the south
of France, the ancient Macedonia, the Crimea, and the
Caucasus.[3]   If we compare the accounts of travellers and
of the authors of floras, it will be seen that towards the
limits of this area there is often a doubt as to the wild
and indigenous (that is to say ancient in the country)
nature of the species.   Sometimes it offers itself as a
shrub which fruits little or not at all ; and sometimes, as
in the Crimea, the plants are rare as though they had
escaped, as an exception, the destructive effects of winters
too severe to allow of a definite establishment.   As
regards Algeria and the south of France, these doubts
have been the subject of a discussion among competent
men in the Botanical Society.[4]   They repose upon the
uncontestable fact that birds often transport the seed of
the olive into uncultivated and sterile places, where the
wild form, the oleaster, is produced and naturalized.

The question is not clearly stated when we ask if
such and such olive trees of a given locality are really
wild.   In a woody species which lives so long and shoots
again from the same stock when cut off by accident, it is
impossible to know the origin of the individuals observed.
They may have been sown by man or birds at a very
early epoch, for olive trees of more than a thousand years
old are known.   The effect of such sowing is a naturalization, which is equivalent to an extension of area.   The
point in question is, therefore, to discover what was the

[1] Aitchison, *Catalogue*, p. 86.
[2] Lowe, *Man. Fl. of Madeira*, ii. p. 20 ; Webb and Berthelot, *Hist.
Nat. des Canaries, Géog. Bot.*, p. 48 ; Ball, *Spicil. Fl. Maroc.*, p. 565.
[3] Cosson, *Bull. Soc. Bot. France*, iv. p. 107, and vii. p. 31 ; Grisebach,
Spicil. Fl. Rumelicœ, ii. p. 71 ; Steven, *Verzeich. der Taurisch. Halbins.*,
p. 248 ; Ledebour, *Fl. Ross.*, p. 38.
[4] *Bulletin*, iv. p. 107.

home of the species in very early prehistoric times, and how this area has grown larger by different modes of transport.

It is not by the study of living olive trees that this question can be answered. We must seek in what countries the cultivation began, and how it was propagated. The more ancient it is in any region, the more probable it is that the species has existed wild there from the time of those geological events which took place before the coming of prehistoric man.

The earliest Hebrew books mention the olive *sait*, or *zeit*,[1] both wild and cultivated. It was one of the trees promised in the land of Canaan. It is first mentioned in Genesis, where it is said that the dove sent out by Noah should bring back a branch of olive. If we take into account this tradition, which is accompanied by miraculous details, it may be added that the discoveries of modern erudition show that the Mount Ararat of the Bible must be to the east of the mountain in Armenia which now bears that name, and which was anciently called Masis. From a study of the text of the Book of Genesis, François Lenormand [2] places the mountain in question in the Hindu Kush, and even near the sources of the Indus. This theory supposes it near to the land of the Aryans, yet the olive has no Sanskrit name, not even in that Sanskrit from which the Indian languages [3] are derived. If the olive had then, as now, existed in the Punjab, the eastern Aryans in their migrations towards the south would probably have given it a name, and if it had existed in the Mazanderan, to the south of the Caspian Sea, as at the present day, the western Aryans would perhaps have known it. To these negative indications, it can only be objected that the wild olive attracts no considerable attention, and that the idea of extracting oil from it perhaps arose late in this part of Asia.

[1] Rosenmüller, *Handbuch der Bibl. Alterth.*, vol. iv. p. 258 ; Hamilton, *Bot. de la Bible*, p. 80, where the passages are indicated.

[2] Fr. Lenormand, *Manuel de l'Hist. Auc. de l'Orient.*, 1869, vol. i. p. 31.

[3] Fick, *Wörterbuch*, Piddington, *Index*, only mentions one Hindu name, *julpai*.

Herodotus [1] tells us that Babylonia grew no olive trees, and that its inhabitants made use of oil of sesame. It is certain that a country so subject to inundation was not at all favourable to the olive. The cold excludes the higher plateaux and the mountains of the north of Persia.

I do not know if there is a name in Zend, but the Semitic word *sait* must date from a remote antiquity, for it is found in modern Persian, *seitun*,[2] and in Arabic, *zeitun, sjetun*.[3] It even exists in Turkish and among the Tartars of the Crimea, *seitun*,[4] which may signify that it is of Turanian origin, or from the remote epoch when the Turanian and Semitic peoples intermixed.

The ancient Egyptians cultivated the olive tree, which they called *tat*.[5] Several botanists have ascertained the presence of branches or leaves of the olive in the sarcophagi.[6] Nothing is more certain, though Hehn [7] has recently asserted the contrary, without giving any proof in support of his opinion. It would be interesting to know to what dynasty belong the most ancient mummy-cases in which olive branches have been found. The Egyptian name, quite different to the Semitic, shows an existence more ancient than the earliest dynasties. I shall mention presently another fact in support of this great antiquity.

Theophrastus says [8] that the olive was much grown, and the harvest of oil considerable in Cyrenaica, but he does not say that the species was wild there, and the quantity of oil mentioned seems to point to a cultivated variety. The low-lying, very hot country between Egypt and the Atlas is little favourable to a naturalization of the olive outside the plantations. Kralik, a very accurate botanist, did not anywhere see on his journey

[1] Herodotus, *Hist.*, bk. i. c. 193.   [2] Boissier, *Fl. Orient.*, iv. p. 36.
[3] Ebn Baithar, Germ. trans., p. 569; Forskal, *Plant. Egypt.*, p. 49.
[4] Boissier, *ibid.* ; Steven, *ibid.*
[5] Unger, *Die Pflanz. der Alten. Ægypt*, p. 45.
[6] De Candolle, *Physiol. Végét.*, p. 696; Pleyte, quoted by Braun and Ascherson, *Sitzber. Naturfor. Ges.*, May 15, 1877.
[7] Hehn, *Kulturpflanzen*, edit. 3, p. 88, line 9.
[8] Theophrastus, *Hist. Plant.*, lib. iv. c. 3.

to Tunis and into Egypt the olive growing wild,[1] although it is cultivated in the oases. In Egypt it is only cultivated, according to Schweinfurth and Ascherson,[2] in their *resumé* of the Flora of the Nile Valley.

Its prehistoric area probably extended from Syria towards Greece, for the wild olive is very common along the southern coast of Asia Minor, where it forms regular woods.[3] It is doubtless here and in the archipelago that the Greeks early knew the tree. If they had not known it on their own territory, had received it from the Semites, they would not have given it a special name, *elaia,* whence the Latin *olea.* The *Iliad* and the *Odyssey* mention the hardness of the olive wood and the practice of anointing the body with olive oil. The latter was in constant use for food and lighting. Mythology attributed to Minerva the planting of the olive in Attica, which probably signifies the introduction of cultivated varieties and suitable processes for extracting the oil. Aristæus introduced or perfected the manner of pressing the fruit.

The same mythical personage carried, it was said, the olive tree from the north of Greece into Sicily and Sardinia. It seems that this may have been early done by the Phœnicians, but in support of the idea that the species, or a perfected variety of it, was introduced by the Greeks, I may mention that the Semitic name *seit* has left no trace in the islands of the Mediterranean. We find the Græco-Latin name here as in Italy,[4] while upon the neighbouring coast of Africa, and in Spain, the names are Egyptian or Arabic, as I shall explain directly.

The Romans knew the olive later than the Greeks. According to Pliny,[5] it was only at the time of Tarquin the Ancient, 627 B.C., but the species probably existed already in Great Greece, as in Greece and Sicily. Besides, Pliny was speaking of the cultivated olive.

A remarkable fact, and one which has not been noted

[1] Kralik, *Bull. Soc. Bot. Fr.,* iv. p. 108.
[2] *Beitrage zur Fl. Æthiopiens,* p. 281.
[3] Balansa, *Bull. Soc. Bot. de Fr.,* iv. p. 107.
[4] Moris, *Fl. Sard.,* iii. p. 9 ; Bertoloni, *Fl. Ital.,* i. p. 46.
[5] Pliny, *Hist.,* lib. xv. cap. 1.

or discussed by philologists, is that the Berber name for the olive, both tree and fruit, has the root *taz* or *tas*, similar to the *tat* of the ancient Egyptians. The Kabyles of the district of Algiers, according to the French-Berber dictionary, published by the French Government, calls the wild olive *tazebboujt, tesettha, ou' zebbouj*, and the grafted olive *tazemmourt, tasettha, ou' zemmour*. The Touaregs, another Berber nation, call it *tamahinet*.[1] These are strong indications of the antiquity of the olive in Africa. The Arabs having conquered this country and driven back the Berbers into the mountains and the desert, having likewise subjected Spain excepting the Basque country, the names derived from the Semitic *zeit* have prevailed even in Spanish. The Arabs of Algiers say *zenboudje* for the wild, *zitoun* for the cultivated olive,[2] *zit* for olive oil. The Andalusians call the wild olive *azebuche*, and the cultivated *aceytuno*.[3] In other provinces we find the name of Latin origin, *olivio*, side by side with the Arabic words.[4] The oil is in Spanish *aceyte*, which is almost the Hebrew name; but the holy oils are called *oleos santos*, because they belong to Rome. The Basques use the Latin name for the olive tree.

Early voyagers to the Canaries, Bontier for instance, in 1403, mention the olive tree in these islands, where modern botanists regard it as indigenous.[5] It may have been introduced by the Phœnicians, if it did not previously exist there. We do not know if the Guanchos had names for the olive and its oil. Webb and Berthelot do not give any in their learned chapter on the language of the aborigines,[6] so the question is open to conjecture. It seems to me that the oil would have played an important part among the Guanchos if they had possessed the olive, and that some traces of it would have remained in the actual speech of the people. From this point of view

---

[1] Duveyrier, *Les Touaregs du Nord* (1864), p. 179.
[2] Munby, *Flore de l'Algerie*, p. 2 ; Debeaux, *Catal. Boghar*, p. 68.
[3] Boissier, *Voyage Bot. en Espagne*, edit. 1, vol. ii. p. 407.
[4] Willkomm and Lange, *Prod. Fl. Hispan.*, ii. p. 672.
[5] Webb and Berthelot, *Hist. Nat. des Canaries, Géog. Bot.*, pp. 47, 48.
[6] Webb and Berthelot, *ibid., Ethnographie*, p 188.

the naturalization in the Canaries is perhaps not more ancient than the Phœnician voyages.

No leaf of the olive has hitherto been found in the tufa of the south of France, of Tuscany, and Sicily, where the laurel, the myrtle, and other shrubs now existing have been discovered. This is an indication, until the contrary is proved, of a subsequent naturalization.

The olive thrives in dry climates like that of Syria and Assyria. It succeeds at the Cape, in parts of America, in Australia, and doubtless it will become wild in these places when it has been more generally planted. Its slow growth, the necessity of grafting or of choosing the shoots of good varieties, and especially the concurrence of other oil-producing species, have hitherto impeded its extension; but a tree which produces in an ungrateful soil should not be indefinitely neglected. Even in the old world, where it has existed for so many thousands of years, its productiveness might be doubled by taking the trouble to graft on wild trees, as the French have done in Algeria.

Star Apple—*Chrysophyllum Caïnito,* Linnæus.

The star apple belongs to the family of the Sapotaceæ, It yields a fruit valued in tropical America, though Europeans do not care much for it. I do not find that any pains have been taken to introduce it into the colonies of Asia or Africa. Tussac gives a good illustration of it in his *Flore des Antilles,* vol. ii. pl. 9.

Seemann [1] saw the star apple wild in several places in the Isthmus of Panama. De Tussac, a San Domingo colonist, considered it wild in the forests of the West India Islands, and Grisebach [2] says it is both wild and cultivated in Jamaica, San Domingo, Antigua, and Trinidad. Sloane considered it had escaped from cultivation in Jamaica, and Jacquin says vaguely, "Inhabits Martinique and San Domingo." [3]

Caimito, or **Abi**—*Lucuma Caïnito,* Alph. de Candolle.

This Peruvian Caïmito must not be confounded with

---

[1] Seemann, *Bot. of the Herald.,* p. 166.

[2] Grisebach, *Flora of Brit. W. Ind. Isl.,* p. 398.

[3] Sloane, *Jamaica,* ii. p. 170; Jacquin, *Amer.,* p. 52.

the *Chrysophyllum Caïnito* of the West Indies. Both belong to the family Sapotaceæ, but the flowers and seeds are different. There is a figure of this one in Ruiz and Pavon, *Flora Peruviana*, vol. iii. pl. 240. It has been transported from Peru, where it is cultivated, to Ega on the Amazon River, and to Para, where it is commonly called *abi* or *abiu*.[1] Ruiz and Pavon say it is wild in the warm regions of Peru, and at the foot of the Andes.

Marmalade Plum, or **Mammee Sapota**—*Lucuma mammosa*, Gærtner.

This fruit tree, of the order Sapotaceæ and a native of tropical America, has been the subject of several mistakes in works on botany.[2] There exists no satisfactory and complete illustration of it as yet, because colonists and travellers think it is too well known to send selected specimens of it, such as may be described in herbaria. This neglect is common enough in the case of cultivated plants. The mammee is cultivated in the West Indies and in some warm regions of America. Sagot tells us it is grown in Venezuela, but not in Cayenne.[3] I do not find that it has been transported into Africa and Asia, the Philippines[4] excepted. This is probably due to the insipid taste of the fruit. Humboldt and Bonpland found it wild in the forests on the banks of the Orinoco.[5] All authors mention it in the West Indies, but as cultivated or without asserting that it is wild. In Brazil it is only a garden species.

Sapodilla—*Sapota achras*, Miller.

The sapodilla is the most esteemed of the order Sapotaceæ, and one of the best of tropical fruits. "An over-ripe sapodilla," says Descourtilz, in his *Flore des Antilles*, "is melting, and has the sweet perfumes of honey, jasmin, and lily of the valley." There is a very good illustration in the *Botanical Magazine*, pls. 3111 and 3112, and in Tussac, *Flore des Antilles*, i. pl. 5. It

[1] *Flora Brasil.*, vol. vii. p. 88.
[2] See the synonyms in the *Flora Brasiliensis*, vol. vii. p. 66.
[3] Sagot, *Journ. Soc. d'Hortic. de France*, 1872, p. 347.
[4] Blanco, *Fl. de Filipinas*, under the name *Achras lucuma.*
[5] *Nova Genera*, iii. p. 240.

has been introduced into gardens in Mauritius, the Malay Archipelago, and India, from the time of Rheede and Rumphius, but no one disputes its American origin. Several botanists have seen it wild in the forests of the Isthmus of Panama, of Campeachy,[1] of Venezuela,[2] and perhaps of Trinidad.[3] In Jamaica, in the time of Sloane, it existed only in gardens.[4] It is very doubtful that it is wild in the other West India Islands, although perhaps the seeds, scattered here and there, may have naturalized it to a certain degree. Tussac says that the young plants are not easy to rear in the plantations.

Aubergine—*Solanum melongena*, Linnæus ; *Solanum esculentum*, Dunal.

The aubergine has a Sanskrit name, *vartta*, and several names, which Piddington in his *Index* considers as both Sanskrit and Bengali, such as *bong, bartakon, mahoti, hingoli*. Wallich, in his edition of Roxburgh's *Indian Flora*, gives *vartta, varttakou, varttaka bunguna*, whence the Hindustani *bungan*. Hence it cannot be doubted that the species has been known in India from a very remote epoch. Rumphius had seen it in gardens in the Sunda Islands, and Loureiro in those of Cochin-China. Thunberg does not mention it in Japan, though several varieties are now cultivated in that country. The Greeks and Romans did not know the species, and no botanist mentions it in Europe before the beginning of the seventeenth century,[5] but its cultivation must have spread towards Africa before the Middle Ages. The Arab physician, Ebn Baithar,[6] who wrote in the thirteenth century, speaks of it, and he quotes Rhasis, who lived in the ninth century. Rauwolf[7] had seen the plant in the gardens of Aleppo at the end of the sixteenth century. It was called *melanzana* and *bedengiam*. This Arabic

---

[1] Dampier and Lussan, in Sloane's *Jamaica*, ii. p. 172; Seemann, *Botany of the Herald.*, p. 166.

[2] Jacquin, *Amer.*, p. 59; Humboldt and Bonpland, *Nova Genera*, iii. p. 239.

[3] Grisebach, *Flora. of Brit. W. Ind.*, p. 399.   [4] Sloane, *ubi supra.*

[5] Dunal, *Hist. des Solanum*, p. 209.

[6] Ebn Baithar, Germ. trans., i. p. 116.

[7] Rauwolf, *Flora Orient.*, ed. Groningue, p. 26.

name, which Forskal writes *badinjan*, is the same as
the Hindustani *badanjan*, which Piddington gives. A
sign of antiquity in Northern Africa is the existence of
a name, *tabendjalts*, among the Berbers or Kabyles of the
province of Algiers,[1] which differs considerably from
the Arab word. Modern travellers have found the
aubergine cultivated in the whole of the Nile Valley and
on the coast of Guinea.[2] It has been transported into
America.

The cultivated form of *Solanum melongena* has not
hitherto been found wild, but most botanists are agreed
in regarding *Solanum insanum*, Roxburgh, and *S.
incanum*, Linnæus, as belonging to the same species.
Other synonyms are sometimes added, the result of a
study made by Nees von Esenbeck from numerous specimens.[3] *S. insanum* appears to have been lately found
wild in the Madras presidency and at Tong-dong in
Burmah. The publication of the article on the Solanaceæ in the *Flora of British India* will probably give
more precise information on this head.

Red Pepper—*Capsicum.* In the best botanical works
the genus Capsicum is encumbered with a number of
cultivated forms, which have never been found wild, and
which differ especially in their duration (which is often
variable), or in the form of the fruit, a character which
is of little value in plants cultivated for that special
organ. I shall speak of the two species most often cultivated, but I cannot refrain from stating my opinion that
no capsicum is indigenous to the old world. I believe
them to be all of American origin, though I cannot
absolutely prove it. These are my reasons.

Fruits so conspicuous, so easily grown in gardens,
and so agreeable to the palate of the inhabitants of hot
countries, would have been very quickly diffused throughout the old world, if they had existed in the south of
Asia, as it has sometimes been supposed. They would
have had names in several ancient languages. Yet

---

[1] *Dict. Fr.-Berbère*, published by the French Government.
[2] Thonning, under the name *S. edule*; Hooker, *Niger Flora*, p. 473.
[3] *Trans. of Linn. Soc.*, xvii. p. 48; Baker, *Fl. of Maurit.*, p. 215.

neither Romans, Greeks, nor even Hebrews were acquainted with them.   They are not mentioned in ancient
Chinese books.[1]   The islanders of the Pacific did not
cultivate them at the time of Cook's voyages,[2] in spite
of their proximity to the Sunda Isles, where Rumphius
mentions their very general use.   The Arabian physician,
Ebn Baithar, who collected in the thirteenth century all
that Eastern nations knew about medicinal plants,
says nothing about it.   Roxburgh knew no Sanskrit
name for the capsicums.   Later, Piddington mentions a
name for *C. frutescens, bran-maricha*,[3] which he says is
Sanskrit; but this name, which may be compared to
that of black pepper (*muricha, murichung*), is probably
not really ancient, for it has left no trace in the Indian
languages which are derived from Sanskrit.[4]   The wild
nature and ancient existence of the capsicum is always
uncertain, owing to its very general cultivation; but
it seems to me to be more often doubtful in Asia than in
South America.   The Indian specimens described by the
most trustworthy authors nearly all come from the herbaria of the East India Company, in which we never
know whether a plant appeared really wild, if it was
found far from dwellings, in forests, etc.   For the
localities in the Malay Archipelago authors often give
rubbish-heaps, hedges, etc.   We pass to a more particular
examination of the two cultivated species.

Annual Capsicum—*Capsicum annuum*, Linnæus.

This species has a number of different names in
European languages,[5] which all indicate a foreign origin
and the resemblance of the taste to that of pepper.   In
French it is often called *poivre de Guinée* (Guinea
pepper), but also *poivre du Brézil, d'Inde* (Indian, Brazilian pepper), etc., denominations to which no importance
can be attributed.   Its cultivation was introduced into
Europe in the sixteenth century.   It was one of the
peppers that Piso and Marcgraf[6] saw grown in Brazil

[1] Bretschneider, *On the Study and Value*, etc., p. 17.
[2] Forster, *De Plantis Escul. Insul.*, etc.   [3] Piddington, *Index.*
[4] Piddington, at the word *Capsicum.*
[5] Nemnich, *Lexicon*, gives twelve French and eight German names.
[6] Piso, p. 107; Marcgraf, p. 39.

under the name *quija* or *quiya*. They say nothing as to its origin. The species appears to have been early cultivated in the West Indies, where it has several Carib names.[1]

Botanists who have most thoroughly studied the genus Capsicum[2] do not appear to have found in herbaria a single specimen which can be considered wild. I have not been more fortunate. The original home is probably Brazil.

C. grossum, Willdenow, seems to be a variety of the same species. It is cultivated in India under the name *kafree murich*, and *kafree chilly*, but Roxburgh did not consider it to be of Indian origin.[3]

Shrubby Capsicum—*Capsicum frutescens*, Willdenow.

This species, taller and with a more woody stock than *C. annuum*, is generally cultivated in the warm regions of both hemispheres. The great part of our so-called Cayenne pepper is made from it, but this name is given also to the product of other peppers. Roxburgh, the author who is most attentive to the origin of Indian plants, does not consider it to be wild in India. Blume says it is naturalized in the Malay Archipelago in hedges.[4] In America, on the contrary, where its culture is ancient, it has been several times found wild in forests, apparently indigenous. De Martius brought it from the banks of the Amazon, Pœppig from the province of Maynas in Peru, and Blanchet from the province of Bahia.[5] So that its area extends from Bahia to Eastern Peru, which explains its diffusion over South America generally.

Tomato—*Lycopersicum esculentum*, Miller.

The tomato, or love apple, belongs to a genus of the Solaneæ, of which all the species are American.[6] It has no name in the ancient languages of Asia, nor even in modern Indian languages.[7] It was not cultivated in Japan in the time of Thunberg, that is to say a century

---

[1] Descourtilz, *Flore Médicale des Antilles*, vi. pl. 423.

[2] Fingerhuth, *Monographia Gen. Capsici*, p. 12; Sendtner, in *Flora Brasil.*, vol. x. p. 147.

[3] Roxburgh, *Fl. Ind.*, edit. Wall, ii. p. 260; edit. 1832, ii. p. 574.

[4] Blume, *Bijdr.*, ii. p. 704.   [5] Sendtner, in *Fl. Bras.*, x. p. 143.

[6] Alph. de Candolle, *Prodr.*, xiii. part 1, p. 26.

[7] Roxburgh, *Fl. Ind.*, edit. 1832, vol. i. p. 565; Piddington, *Index*.

ago, and the silence of ancient writers on China on this head shows that it is of recent introduction there. Rumphius [1] had seen it in gardens in the Malay Archipelago. The Malays called it *tomatte,* but this is an American name, for C. Bauhin calls the species *tumatle Americanorum.* Nothing leads us to suppose it was known in Europe before the discovery of America.

The first names given to it by botanists in the sixteenth century indicate that they received the plant from Peru.[2] It was cultivated on the continent of America before it was grown in the West India Islands, for Sloane does not mention it in Jamaica, and Hughes [3] says it was brought to Barbados from Portugal hardly more than a century ago. Humboldt considered that the cultivation of the tomato was of ancient date in Mexico.[4] I notice, however, that the earliest work on the plants of this country (Hernandez, *Historia*) makes no mention of it. Neither do the early writers on Brazil, Piso and Marcgraf, speak of it, although the species is now cultivated throughout tropical America. Thus by the process of exhaustion we return to the idea of a Peruvian origin, at least for its cultivation.

De Martius [5] found the plant wild in the neighbourhood of Rio de Janeiro and Para, but it had perhaps escaped from gardens. I do not know of any botanist who has found it really wild in the state in which it is familiar to us, with the fruit more or less large, lumpy, and with swelled sides ; but this is not the case with the variety with small spherical fruit, called *L. cerasiforme* in some botanical works, and considered in others (and rightly so, I think [6]) as belonging to the same species. This variety is wild on the sea-shore of

---

[1] Rumphius, *Amboin,* v. p. 416.
[2] *Mala Peruviana, Pomi del Peru,* in Bauhin's *Hist.,* iii. p. 621.
[3] Hughes, *Barbados,* p. 148.
[4] Humboldt, *Espagne,* edit. 2, vol. ii. p. 472.
[5] *Fl. Brasil.,* vol. x. p. 126.
[6] The proportions of the calyx and the corolla are the same as those of the cultivated tomato, but they are different in the allied species *S. Humboldtii,* of which the fruit is also eaten, according to Humboldt, who found it wild in Venezuela.

Peru,[1] at Tarapoto, in Eastern Peru,[2] and on the frontiers of Mexico and of the United States towards California.[3] It is sometimes naturalized in clearings near gardens.[4] It is probably in this manner that its area has extended north and south from Peru.

Avocado, or **Alligator Pear** — *Persea gratissima,* Gærtner.

The avocado pear is one of the most highly prized of tropical fruits. It belongs to the order Laurineæ. It is like a pear containing one large stone, as is well shown in Tussac's illustrations, *Flore des Antilles,* iii. pl. 3, and in the *Botanical Magazine,* pl. 4580. The common names are absurd. The origin of that of *alligator* is unknown; *avocado* is a corruption of the Mexican *ahuaca,* or *aguacate.* The botanical name *Persea* has nothing to do with the *persea* of the Greeks, which was a *Cordia.* Clusius,[5] writing in 1601, says that the avocado pear is an American fruit tree introduced into a garden in Spain ; but as it is widely spread in the colonies of the old world, and has here and there become almost wild,[6] it is possible to make mistakes as to its origin. This tree did not exist in the gardens of British India at the beginning of the nineteenth century. It had been introduced into the Sunda Isles [7] in the middle of the eighteenth century, and in 1750 into Mauritius and Bourbon.[8]

In America its actual area in a wild state is of uncommon extent. The species has been found in forests, on the banks of rivers, and on the sea-shore from Mexico and the West Indies as far as the Amazon.[9] It has not

---

[1] Ruiz and Pavon, *Flor. Peruv.,* ii. p. 37.

[2] Spruce, n. 4143, in Boissier's herbarium.

[3] Asa Gray, *Bot. of Califor.,* i. p. 538.

[4] Baker, *Fl. of Maurit.,* p. 216.　　　[5] Clusius, *Historia,* p. 2.

[6] For instance in Madeira, according to Grisebach, *Fl. of Brit. W. Ind.,* p. 280; in Mauritius, the Seychelles and Rodriguez, according to Baker, *Flora of Mauritius,* p. 290.

[7] It is not in Rumphius.　　　[8] Aublet, *Guyane,* i. p. 364.

[9] Meissner, in de Candolle, *Prodromus,* vol. xv. part 1, p. 52 ; and *Flora Brasil.,* vol. v. p. 158. For Mexico, Hernandez, p. 89; for Venezuela and Para, Nees, *Laurineæ,* p. 129; for Eastern Peru, Pœppig, *Exsicc.,* seen by Meissner.

always occupied this vast region. P. Browne says distinctly that the avocado pear was introduced from the Continent into Jamaica, and Jacquin held the same opinion as regards the West India Islands generally.[1] Piso and Marcgraf do not mention it for Brazil, and Martius gives no Brazilian name.

At the time of the discovery of America, the species was certainly wild and cultivated in Mexico, according to Hernandez. Acosta [2] says it was cultivated in Peru under the name of *palto*, which was that of a people of the eastern part of Peru, among whom it was abundant.[3] I find no proof that it was wild upon the Peruvian littoral.

Papaw—*Carica Papaya*, Linnæus ; *Papaya vulgaris*, de Candolle.

The papaw is a large herbaceous plant rather than a tree. It has a sort of juicy trunk terminated by a tuft of leaves, and the fruit, which is like a melon, hangs down under the leaves.[4] It is now grown in all tropical countries, even as far as thirty to thirty-two degrees of latitude. It is easily naturalized outside plantations. This is one reason why it has been said, and people still say that it is a native of Asia or of Africa, whereas Robert Brown and I proved in 1848 and 1855 its American origin.[5] I repeat the arguments against its supposed origin in the eastern hemisphere.

The species has no Sanskrit name. In modern Indian languages it bears names derived from the American word *papaya*, itself a corruption of the Carib *ababai*.[6] Rumphius [7] says that the inhabitants of the Malay Archipelago considered it as an exotic plant introduced by the Portuguese, and gave it names expressing its likeness to

[1] P. Browne, *Jamaica*, p. 214; Jacquin, *Obs.*, i. p. 38.

[2] Acosta, *Hist. Nat. des Indes.*, edit. 1598, p. 176.

[3] Laet, *Hist. Nouv. Monde*, i. pp. 325, 341.

[4] See the fine plates in Tussac's *Flore des Antilles*, iii. p. 45, pls. 10 and 11.   The papaw belongs to the small family of the *Papayaceæ*, fused by some botanists into the *Passifloræ*, and by others into the *Bixaceæ*.

[5] R. Brown, *Bot. of Congo*, p. 52 ; A. de Candolle, *Géogr. Bot. Rais.*, p. 917.

[6] Sagot, *Journ. de la Soc. Centr. d'Hortic. de France*, 1872.

[7] Rumphius, *Amboin*, i. p. 147.

other species or its foreign extraction. Sloane,[1] in the beginning of the eighteenth century, quotes several of his contemporaries, who mention that it was taken from the West Indies into Asia and Africa. Forster had not seen it in the plantations of the Pacific Isles at the time of Cook's voyages. Loureiro,[2] in the middle of the eighteenth century, had seen it in cultivation in China, Cochin-China, and Zanzibar. So useful and so striking a plant would have been spread throughout the old world for thousands of years if it had existed there. Everything leads to the belief that it was introduced on the coasts of Africa and Asia after the discovery of America.

All the species of the family are American. This one seems to have been cultivated from Brazil to the West Indies, and in Mexico before the arrival of the Europeans, since the earliest writers on the productions of the new world mention it.[3]

Marcgraf had often seen the male plant (always commoner than the female) in the forests of Brazil, while the female plants were in gardens. Clusius, who was the first to give an illustration of the plant, says [4] that his drawing was made in 1607, in the bay of Todos Santos (province of Bahia). I know of no modern author who has confirmed the habitation in Brazil. Martius does not mention the species in his dictionary of the names of fruits in the language of the Tupis.[5] It is not given as wild in Guiana and Columbia. P. Browne [6] asserts, on the other hand, that it is wild in Jamaica, and before his time Ximenes and Hernandez said the same for St. Domingo and Mexico. Oviedo [7] seems to have seen the papaw in Central America, and he gives the common

---

[1] Sloane, *Jamaica*, p. 165.   [2] Loureiro, *Fl. Coch.*, p. 772.

[3] Marcgraf, *Brasil.*, p. 103, and Piso, p. 159, for Brazil; Ximenes in Marcgraf and Hernandez, *Thesaurus*, p. 99, for Mexico; and the last for St. Domingo and Mexico.

[4] Clusius, *Curæ Posteriores*, pp. 79, 80.

[5] Martius, *Beitr. z. Ethnogr.*, ii. p. 418.

[6] P. Browne, *Jamaica*, edit. 2, p. 360. The first edition is of 1756.

[7] The passage of Oviedo is translated into English by Correa de Mello and Spruce, in their paper on the *Proceedings of the Linnæan Society*, x. p. 1.

name *olocoton* for Nicaragua. Yet Correa de Mello and
Spruce, in their important article on the *Papayaceœ*, after
having botanized extensively in the Amazon region, in
Peru and elsewhere, consider the papaw as a native of
the West Indies, and do not think it is anywhere wild
upon the Continent. I have seen [1] specimens from the
mouth of the river Manatee in Florida, from Puebla in
Mexico, and from Columbia, but the labels had no remark
as to their wild character. The indications, it will be
noticed, are numerous for the shores of the Gulf of Mexico
and for the West Indies. The habitation in Brazil which
lies apart is very doubtful.

Fig—*Ficus carica*, Linnæus.

The history of the fig presents a close analogy with
that of the olive in point of origin and geographical
limits. Its area as a wild species may have been extended
by the dispersal of the seeds as cultivation spread. This
seems probable, as the seeds pass intact through the
digestive organs of men and animals. However, countries
may be cited where the fig has been cultivated for a
century at least, and where no such naturalization has
taken place. I am not speaking of Europe north of the
Alps, where the tree demands particular care and the
fruit ripens with difficulty, even the first crop, but of
India for instance, the Southern States of America,
Mauritius, and Chili, where, to judge from the silence of
compilers of floras, the instances of quasi-wildness are
rare. In our own day the fig tree grows wild, or nearly
wild, over a vast region of which Syria is about the
centre; that is to say, from the east of Persia, or even
from Afghanistan, across the whole of the Mediterranean
region as far as the Canaries.[2] From north to south this
zone varies in width from the 25th to the 40th or 42nd
parallel, according to local circumstances. As a rule, the
fig stops like the olive at the foot of the Caucasus and
the mountains of Europe which limit the Mediterranean

---

[1] De Candolle, *Prodr.*, xv. part 1, p. 414.

[2] Boissier, *Fl. Orient.*, iv. p. 1154; Brandis, *Forest Flora of India*,
p. 418; Webb and Berthelot, *Hist. Nat. des Canaries, Botanique*, iii.
p. 257.

basin, but it grows nearly wild on the south-west coast of France, where the winter is very mild.[1]

We turn to historical and philological records to see whether the area was more limited in antiquity. The ancient Egyptians called the fig *teb*,[2] and the earliest Hebrew books speak of the fig, whether wild or cultivated, under the name *teenah*,[3] which leaves its trace in the Arabic *tin*.[4] The Persian name is quite different, *unjir;* but I do not know if it dates from the Zend. Piddington's *Index* has a Sanskrit name, *udumvara*, which Roxburgh, who is very careful in such matters, does not give, and which has left no trace in modern Indian languages, to judge from four names quoted by authors. The antiquity of its existence east of Persia appears to me doubtful, until the Sanskrit name is verified. The Chinese received the fig tree from Persia, but only in the eighth century of our era.[5]   Herodotus [6] says the Persians did not lack figs, and Reynier, who has made careful researches into the customs of this ancient people, does not mention the fig tree. This only proves that the species was not utilized and cultivated, but it perhaps existed in a wild state.

The Greeks called the wild fig *erineos*, and the Latins *caprificus*. Homer mentions a fig tree in the *Iliad* which grew near Troy.[7] Hehn asserts [8] that the cultivated fig cannot have been developed from the wild fig, but all

---

[1] Count Solms Laubach, in a learned discussion (*Herkunft, Domestication, etc., des Feigenbaums*, in 4to, 1882), has himself observed facts of this nature already indicated by various authors. He did not find the seed provided with embryos (p. 64), which he attributes to the absence of the insect (*Blastophaga*), which generally lives in the wild fig, and facilitates the fertilization of one flower by another in the interior of the fruit.   It is asserted, however, that fertilization occasionally takes place without the intervention of the insect.

[2] Chabas, *Mélanges Egyptol.*, 3rd series (1873), vol. ii. p. 92.

[3] Rosenmuller, *Bibl. Alterth.*, i. p. 285 ; Reynier, *Écon. Publ. des Arabes et des Juifs*, p. 470.

[4] Forskal, *Fl. Ægypto-Arab.*, p. 125.   Lagarde (*Revue Critique d'Histoire*, Feb. 27, 1882) says that this Semitic name is very ancient.

[5] Bretschneider, in Solms, *ubi supra*, p. 51.   [6] Herodotus, i. 71.

[7] Lenz, *Botanik der Griechen*, p. 421, quotes four lines of Homer. See also Hehn, *Culturpflanzen*, edit. 3, p. 84.

[8] Hehn, *Culturpflanzen*, edit. 3, p. 513.

botanists hold a contrary opinion;[1] and, without speaking of floral details on which they rely, I may say that Gussone obtained from the same seeds plants of the form *caprificus*, and other varieties.[2] The remark made by several scholars as to the absence of all mention of the cultivated fig *sukai* in the *Iliad*, does not therefore prove the absence of the fig tree in Greece at the time of the Trojan war. Homer mentions the sweet fig in the *Odyssey*, and that but vaguely. Hesiod, says Hehn, does not mention it, and Archilochus (700 B.C.) is the first to mention distinctly its cultivation by the Greeks of Paros. According to this, the species grew wild in Greece, at least in the Archipelago, before the introduction of cultivated varieties of Asiatic origin. Theophrastus and Dioscorides mention wild and cultivated figs.[3]

Romulus and Remus, according to tradition, were nursed at the foot of a fig tree called *ruminalis*, from *rumen*, breast or udder.[4] The Latin name, *ficus*, which Hehn derives, by an effort of erudition, from the Greek *sukai*,[5] also argues an ancient existence in Italy, and Pliny's opinion is positive on this head. The good cultivated varieties were of later introduction. They came from Greece, Syria, and Asia Minor. In the time of Tiberius, as now, the best figs came from the East.

We learnt at school how Cato exhibited to the assembled senators Carthaginian figs, still fresh, as a proof of the proximity of the hated country. The Phœnicians must have transported good varieties to the coast of Africa and their other colonies on the Mediterranean, even as far as the Canaries, where, however, the wild fig may have already existed.

For the Canaries we have a proof in the Guanchos

---

[1] No importance should be attached to the exaggerated divisions made by Gasparini in *Ficus carica*, Linnæus. Botanists who have studied the fig tree since his time retain a single species, and name several varieties of the wild fig. The cultivated forms are numberless.

[2] Gussone, *Enum. Plant. Inarimensium*, p. 301.

[3] For the history of the fig tree and an account of the operation (of doubtful utility) which consists in planting insect-bearing *Caprifici* among the cultivated trees (caprification), see Solms' work.

[4] Pliny, *Hist.*, lib. xv. cap. 18. [5] Hehn, *Culturpflanzen*, edit. 3, p. 513.

words, *arahormaze* and *achormaze*, green figs; *tahare-menen* and *tehahunemen*, dried figs. Webb and Ber-thelot,[1] who quote these names, and who admit the common origin of the Guanchos and Berbers, would have noted with pleasure the existence among the Touaregs, a Berber people, of the word *tahart*, fig tree,[2] and in the French-Berber dictionary, published since their time, the names *tabeksist*, green fig, and *tagrourt*, fig tree. These old names, of more ancient and local origin than Arabic, bear witness to a very ancient habitation in the north of Africa as far as the Canaries.

The result of our inquiry shows, then, that the prehistoric area of the fig tree covered the middle and southern part of the Mediterranean basin from Syria to the Canaries.

We may doubt the antiquity of the fig in the south of France, but a curious fact deserves mention. Plan-chon found in the quaternary tufa of Montpellier, and de Saporta[3] in those of Aygalades near Marseilles, and in the quaternary strata of La Celle near Paris, leaves and even fruit of the wild *Ficus carica*, with teeth of *Elephas primigenius*, and leaves of plants of which some no longer exist, and others, like *Laurus canariensis*, have survived in the Canaries. So that the fig tree perhaps existed in its modern form in this remote epoch. It is possible that it perished in the south of France, as it certainly did at Paris, and re-appeared later in a wild state in the southern region. Perhaps the fig trees which Webb and Berthelot had seen as old plants in the wildest part of the Canaries were descended from those which existed in the fourth epoch.

Bread-Fruit—*Artocarpus incisa*, Linnæus.

The bread-fruit tree was cultivated in all the islands of the Asiatic Archipelago, and of the great oceans near

---

[1] Webb and Berthelot, *Hist. Nat. des Canaries Ethnogr.*, p. 186; *Phytogr.*, iii. p. 257.

[2] Duveyrier, *Les Touaregs du Nord.*, p. 193.

[3] Planchon, *Étude sur les tufs de Montpellier*, p. 63; de Saporta, *La flore des tufs quaternaires en Provence*, in *Comptes rendus de la 32e Session du Congrès Scientifique de France; Bull. Soc. Geolog.*, 1873–74, p. 442.

the equator, from Sumatra to the Marquesas Isles, when first Europeans began to visit them. Its fruit is constituted, like the pine-apple, of an assemblage of bracts and fruits welded into a fleshy mass, more or less spherical; and as in the pine-apple, the seeds come to nothing in the most productive cultivated varieties.[1]

Sonnerat[2] carried the bread-fruit tree to Mauritius, where the Intendant Poivre took care to spread it. Captain Bligh was commissioned to introduce it into the English West Indian Isles. The mutiny of his crew prevented his succeeding the first time, but a second attempt proved more fortunate. In January, 1793, he landed 153 plants at St. Vincent, whence the species has been diffused into several parts of tropical America.[3]

Rumphius[4] saw the species wild in several of the Sunda Isles. Modern authors, less careful, or acquainted only with cultivated species, say nothing on this head. Seemann[5] says for the Fiji Isles, " cultivated, and to all appearance wild in some places." On the continent of Asia it is not even cultivated, as the climate is not hot enough.

The bread-fruit is evidently a native of Java, Amboyna, and the neighbouring islands; but the antiquity of its cultivation in the whole of the archipelago, proved by the number of varieties, and the facility of propagating it by buds and suckers, prevent us from knowing its history accurately. In the islands to the extreme east, like Otahiti, certain fables and traditions point to an introduction which is not very ancient, and the absence of seeds confirms this.[6]

Jack-Fruit—*Artocarpus integrifolia*, Linnæus.

The jack-fruit, larger than the bread-fruit, for it sometimes weighs as much as eighty pounds, hangs from

---

[1] See the fine plates published in Tussac's *Flore des Antilles*, vol. ii. pls. 2 and 3 ; and Hooker, *Bot. Mag.*, t. 2869–2871.
[2] *Voyages à la Nouvelle Guinée*, p. 100.   [3] Hooker, *ubi supra*.
[4] Rumphius, *Herb. Amboin*, i. p. 112, pl. 33.
[5] *Flora Vitiensis*, p. 255.
[6] Seemann, *Fl. Vit.*, p. 255 ; Nadeaud, *Enum. des Pl. Indig. de Taiti*, p. 44; Idem, *Pl. usuelles des Taitiens*, p. 24.

the branches of a tree thirty to fifty feet high.[1] The
common name is derived from the Indian names *jaca*, or
tsjaka.

The species has long been cultivated in southern
Asia, from the Punjab to China, from the Himalayas to
the Moluccas. It has not spread into the small islands
more to the east, such as Otahiti, which leads us to suppose it has not been so long in the archipelago as upon
the continent. In the north-west of India, also, its
cultivation does not perhaps date from a very remote
epoch, for the existence of a Sanskrit name is not absolutely certain. Roxburgh mentions one, *punusa*, but
Piddington does not admit it into his *Index*. The Persians and the Arabs do not seem to have known the
species. Its enormous fruit must, however, have struck
them if the species had been cultivated near their frontiers. Dr. Bretschneider does not speak of any Artocarpus in his work on the plants known to the ancient
Chinese, whence it may be inferred that towards China,
as in other directions, the jack-fruit was not diffused at
a very early epoch. The first statement as to its existence in a wild state is given by Rheede in ambiguous
terms : "This tree grows everywhere in Malabar and
throughout India." He perhaps confounded the planted
tree with the wild one. After him, however, Wight
found the species several times in the Indian Peninsula,
notably in the Western Ghauts, with every appearance
of a wild and indigenous tree. It has been extensively
planted in Ceylon ; but Thwaites, the best authority for
the flora of this island, does not recognize it as wild.
Neither is it wild in the archipelago to the south of
India, according to the general opinion. Lastly, Brandis
found it growing in the forests of the district of Attaran,
in Burmah, but, he adds, always in the neighbourhood of
abandoned settlements. Kurz did not find it wild in
British Burmah.[2]

[1] See Tussac's plates, *Flore des Antilles*, pl. 4; and Hooker, *Bot. Mag.*,
t. 2833, 2834.
[2] Rheede, *Malabar*, iii. p. 18; Wight, *Icones*, ii. No. 678; Brandis,
Forest Flora of India, p. 426; Kurz, *Forest Flora of Brit. Burmah*, p. 432.

The species is, therefore, a native of the region lying at the foot of the western mountains of the Indian Peninsula, and its cultivation in the neighbourhood is probably not earlier than the Christian era. It was introduced into Jamaica by Admiral Rodney in 1782, and thence into San Domingo.[1] It has also been introduced into Brazil, Mauritius, the Seychelles, and Rodriguez Island.[2]

Date-Palm—*Phœnix dactylifera*, Linnæus.

The date-palm has existed from prehistoric times in the warm dry zone, which extends from Senegal to the basin of the Indus, principally between parallels 15 and 30. It is seen here and there further to the north, by reason of exceptional circumstances and of the aim which is proposed in its cultivation. For beyond the limit within which the fruit ripens every year, there is a zone in which they ripen ill or seldom, and a further region within which the tree can live, but without fruiting or even flowering. These limits have been traced by de Martius, Carl Ritter, and myself.[3] It is needless to reproduce them here, the aim of the present work being to study questions of origin.

As regards the date-palm, we can hardly rely on the more or less proved existence of really wild indigenous individuals. Dates are easily transported; the stones germinate when sown in damp soil near the source of a river, and even in the fissures of rocks. The inhabitants of oases have planted or sown date-palms in favourable localities where the species perhaps existed before man, and when the traveller comes across isolated trees, at a distance from dwellings, he cannot know that they did not spring from stones thrown away by caravans. Botanists admit a variety, *sylvestris*, that is to say wild, with small and sour fruit; but it is perhaps the result of recent naturalization in an unfavourable soil. Historical and philological data are of more value here, though doubtless from the antiquity of cultivation they can only establish probabilities.

---

[1] Tussac, *Flore des Antilles*, pl. 4.   [2] Baker, *Fl. of Maurit.*, p. 282.
[3] Martius, *Gen. et Spec. Palmarum*, in folio, vol. iii. p. 257 ; C. Ritter, *Erdkunde*, xiii. p. 760 ; Alph. de Candolle, *Géog. Bot. Rais.*, p. 343.

From Egyptian and Assyrian remains, as well as from
tradition and the most ancient writings, we find that the
date-palm grew in abundance in the region lying between
the Euphrates and the Nile. Egyptian monuments contain fruits and drawings of the tree.[1] Herodotus, in a
more recent age (fifth century before Christ), mentions
the wood of the date-palms of Babylonia, and still later
Strabo used similar expressions about those of Arabia,
whence it seems that the species was commoner than it
is now, and more in the condition of a natural forest
tree. On the other hand, Carl Ritter makes the ingenious
observation that the earliest Hebrew books do not speak
of the date-palm as producing a fruit valued as a food
for man. David, about one thousand years before Christ,
and about seven centuries after Moses, does not mention
the date palm in his list of trees to be planted in his
gardens. It is true that except at Jericho dates seldom
ripen in Palestine. Later, Herodotus says of the Baby-
lonian date-palms that only the greater part produced
good fruit which was used for food. This seems to indicate the beginning of a cultivation perfected by the
selection of varieties and of the transport of male flowers
into the middle of the branches of female trees, but it
perhaps signifies also that Herodotus was ignorant of the
existence of the male plant.

To the west of Egypt the date-palm had probably
existed for centuries or for thousands of years when
Herodotus mentioned them. He speaks of Libya.
There is no historical record with respect to the oases in
the Sahara, but Pliny[2] mentions the date-palm in the
Canaries.

The names of the species bear witness to its great
antiquity both in Asia and in Africa, seeing they are numerous and very different. The Hebrews called the date-
palm *tamar*, and the ancient Egyptians *beq*.[3] The complete difference between these words, both very ancient,
shows that these peoples found the species indigenous
and perhaps already named in Western Asia and in

---

[1] Unger, *Pflanzen d. Alt. Ægypt.*, p. 38.
[2] Pliny, *Hist.*, lib. vi. cap. 37.   [3] Unger, *ubi supra.*

Egypt. The number of Persian, Arabic, and Berber names is incredible.[1] Some are derived from the Hebrew word, others from unknown sources. They often apply to different states of the fruit, or to different cultivated varieties, which again shows ancient cultivation in different countries. Webb and Berthelot have not discovered a name for the date-palm in the language of the Guanchos, and this is much to be regretted. The Greek name, *phœnix*, refers simply to Phœnicia and the Phœnicians, possessors of the date-palm.[2] The names *dactylus* and *date* are derivations of *dachel* in a Hebrew dialect.[3] No Sanskrit name is known, whence it may be inferred that the plantations of the date-palm in Western India are not very ancient. The Indian climate does not suit the species.[4] The Hindustani name *khurma* is borrowed from the Persian.

Further to the East the date-palm remained long unknown. The Chinese received it from Persia, in the third century of our era, and its cultivation was resumed at different times, but they have now abandoned it.[5] As a rule, beyond the arid region which lies between the Euphrates and the south of the Atlas and the Canaries, the date-palm has not succeeded in similar latitudes, or at least it has not become an important culture. It might be grown with success in Australia and at the Cape, but the Europeans who have colonized these regions are not satisfied, like the Arabs, with figs and dates for their staple food. I think, in fine, that in times anterior to the earliest Egyptian dynasties the date-palm already existed, wild or sown here and there by wandering tribes, in a narrow zone extending from the Euphrates to the Canaries, and that its cultivation began later as far as the north-west of India on the one hand and the Cape de Verde Islands [6] on the other, so that the natural area

---

[1] See C. Ritter, *ubi supra*.   [2] Hehn, *Culturpflanzen*, edit. 3, p. 234.
[3] C. Ritter, *ibid.*, p. 828.   [4] According to Roxburgh, Royle, etc.
[5] Bretschneider, *Study and Value*, etc., p. 31.
[6] According to Schmidt, *Fl. d. Cap.-Verd. Isl.*, p. 168, the date-palm is rare in these islands, and is certainly not wild. Webb and Berthelot, on the contrary, assert that in some of the Canaries it is apparently indigenous (*Hist. Nat. des Canaries, Botanique,* iii. p. 289).

has remained very nearly the same for about five thousand years. What it was previously, palæontological discoveries may one day reveal.

Banana—*Musa sapientum* and *M. paradisiaca,* Linnæus; *M. sapientum,* Brown.

The banana or bananas were generally considered to be natives of Southern Asia, and to have been carried into America by Europeans, till Humboldt threw doubts upon their purely Asiatic origin. In his work on New Spain[1] he quoted early authors who assert that the banana was cultivated in America before the conquest.

He admits, on Oviedo's authority,[2] its introduction by Father Thomas of Berlangas from the Canaries into San Domingo in 1516, whence it was introduced into other islands and the mainland.[3] He recognizes the absence of any mention of the banana in the accounts of Columbus, Alonzo Negro, Pinzon, Vespuzzi, and Cortez. The silence of Hernandez, who lived half a century after Oviedo, astonishes him and appears to him a remarkable carelessness; "for," he says,[4] "it is a constant tradition in Mexico and on the whole of the mainland that the *platano arton,* and the *dominico* were cultivated long before the Spanish conquest." The author who has most carefully noted the different epochs at which American agriculture has been enriched by foreign products, the Peruvian Garcilasso de la Vega,[5] says distinctly that at the time of the Incas, maize, quinoa, the potato, and, in the warm and temperate regions, bananas formed the staple food of the natives. He describes the *Musa* of the valleys in the Andes; he even distinguishes the rarer species, with a small fruit and a sweet aromatic flavour, the *dominico,* from the common banana or *arton.*

[1] Humboldt, *Nouvelle Espagne,* 1st edit., ii. p. 360.
[2] Oviedo, *Hist. Nat.,* 1556, p. 112. Oviedo's first work is of 1526. He is the earliest naturalist quoted by Dryander (*Bibl. Banks*) for America.
[3] I have also seen this passage in the translation of Oviedo by Ramusio, vol. iii. p. 115.
[4] Humboldt, *Nouvelle Espagne,* 2nd edit., p. 385.
[5] Garcilasso de la Vega, *Commentarios Reales,* i. p. 282.

Father Acosta[1] asserts also, although less positively, that the *Musa* was cultivated by the Americans before the arrival of the Spaniards. Lastly, Humboldt adds from his own observation, " On the banks of the Orinoco, of the Cassiquaire or of the Beni, between the mountains of Esmeralda and the banks of the river Carony, in the midst of the thickest forests, almost everywhere that Indian tribes are found who have had no relations with European settlements, we meet with plantations of Manioc and bananas." Humboldt suggests the hypothesis that several species or constant varieties of the Banana have been confounded, some of which are indigenous to the new world.

Desvaux studied the specific question, and in a really remarkable work, published in 1814,[2] he gives it as his opinion that all the bananas cultivated for their fruits are of the same species. In this species he distinguishes forty-four varieties, which he arranges in two groups; the large-fruited bananas (seven to fifteen inches long), and the small-fruited bananas (one to six inches), commonly called fig bananas. R. Brown, in 1818, in his work on the *Plants of the Congo,* p. 51, maintains also that no structural difference in the bananas cultivated in Asia and those in America prevents us from considering them as belonging to the same species. He adopts the name *Musa sapientum,* which appears to me preferable to that of *M. paradisiaca* adopted by Desvaux, because the varieties with small fertile fruit appear to be nearer the condition of the wild *Musæ* found in Asia.

Brown remarks on the question of origin that all the other species of the genus *Musa* belong to the old world; that no one pretends to have found in America, in a wild state, varieties with fertile fruit, as has happened in Asia; lastly, that Piso and Marcgraf considered that the banana was introduced into Brazil from Congo. In spite of the force of these three arguments, Humboldt, in his second edition of his essay upon New Spain (ii. p. 397), does not entirely renounce his opinion. He

---

[1] Acosta, *Hist. Nat. De Indias,* 1608, p. 250.
[2] Desvaux, *Journ. Bot.,* iv. p. 5.

says that the traveller Caldcleugh [1] found among the
Puris the tradition that a small species of banana was
cultivated on the borders of the Prato long before they
had any communications with the Portuguese. He adds
that words which are not borrowed ones are found in
American languages to distinguish the fruit of the *Musa;*
for instance, *paruru* in Tamanac, etc., *arata* in Maypur.
I have also read in Stevenson's travels [2] that beds of
the leaves of the two bananas commonly cultivated in
America have been found in the *huacas* or Peruvian
tombs anterior to the conquest; but as this traveller
also says that he saw beans [3] in these *huacas*, a plant
which undoubtedly belongs to the old world, his assertions are not very trustworthy.

Boussingault [4] thought that the *platano arton* at
least was of American origin, but he gives no proof.
Meyen, who had also been in America, adds no argument
to those which were already known; [5] nor does the
geographer Ritter, [6] who simply reproduces the facts
about America, given by Humboldt.

On the other hand, the botanists who have more
recently visited America have no hesitation as to the
Asiatic origin. I may name Seemann for the Isthmus of
Panama, Ernst for Venezuela, and Sagot for Guiana. [7]
The two first insist upon the absence of names for the
banana in the languages of Peru and Mexico. Piso
knew no Brazilian name. Martius [8] has since indicated,
in the Tupi language of Brazil, the names *pacoba* or
bacoba. This same word *bacove* is used, according to
Sagot, by the French in Guiana. It is perhaps derived
from the name *bala,* or *palan,* of Malabar, from an introduction by the Portuguese, subsequent to Piso's voyage.

The antiquity and wild character of the banana in
Asia are incontestable facts. There are several Sanskrit

[1] Caldcleugh, *Trav. in S. Amer.*, 1825, i. p. 23.
[2] Stevenson, *Trav. in S. Amer.*, i. p. 328.
[3] *Ibid.*, p. 363.   [4] Boussingault, *C. r. Acad. Sc. Paris*, May 9, 1836.
[5] Meyen, *Pflanzen Geog.*, 1836, p. 383.   [6] Ritter, *Erdk.*, iv. p. 870.
[7] Seemann, *Bot. of the Herald*, p. 213; Ernst, in Seemann's *Journ.
of Bot.*, 1867, p. 289; Sagot, *Journ. de la Soc. d'Hort. de Fr.*, 1872, p. 226.
[8] Martius, *Eth. Sprachenkunde Amer.*, p. 123.

names.[1] The Greeks, Latins, and Arabs have mentioned
it as a remarkable Indian fruit tree. Pliny[2] speaks of
it distinctly. He says that the Greeks of the expedition of Alexander saw it in India, and he quotes the
name *pala* which still persists in Malabar. Sages re-
posed beneath its shade and ate of its fruit. Hence
the botanical name *Musa sapientum*. *Musa* is from the
Arabic *mouz* or *mauwz*, which we find as early as the
thirteenth century in Ebn Baithar. The specific name
paradisiaca comes from the ridiculous hypothesis which
made the banana figure in the story of Eve and of
Paradise.

It is a curious fact that the Hebrews and the ancient
Egyptians[3] did not know this Indian plant. It is a
sign that it did not exist in India from a very remote
epoch, but was first a native of the Malay Archipelago.

There is an immense number of varieties of the
banana in the south of Asia, both on the islands and on
the continent; the cultivation of these varieties dates
in India, in China, and in the archipelago, from an epoch
impossible to realize; it even spread formerly into the
islands of the Pacific[4] and to the west coast of Africa;[5]
lastly, the varieties bore distinct names in the most
separate Asiatic languages, such as Chinese, Sanskrit,
and Malay.   All this indicates great antiquity of culture,
consequently a primitive existence in Asia, and a diffusion contemporary with or even anterior to that of the
human races.

The banana is said to have been found wild in several
places. This is the more worthy of attention since the
cultivated varieties seldom produce seed, and are
multiplied by division, so that the species can hardly
have become naturalized from cultivation by sowing itself.
Roxburgh had seen it in the forests of Chittagong,[6] in

---

[1] Roxburgh and Wallich, *Fl. Ind.*, ii. p. 485 ; Piddington, *Index*.

[2] Pliny, *Hist.*, lib. xii. cap. 6.

[3] Unger, *ubi supra*, and Wilkinson, **ii.** p. 403, do not mention it.   The
banana is now cultivated in Egypt.

[4] Forster, *Plant. Esc.*, p. 28.

[5] Clusius, *Exot.*, p. 229 ; Brown, *Bot. Congo*, p. **51.**

[6] Roxburgh, *Corom.*, tab. 275 ; *Fl. Ind.*

the form of *Musa sapientum.* Rumphius [1] describes a wild variety with small fruits in the Philippine Isles. Loureiro [2] probably speaks of the same form by the name *M seminifera agrestis,* which he contrasts with *M. seminifera domestica,* which is wild in Cochin-China. Blanco also mentions a wild banana in the Philippines, [4] but his description is vague. Finlayson [5] found the banana wild in abundance in the little island of Pulo Ubi at the southern extremity of Siam. Thwaites [6] saw the variety *M. sapientum* in the rocky forests of the centre of Ceylon, and does not hesitate to pronounce it the original stock of the cultivated bananas. Sir Joseph Hooker and Thomson [7] found it wild at Khasia.

The facts are quite different in America. The wild banana has been seen nowhere except in Barbados, [8] but here it is a tree of which the fruit does not ripen, and which is, consequently, in all probability the result of cultivated varieties of which the seed is not abundant. Sloane's *wild plantain* [9] appears to be a plant very different to the *musa.* The varieties which are supposed to be possibly indigenous in America are only two, and as a rule far fewer varieties are grown than in Asia. The culture of the banana may be said to be recent in the greater part of America, for it dates from but little more than three centuries. Piso [10] says positively that it was imported into Brazil, and has no Brazilian name. He does not say whence it came. We have seen that, according to Oviedo, the species was brought to San Domingo from the Canaries. This fact and the silence of Hernandez, generally so accurate about the useful plants, wild or cultivated, in Mexico, convince me that at the time of the discovery of America the banana did not exist in the whole of the eastern part of the continent.

[1] Rumphius, *Amb.,* **v.** p. 139.　　[2] Loureiro, *Fl. Coch.,* p. 791.
[3] Loureiro, *Fl. Coch.,* p. 791.　　[4] Blanco, *Flora,* 1st edit., p. 247.
[5] Finlayson, *Journey to Siam,* 1826, p. 86, according to Ritter, *Erdk.,* iv. p. 878.
[6] Thwaites, *Enum. Pl. Cey.,* p. 321.
[7] Aitchison, *Catal. of Punjab,* p. 147.
[8] Hughes, *Barb.,* p. 182; Maycock, *Fl. Barb.,* p 396.
[9] Sloane, *Jamaica,* ii. p. 118.　　[10] Piso, edit. 1648, *Hist. Nat.,* p. 75.

Did it exist, then, in the western part on the shores of the Pacific ? This seems very unlikely when we reflect that communication was easy between the two coasts towards the isthmus of Panama, and that before the arrival of the Europeans the natives had been active in diffusing throughout America useful plants like the manioc, maize, and the potato. The banana, which they have prized so highly for three centuries, which is so easily multiplied by suckers, and whose appearance must strike the least observant, would not have been forgotten in a few villages in the depths of the forest or upon the littoral.

I admit that the opinion of Garcilasso, descendant of the Incas, an author who lived from 1530 to 1568, has a certain importance when he says that the natives knew the banana before the conquest. However, the expressions of another writer, extremely worthy of attention, Joseph Acosta, who had been in Peru, and whom Humboldt quotes in support of Garcilasso, incline me to adopt the contrary opinion.[1] He says,[2] "The reason the Spaniards called it *plane* (for the natives had no such name) was that, as in the case of their trees, they found some resemblance between them." He goes on to show how different was the *plane* (*Platanus*) of the ancients. He describes the banana very well, and adds that the tree is very common in the Indies (*i.e.* America), " although they (the Indians) say that its origin is Ethiopia. . . . There is a small white species of plantain (banana), very delicate, which is called in Espagnolle [3] *dominico*. There are others coarser and larger, and of a red colour. There are none in Peru, but they are imported thither from the Indies,[4] as

---

[1] Humboldt quotes the Spanish edition of 1608. The first edition is of 1591. I have only been able to consult the French translation of Regnault, published in 1598, and which is apparently accurate.

[2] Acosta, trans., lib. iv. cap. 21.

[3] That is probably Hispaniola or San Domingo ; for if he had meant the Spanish language, it would have been translated by *castillan* and without the capital letter.

[4] This is probably a misprint for *Andes*, for the word *Indes* has no sense. The work says (p. 166) that pine-apples do not grow in Peru, but that they are brought thither from the Andes, and (p. 173) that the cacao comes from the Andes. It seems to have meant hot regions. The word

into Mexico from Cuernavaca and the other valleys. On
the continent and in some of the islands there are great
plantations of them which form dense thickets." Surely it
is not thus that the author would express himself were
he writing of a fruit tree of American origin. He would
quote American names and customs; above all, he would
not say that the natives regarded it as a plant of foreign
origin. Its diffusion in the warm regions of Mexico may
well have taken place between the epoch of the conquest
and the time when Acosta wrote, since Hernandez, whose
conscientious researches go back to the earliest times of
the Spanish dominion in Mexico (though published later
in Rome), says not a word of the banana.[1] Prescott the
historian saw ancient books and manuscripts which assert
that the inhabitants of Tumbez brought bananas to
Pizarro when he disembarked upon the Peruvian coast,
and he believes that its leaves were found in the *huacas,*
but he does not give his proofs.[2]

As regards the argument of the modern native
plantations in regions of America, remote from European
settlements, I find it hard to believe that tribes have
remained absolutely isolated, and have not received so
useful a tree from colonized districts.

Briefly, then, it appears to me most probable that the
species was early introduced by the Spanish and Portuguese into San Domingo and Brazil, and I confess that
this implies that Garcilasso was in error with regard to
Peruvian traditions. If, however, later research should
prove that the banana existed in some parts of America
before the advent of the Europeans, I should be inclined
to attribute it to a chance introduction, not very ancient,
the effect of some unknown communication with the
islands of the Pacific, or with the coast of Guinea, rather
than to believe in the primitive and simultaneous existence

Andes has since been applied to the chain of mountains by a strange
and unfortunate transfer.
[1] I have read through the entire work, to make sure of this fact.
[2] Prescott, *Conquest of Peru.* The author has consulted valuable
records, among others a manuscript of Montesinos of 1527; but he
does not quote his authorities for each fact, and contents himself with
vague and general indications, which are very insufficient.

of the species in both hemispheres. The whole of geographical botany renders the latter hypothesis improbable, I might almost say impossible, to admit, especially in a genus which is not divided between the two worlds.

In conclusion, I would call attention to the remarkable way in which the distribution of varieties favours the opinion of a single species—an opinion adopted, purely from the botanical point of view, by Roxburgh, Desvaux, and R. Brown. If there were two or three species, one would probably be represented by the varieties suspected to be of American origin, the other would belong, for instance, to the Malay Archipelago or to China, and the third to India. On the contrary all the varieties are geographically intermixed, and the two which are most widely diffused in America differ sensibly the one from the other, and each is confounded with or approaches very nearly to Asiatic varieties.

Pine-Apple — *Ananassa sativa*, Lindley; *Bromelia Ananas*, Linnæus.

In spite of the doubts of a few writers, the pine-apple must be an American plant, early introduced by Europeans into Asia and Africa.

Nana was the Brazilian name,[1] which the Portuguese turned into *ananas*. The Spanish called it *pinas*, because the shape resembles the fruit of a species of pine.[2] All early writers on America mention it.[3] Hernandez says that the pine-apple grows in the warm regions of Haïti and Mexico. He mentions a Mexican name, *matzatli*. A pine-apple was brought to Charles V., who mistrusted it, and would not taste it.

The works of the Greeks, Romans, and Arabs make no allusion to this species, which was evidently introduced into the old world after the discovery of America. Rheede [4] in the seventeenth century was persuaded of this; but Rumphius [5] disputed it later, because he said

---

[1] Marcgraf, *Brasil.*, p. 33.
[2] Oviedo, Ramusio's trans., iii. p. 113 ; Jos. Acosta, *Hist. Nat. des Indes*, French trans., p. 166.
[3] Thevet, Piso, etc. ; Hernandez, *Thes.*, p. 341.
[4] Rheede, *Hort. Malab.*, xi. p. 6. [5] Rumphius, *Amboin*, v. p. 228.

the pine-apple was cultivated in his time in every part of India, and was found wild in Celebes and elsewhere. He notices, however, the absence of an Asiatic name. That given by Rheede for Malabar is evidently taken from a comparison with the jack-fruit, and is in no sense original. It is doubtless a mistake on the part of Piddington to attribute a Sanskrit name to the pine-apple, as the name *anarush* seems to be a corruption of *ananas.* Roxburgh knew of none, and Wilson's dictionary does not mention the word *anarush.* Royle [1] says that the pine-apple was introduced into Bengal in 1594. Kircher [2] says that the Chinese cultivated it in the seventeenth century, but it was believed to have been brought to them from Peru.

Clusius [3] in 1599 had seen leaves of the pine-apple brought from the coast of Guinea. This may be explained by an introduction there subsequent to the discovery of America. Robert Brown speaks of the pine-apple among the plants cultivated in Congo; but he considers the species to be an American one.

Although the cultivated pine-apple bears few seeds or none at all, it occasionally becomes naturalized in hot countries. Examples are quoted in Mauritius, the Seychelles, and Rodriguez Island,[4] in India,[5] in the Malay Archipelago, and in some parts of America, where it was probably not indigenous—the West Indies, for instance.

It has been found wild in the warm regions of Mexico (if we may trust the phrase used by Hernandez), in the province of Veraguas [6] near Panama, in the upper Orinoco valley,[7] in Guiana [8] and the province of Bahia.[9]

---

[1] Royle, *Ill.*, p. 376.
[2] Kircher, *Chine Illustrée*, trans. of 1670, p. 253.
[3] Clusius, *Exotic.*, cap. 44.   [4] Baker, *Fl. of Maurit.*
[5] Royle, *ubi supra.*   [6] Seemann, *Bot. of the Herald*, p. 215.
[7] Humboldt, *Nouv. Esp.*, 2nd edit., ii. p. 478.
[8] *Gardeners' Chronicle*, 1881, vol. i. p. 657.
[9] Martius, letter to A. de Candolle, *Géogr. Bot. Rais.*, p. 927.

CHAPTER V.

Article I.—Seeds used for Food.

Cacao—*Theobroma Cacao,* Linnæus.

The genus *Theobroma,* of the order *Byttneriaceæ,* allied to the *Malvaceæ,* consists of fifteen to eighteen species, all belonging to tropical America, principally in the hotter parts of Brazil, Guiana, and Central America.

The common cacao, *Theobroma Cacao,* is a small tree wild in the forests of the Amazon and Orinoco basins [1] and of their tributaries up to four hundred feet of altitude. It is also said to grow wild in Trinidad, which lies near the mouth of the Orinoco.[2] I find no proof that it is indigenous in Guiana, although it seems probable. Many early writers indicate that it was both wild and cultivated at the time of the discovery of America from Panama to Guatemala and Campeachy; but from the numerous quotations collected by Sloane,[3] it is to be feared that its wild character was not sufficiently verified. Modern botanists are not very explicit on this head, and in general they only mention the cacao as cultivated in these regions and in the West India Islands. G. Bernoulli,[4] who had resided in Guatemala, only says, " wild

---

[1] Humboldt, *Voy.,* ii. p. 511; Kunth, in Humboldt and Bonpland, *Nova Genera,* v. p. 316 ; Martius, *Ueber den Cacao,* in Büchner, *Repert. Pharm.*

[2] Schach, in Grisebach, *Flora of Brit. W. Ind. Is.,* p. 91.

[3] Sloane, *Jamaica,* ii. p. 15.

[4] G. Bernoulli, *Uebersicht der Arten von Theobroma,* p. 5.

and cultivated throughout tropical America;" and Hemsley,[1] in his review of the plants of Mexico and Central America, made in 1879 from the rich materials of the Kew herbarium, gives no locality where the species is indigenous. It was perhaps introduced into Central America and into the warm regions of Mexico by the Indians before the discovery of America. Cultivation may have naturalized it here and there, as is said to be the case in Jamaica.[2] In support of this hypothesis, it must be observed that Triana[3] indicates the cacao as only cultivated in the warm regions of New Granada, a country situated between Panama and the Orinoco valley.

However this may be, the species was grown in Central America and Yucatan at the time of the discovery of America. The seeds were sent into the highlands of Mexico, and were even used as money, so highly were they valued. The custom of drinking chocolate was general. The name of this excellent drink is Mexican. The Spaniards carried the cacao from Acapulco to the Philippine Isles in 1674 and 1680,[4] where it succeeded wonderfully. It is also cultivated in the Sunda Isles. I imagine it would succeed on the Guinea and Zanzibar coasts, but it is of no use to attempt to grow it in countries which are not very hot and very damp.

Another species, *Theobroma bicolor*, Humboldt and Bonpland, is found growing with the common cacao in American plantations. It is not so much prized. On the other hand, it does not require so high a temperature, and can live at an altitude of nearly three thousand feet in the valley of the Magdalena. It abounds in a wild state in New Granada.[5] Bernoulli asserts that it is only cultivated in Guatemala, though the inhabitants call it mountain cacao.

Litchi—*Nephelium Litchi*, Cambessides.

The seed of this species and of the two following is

---

[1] Hemsley, *Biologia Centrali Americana*, part ii. p. 133.
[2] Grisebach, *ubi supra.*
[3] Triana and Planchon, *Prodr. Fl. Novo Granatensis*, p. 208.
[4] Blanco, *Fl. de Filipinas*, edit. 2, p. 420.
[5] Kunth, in Humboldt and Bonpland, *ubi supra ;* Triana, *ubi supra.*

covered with a fleshy excrescence, very sweet and scented, which is eaten with tea.

Like most of the *Sapindaceæ,* the nepheliums are trees. This one has been cultivated in the south of China, India, and the Malay Archipelago from a date of which we cannot be certain. Chinese authors living at Pekin only knew the *Litchi* late in the third century of our era.[1] Its introduction into Bengal took place at the end of the eighteenth century.[2] Every one admits that the species is a native of the south of China, and, Blume[3] adds, of Cochin-China and the Philippine Isles, but it does not seem that any botanist has found it in a truly wild state. This is probably because the southern part of China towards Siam has been little visited. In Cochin-China and in Burmah and at Chittagong the *Litchi* is only cultivated.[4]

Longan—*Nephelium longana,* Cambessides.

This second species, very often cultivated in Southern Asia, like the *Litchi,* is wild in British India, from Ceylon and Concan as far as the mountains to the east of Bengal, and in Pegu.[5] The Chinese introduced it into the Malay Archipelago some centuries ago.

Rambutan—*Nephelium lappaceum,* Linnæus.

It is said to be wild in the Indian Archipelago, where it must have been long cultivated, to judge from the number of its varieties. A Malay name, given by Blume, signifies wild tree. Loureiro says it is wild in Cochin-China and Java. Yet I find no confirmation for Cochin-China in modern works, nor even for the islands. The new flora of British India[6] indicates it at Singapore and Malacca without affirming that it is indigenous, on which head the labels in herbaria commonly tell us nothing. Certainly the species is not wild on the continent of Asia, in spite of the vague expressions of Blume and

---

[1] Bretschneider, letter of Aug. 23, 1881.
[2] Roxburgh, *Fl. Indica,* ii. p. 269.   [3] Blume, *Rumphia,* iii. p. 106.
[4] Loureiro, *Flora Coch.,* p. 233 ; Kurz, *Forest Fl. of Brit. Burmah,* p. 293.
[5] Roxburgh, *Fl. Ind.,* ii. p. 271 ; Thwaites, *Enum. Zeyl.,* p. 58 ; Hiern, in *Fl. of Brit. Ind.,* i. p. 688.
[6] Hiern, in *Fl. of Brit. Ind.,* i. p. 687.

Miquel,[1] but it is more probably a native of the Malay Archipelago.

In spite of the reputation of the nepheliums, of which the fruit can be exported, it does not appear that these trees have been introduced into the tropical colonies of Africa and America except into a few gardens as curiosities.

Pistachio Nut—*Pistacia vera*, Linnæus.

The pistachio, a shrub belonging to the order *Anacardiaceæ*, grows naturally in Syria. Boissier[2] found it to the north of Damascus in Anti-Lebanon, and he saw specimens of it brought from Mesopotamia, but he could not be sure that they were found wild. There is the same doubt about branches gathered in Arabia, which have been mentioned by some writers. Pliny and Galen[3] knew that the species was a Syrian one. The former tells us that the plant was introduced into Italy by Vitellius at the end of the reign of Tiberius, and thence into Spain by Flavius Pompeius.

There is no reason to believe that the cultivation of the pistachio was ancient even in its primitive country, but it is practised in our own day in the East, as well as in Sicily and Tunis. In the south of France and Spain it is of little importance.

Broad Bean—*Faba vulgaris*, Mœnch ; *Vicia faba*, Linnæus.

Linnæus, in his best descriptive work, *Hortus cliffortianus*, admits that the origin of this species is obscure, like that of most plants of ancient cultivation. Later, in his *Species*, which is more often quoted, he says, without giving any proof, that the bean "inhabits Egypt." Lerche, a Russian traveller at the end of the last century, found it wild in the Mungan desert of the Mazanderan, to the south of the Caspian Sea.[4] Travellers

---

[1] Blume, *Rumphia*, iii. p. 103 ; Miquel, *Fl. Indo-Batava*, i. p. 554.

[2] Boissier, *Fl. Orient.*, ii. p. 5.

[3] Pliny, *Hist. Nat.*, lib. xiii. cap. 15; lib. xv. cap. 22; Galen, *De Alimentis*, lib. ii. cap. 30.

[4] Lerche, *Nova Acta Acad. Cesareo-Leopold*, vol. v., appendix, p. 203, published in 1773. Maximowicz, in a letter of Feb. 24, 1882, tells me that Lerche's specimen exists in the herbarium of the Imperial Garden

who have collected in this region have sometimes come across it,[1] but they do not mention it in their writings,[2] excepting Ledebour,[3] and the quotation on which he relies is not correct. Bosc[4] says that Olivier found the bean wild in Persia; I do not find this confirmed in Olivier's *Voyage*, and as a rule Bosc seems to have been too ready to believe that Olivier found a good many of our cultivated plants in the interior of Persia. He says it of buckwheat and of oats, which Olivier does not mention

The only indication besides that of Lerche which I find in floras is a very different locality. Munby mentions the bean as wild in Algeria, at Oran. He adds that it is rare. No other author, to my knowledge, has spoken of it in northern Africa. Cosson, who knows the flora of Algeria better than any one, assures me he has not seen or received any specimen of the wild bean from the north of Africa. I have ascertained that there is no specimen in Munby's[5] herbarium, now at Kew. As the Arabs grow the bean on a large scale, it may perhaps be met with accidentally outside cultivated plots. It must not be forgotten, however, that Pliny (lib. xviii. c. 12) speaks of a wild bean in Mauritania, but he adds that it is hard and cannot be cooked, which throws doubt upon the species. Botanists who have written upon Egypt and Cyrenaïca, especially the more recent,[6] give the bean as cultivated.

This plant alone constitutes the genus *Faba*. We cannot, therefore, call in the aid of any botanical analogy

at St. Petersburgh. It is in flower, and resembles the cultivated bean in all points excepting height, which is about half a foot. The label mentions the locality and its wild character without other remarks.

[1] There are Transcaucasian specimens in the same herbarium, but taller, and they are not said to be wild.

[2] Marschall Bieberstein, *Flora Caucaso-Taurica;* C. A. Meyer, *Verzeichniss;* Hohenacker, *Enum. Plant. Talysch ;* Boissier, *Fl. Orient.*, p. 578, Buhse and Boissier, *Plant. Transcaucasiæ.*

[3] Ledebour, *Fl. Ross.*, i. p. 664, quotes de Candolle, *Prodromus*, ii. p. 354; now Seringe wrote the article *Faba* in *Prodromus*, in which the south of the Caspian is indicated, probably on Lerche's authority.

[4] *Dict. d'Agric.*, v. p. 512.

[5] Munby, *Catal. Plant. in Alger. sponte nascent.*, edit. 2, p. 12.

[6] Schweinfurth and Ascherson, *Aufzählung*, p. 256 ; Rohlfs, *Kufra.*

to discover its origin. We must have recourse to the history of its cultivation and to the names of the species to find out the country in which it was originally indigenous.

We must first eliminate an error which came from a wrong interpretation of Chinese works. Stanislas Julien believed that the bean was one of the five plants which the Emperor Chin-nong commanded, 4600 years ago, to be sown every year with great solemnity.[1] Now, according to Dr. Bretschneider,[2] who is surrounded at Pekin with every possible resource for arriving at the truth, the seed similar to a bean which the emperors sow in the enjoined ceremony is that of *Dolichos soja*, and the bean was only introduced into China from Western Asia a century before the Christian era, at the time of Chang-kien's embassy. Thus falls an assertion which it is hard to reconcile with other facts, for instance with the absence of an ancient cultivation of the bean in India, and of a Sanskrit name, or even of any modern Indian name.

The ancient Greeks were acquainted with the bean, which they called *kuamos*, and sometimes *kuamos ellenikos*, to distinguish it from that of Egypt, which was the seed of a totally different aquatic species, *Nelumbium*. The *Iliad*[3] already mentions the bean as a cultivated plant, and Virchow found some beans in the excavations at Troy.[4] The Latins called it *faba*. We find nothing in the works of Theophrastus, Dioscorides, Pliny, etc., which leads us to believe the plant indigenous in Greece or Italy. It was early known, because it was an ancient Roman rite to put beans in the sacrifices to the goddess Carna, whence the name *Fabariæ Calendæ*.[5] The Fabii perhaps took their name from *faba*, and the twelfth chapter of the eighteenth book of Pliny shows, without the possibility of a doubt, the antiquity and importance of the bean in Italy.

---

[1] Loiscleur Deslongchamps, *Consid. sur les Céréales*, part i. p. 29.

[2] Bretschneider, *Study and Value*, etc., pp. 7, 15.

[3] *Iliad*, 13, v. 589.

[4] Wittmack, *Sitz. bericht Vereins*, Brandenburg, 1879.

[5] *Novitius Dictionnarium*, at the word *Faba*.

The word *faba* recurs in several of the Aryan languages of Europe, but with modifications which philologists alone can recognize. We must not forget, however, Adolphe Pictet's very just remark,[1] that in the cases of the seeds of cereals and leguminous plants the names of one species are often transferred to another, or that certain names were sometimes specific and sometimes generic. Several seeds of like form were called *kuamos* by the Greeks; several different kinds of haricot bean (*Phaseolus, Dolichos*) bear the same name in Sanskrit, and *faba* in ancient Slav, *bobu* in ancient Prussian, *babo* in Armorican, *fav*, etc., may very well have been used for peas, haricot beans, etc. In our own day the phrase coffee-*bean* is used in the trade. It has been rightly supposed that when Pliny speaks of *fabariæ* islands, where beans were found in abundance, he alludes to a species of wild pea called botanically *Pisum maritimum*.

The ancient inhabitants of Switzerland and of Italy in the age of bronze cultivated a small-fruited variety of *Faba vulgaris.*[2] Heer calls it *Celtica nana*, because it is only six to nine millimetres long, whereas our modern field bean is ten to twelve millimetres. He has compared the specimens from Montelier on Lake Morat, and St. Peter's Islands on Lake Bienne, with others of the same epoch from Parma. Mortellet found, in the contemporary lake-dwellings on the Lake Bourget, the same small bean, which is, he says, very like a variety cultivated in Spain at the present day.[3]

The bean was cultivated by the ancient Egyptians.[4] It is true that hitherto no beans have been found in the sarcophagi, or drawings of the plant seen on the monuments. The reason is said to be that the plant was reckoned unclean.[5] Herodotus[6] says, "The Egyptians

---

[1] *Origines Indo-Européennes*, edit. 2, vol. i. p. 353.

[2] Heer, *Pflanzen der Pfahlbauten*, p. 22, figs. 44–47.

[3] Perrin, *Étude Préhistorique sur la Savoie*, p. 2.

[4] Delile, *Plant. Cult. en Égypte*, p. 12; Reynier, *Économie des Egyptiens et Carthaginois*, p. 340; Unger, *Pflan. d. Alt. Ægyp.*, p. 64; Wilkinson, *Man. and Cus. of Anc. Egyptians*, p. 402.

[5] Reynier, *ubi supra*, tries to discover the reason of this.

[6] Herodotus, *Histoire*, Larcher's trans., vol. ii. p. 32.

never sow the bean in their land, and if it grows they do
not eat it either cooked or raw.   The priests cannot even
endure the sight of it; they imagine that this vegetable is
unclean." The bean existed then in Egypt, and probably
in cultivated places, for the soil which would suit it was
as a rule under cultivation.   Perhaps the poor population
and that of certain districts did not share the prejudices
of the priests; we know that the superstitions varied
with the *nomes*.   Plutarch and Diodorus Siculus mention
the cultivation of the bean in Egypt, but they wrote
five hundred years later than Herodotus.

The word *pol* occurs twice in the Old Testament; [1] it
has been translated bean because of the traditions preserved by the Talmud, and of the Arabic name *foul, fol,*
or *ful*, which is that of the bean.   The first of the two
verses shows that the Hebrews were acquainted with the
bean one thousand years before Christ.

Lastly, I shall mention a sign of the ancient existence
of the bean in the north of Africa.   This is the Berber
name *ibiou*, in the plural *iabouen*, used by the Kabyles of
the province of Algiers.[2]   It has no resemblance to the
Semitic name, and dates perhaps from a remote antiquity.
The Berbers formerly inhabited Mauritania, where Pliny
asserts that the species was wild.   It is not known
whether the Guanchos (the Berber people of the Canaries)
knew the bean.   I doubt whether the Iberians had it, for
their supposed descendants, the Basques, use the name
baba,[3] answering to the Roman *faba*.

We judge from these facts that the bean was cultivated in Europe in prehistoric terms.   It was introduced
into Europe probably by the western Aryans at the time
of their earliest migrations (Pelasgians, Kelts, Slavs).   It
was taken to China later, a century before the Christian
era, and still later into Japan, and quite recently into
India.

Its wild habitat was probably twofold some thousands
of years ago, one of the centres being to the south of the

[1] 2 Sam. xvii. 28; Ezek. iv. 9.
[2] *Dict. Français-Berbère*, published by the French government.
[3] Note communicated to M. Clos by M. d'Abadie.

Caspian, the other in the north of Africa. This kind of area, which I have called disjunctive, and to which I formerly paid a good deal of attention,[1] is rare in dicotyledons, but there are examples in those very countries of which I have just spoken.[2] It is probable that the area of the bean has long been in process of diminution and of extinction. The nature of the plant is in favour of this hypothesis, for its seed has no means of dispersing itself, and rodents or other animals can easily make prey of it. Its area in Western Asia was probably less limited at one time, and that in Africa in Pliny's day was more or less extensive. The struggle for existence which was going against this plant, as against maize, would have gradually isolated it and caused it to disappear, if man had not saved it by cultivation.

The plant which most nearly resembles the bean is *Vicia narbonensis*. Authors who do not admit the genus *Faba*, of which the characters are not very distinct from those of *Vicia*, place these two species in the same section. Now, *Vicia narbonensis* is wild in the Mediterranean basin and in the East as far as the Caucasus, in the north of Persia, and in Mesopotamia.[3] Its area is continuous, but this renders the hypothesis I mentioned above probable by analogy.

Lentil—*Ervum lens*, Linnæus; *Lens esculenta*, Mœnch.

The plants which most nearly resemble the lentil are classed by authors now in the genus *Ervum*, now in a distinct genus *Lens*, and sometimes in the genus *Cicer*; but the species of these ill-defined groups all belong to the Mediterranean basin or to Western Asia. This throws some light on the origin of the cultivated plant. Unfortunately, the lentil is no longer to be found in a wild state, at least with certainty. The floras of the south of Europe, of Northern Africa, of the East, and of India always mention it as cultivated, or as growing in fields after or with other cultivated species. A botanist[4]

[1] A. de Candolle, *Géogr. Bot. Rais.*, chap. x.

[2] *Rhododendron ponticum* now exists only in Asia Minor and in the south of the Spanish ｒeninsula.

[3] Boissier, *Fl. Orient.*, ii. p. 577.

[4] C. A. Meyer, *Verzeichniss Fl. Caucas.*, p. 147.

saw it in the provinces to the south of the Caucasus, "cultivated and nearly wild here and there round villages." Another[1] indicates it vaguely in the south of Russia, but more recent floras fail to confirm this.

The history and names of this plant may give clearer indications of its origin. It has been cultivated in the East, in the Mediterranean basin and even in Switzerland, from prehistoric time. According to Herodotos, Theophrastus, etc., the ancient Egyptians used it largely. If their monuments give no proof of this, it was probably because the lentil was, like the bean, considered common and coarse. The Old Testament mentions it three times, by the name *adaschum* or *adaschim*, which must certainly mean lentil, for the Arabic name is *ads*,[2] or *adas*.[3] The *red* colour of Esau's famous mess of pottage has not been understood by most authors. Reynier,[4] who had lived in Egypt, confirms the explanation given formerly by Josephus; the lentils were red because they were hulled. It is still the practice in Egypt, says Reynier, to remove the husk or outer skin from the lentil, and in this case they are a pale red. The Berbers have the Semitic name *adès* for the lentil.[5]

The Greeks cultivated the species—*fakos* or *fakai*. Aristophanes mentions it as an article of food of the poor.[6] The Latins called it *lens*, a name whose origin is unknown, which is evidently allied to the ancient Slav *lesha*, Illyrian *lechja*, Lithuanian *lenszic*.[7] The difference between the Greek and Latin names shows that the species perhaps existed in Greece and Italy before it was cultivated. Another proof of ancient existence in Europe is the discovery of lentils in the lake-dwellings of St. Peter's Island, Lake of Bienne,[8] which are of the age of

[1] Georgi, in Ledebour, *Fl. Ross.*

[2] Forskal, *Fl. Ægypt.*; Delile, *Plant. Cult. en Égypte*, p. 13.

[3] Ebn Baithar, ii. p. 134.

[4] Reynier, *Économie publique et rurale des Arabes et des Juifs*, Genève, 1820, p. 429.

[5] *Dict. Franç.-Berbère*, in 8vo, 1844.

[6] Hehn, *Culturpflanzen*, etc., edit. 3, vol. ii. p. 188.

[7] Ad. Pictet, *Origines Indo-Européennes*, edit. 2, vol. i. p. 364; Hehn, *ubi supra.*

[8] Heer, *Pflanzen der Pfahlbauten*, p. 23, fig. 49.

bronze. The species may have been introduced from Italy.

According to Theophrastus,[1] the inhabitants of Bactriana (the modern Bokkara) did not know the *fakos* of the Greeks. Adolphe Pictet quotes a Persian name, *mangu* or *margu*, but he does not say whether it is an ancient name, existing, for instance, in the Zend Avesta. He admits several Sanskrit names for the lentil, *masura, renuka, mangalya*, etc., while Anglo-Indian botanists, Roxburgh and Piddington, knew none.[2] As these authors mention an analogous name in Hindustani and Bengali, *mussour*, we may suppose that *masura* signifies lentil, while *mangu* in Persian recalls the other name *mangalya*. As Roxburgh and Piddington give no name in other Indian languages, it may be supposed that the lentil was not known in this country before the invasion of the Sanskrit-speaking race. Ancient Chinese works do not mention the species; at least, Dr. Bretschneider says nothing of them in his work published in 1870, nor in the more detailed letters which he has since written to me.

The lentil appears to have existed in western temperate Asia, in Greece, and in Italy, where its cultivation was first undertaken in very early prehistoric time, when it was introduced into Egypt. Its cultivation appears to have been extended at a less remote epoch, but still hardly in historic time, both east and west, that is into Europe and India.

Chick-Pea—*Cicer arietinum*, Linnæus.

Fifteen species of the genus *Cicer* are known, all of Western Asia or Greece, except one, which is Abyssinian. It seems, therefore, most probable that the cultivated species comes from the tract of land lying between Greece and the Himalayas, vaguely termed the East. The species has not been found undoubtedly wild. All the floras of the south of Europe, of Egypt, and of Western Asia as far as the Caucasus and India, give it as a cultivated species, or growing in fields and cultivated

---

[1] Theophrastus, *Hist.*, lib. iv. cap. 5.
[2] Roxburgh, *Fl. Ind.*, edit. 1832, vol. iii. p. 324; Piddington, *Index.*

grounds. It has sometimes[1] been indicated in the Crimea, and to the north, and especially to the south of the Caucasus, as nearly wild ; but well-informed modern authors do not think so.[2] This quasi-wildness can only point to its origin in Armenia and the neighbouring countries. The cultivation and the names of the species may perhaps throw some light on the question.

The Greeks cultivated this species of pea as early as Homer's time, under the name of *erebinthos*,[3] and also of *krios*,[4] from the resemblance of the pea to the head of a ram. The Latins called it *cicer*, which is the origin of all the modern names in the south of Europe. The name exists also among the Albanians, descendants of the Pelasgians, under the form *kikere*.[5] The existence of such widely different names shows that the plant was very early known, and perhaps indigenous, in the south-east of Europe.

The chick-pea has not been found in the lake-dwellings of Switzerland, Savoy, and Italy. In the first-named locality its absence is not singular; the climate is not hot enough. A common name among the peoples of the south of the Caucasus and of the Caspian Sea is, in Georgian, *nachuda ;* in Turkish and Armenian, *nachius, nachunt ;* in Persian, *nochot*.[6] Philologists can tell if this is a very ancient name, and if it has any connection with the Sanskrit *chennuka*.

The chick-pea is so frequently cultivated in Egypt from the earliest times of the Christian era,[7] that it is supposed to have been also known to the ancient Egyptians. There is no proof to be found in the drawings or stores of grain in their monuments, but it may be supposed that this pea, like the bean and the lentil, was

---

[1] Ledebour, *Fl. Ross.*, i. p. 660, according to Pallas, Falk, and Koch.

[2] Boissier, *Fl. Orient.*, ii. p. 560; Steven, *Verzeichniss des Taurischen Hablinseln*, p. 134.

[3] *Iliad*, bk. 13, verse 589 ; Theophrastus, *Hist.*, lib. viii. c. 3.

[4] Dioscorides, lib. ii. c. 126.

[5] Heldreich, *Nutzpflanzen Griechenlands*, p. 71.

[6] Nemnich, *Polyglott. Lex.*, i. p. 1037 ; Bunge, in *Goebels Reise*, ii. p. 328.

[7] Clément d'Alexandrie, *Strom.*, lib. i., quoted from Reynier, *Écon. des Égyp. et Carthag.*, p. 343.

considered common or unclean. Reynier[1] thought that the *ketsech*, mentioned by Isaiah in the Old Testament, was perhaps the chick-pea; but this name is generally attributed, though without certainty, to *Nigella sativa* or *Vicia sativa*.[2] As the Arabs have a totally different name for the chick-pea, *omnos*, *homos*, which recurs in the Kabyl language as *hammez*,[3] it is not likely that the *ketsech* of the Jews was the same plant. These details lead me to suspect that the species was unknown to the ancient Egyptians and to the Hebrews. It was perhaps introduced among them from Greece or Italy towards the beginning of our era.

It is of more ancient introduction into India, for there is a Sanskrit name, and several others, analogous or different, in modern Indian languages.[4] Bretschneider does not mention the species in China.

I do not know of any proof of antiquity of culture in Spain, yet the Castilian name *garbanzo*, used also by the Basques under the form *garbantzua*, and by the French as *garvance*, being neither Latin nor Arabic, may date from an epoch anterior to the Roman conquest.

Botanical, historical, and philological data agree in indicating a habitation anterior to cultivation in the countries to the south of the Caucasus and to the north of Persia. The western Aryans (Pelasgians, Hellenes) perhaps introduced the plant into Southern Europe, where, however, there is some probability that it was also indigenous. The western Aryans carried it into India. Its area perhaps extended from Persia to Greece, and the species now exists only in cultivated ground, where we do not know whether it springs from a stock originally wild or from cultivated plants.

Lupin—*Lupinus albus*, Linnæus.

The ancient Greeks and Romans cultivated this leguminous plant to bury it as a green manure, and also

---

[1] Reynier, *Écon. des Arabes et Juifs*, p. 430.

[2] Rosenmüller, *Bibl. Alterth.*, i. p. 100; Hamilton, *Bot. de la Bible*, p. 180.

[3] Rauwolf, *Fl. Orient.*, No. 220; Forskal, *Fl. Ægypt.*, p. 81; *Dict. Franç.-Berbère.*

[4] Roxburgh, *Fl. Ind.*, iii. p. 324; Piddington, *Index.*

for the sake of the seeds, which are a good fodder for cattle, and which are also used by man. The expressions of Theophrastus, Dioscorides, Cato, Varro, Pliny, etc., quoted by modern writers, refer to the culture or to the medical properties of the seeds, and do not show whether the species was the white lupin, *L. albus*, or the blue-flowered lupin, *L. hirsutus*, which grows wild in the south of Europe. Fraas says [1] that the latter is grown in the Morea at the present day; but Heldreich says [2] that *L. albus* grows in Attica. As this is the species which has been long cultivated in Italy, it is probable that it is the lupin of the ancients. It was much grown in the eighteenth century, especially in Italy,[3] and de l'Ecluse settles the question of the species, as he calls it *Lupinus sativus albo flore*.[4] The antiquity of its cultivation in Spain is shown by the existence of four different common names, according to the province; but the plant is only found cultivated or nearly wild in fields and sandy places.[5] The species is indicated by Bertoloni in Italy, on the hills of Sarzana. Yet Caruel does not believe it to be wild here, any more than in other parts of the peninsula.[6] Gussone [7] is very positive for Sicily—" on barren and sandy hills, and in meadows (*in herbidis*)." Lastly, Grisebach [8] found it in Turkey in Europe, near Ruskoï, and d'Urville [9] saw it in abundance, in a wood near Constantinople. Castagne confirms this in a manuscript catalogue in my possession. Boissier does not mention any locality in the East; the species does not exist in India, but Russian botanists have found it to the south of the Caucasus, though we do not know with certainty if it was really wild.[10] Other localities will perhaps be found between Sicily, Macedonia, and the Caucasus.

---

[1] See Fraas, *Fl. Class.*, p. 51 ; Lenz., *Bot. der Alten*, p. 73.

[2] Heldreich, *Nutzpflanzen Griechenlands*, p. 69.

[3] Olivier de Serres, *Théâtre de l'Agric.*, edit. 1529, p. 88.

[4] Clusius, *Hist. Plant.*, ii. p. 228.

[5] Willkomm and Lange, *Fl. Hisp.*, iii. p. 466.

[6] Caruel, *Fl. Toscana*, p. 136.

[7] Gussone, *Fl. Siculæ Syn.*, edit. 2, vol. ii. p. 436.

[8] Grisebach, *Spicil. Fl. Rumel.*, p. 11.   [9] D'Urville, *Enum.*, p. 86.

[10] Ledebour, *Fl. Ross.*, i. p. 510.

Egyptian Lupin—*Lupinus termis*, Forskal.

This species of lupin, so nearly allied to *L. albus* that it has sometimes been proposed to unite them,[1] is largely cultivated in Egypt and even in Crete. The most obvious difference is that the upper part of the flowers of *L. termis* is blue. The stem is taller than that of *L. albus*. The seeds are used like those of the common lupin, after they have been steeped to get rid of their bitterness.

L. termis is wild in sandy soil and mountainous districts, in Sicily, Sardinia, and Corsica;[2] in Syria and Egypt, according to Boissier;[3] but Schweinfurth and Ascherson[4] say that it is only cultivated in Egypt. Hartmann saw it wild in Upper Egypt.[5] Unger[6] mentions it among the cultivated specimens of the ancient Egyptians, but he gives neither specimen nor drawing. Wilkinson[7] says only that it has been found in the tombs.

No lupin is grown in India, nor is there any Sanskrit name; its seeds are sold in bazaars under the name *tourmus* (Royle, *Ill.*, p. 194).

The Arabic name, *termis* or *termus*, is also that of the Greek lupin, *termos*. It may be inferred that the Greeks had it from the Egyptians. As the species was known to the ancient Egyptians, it seems strange that it has no Hebrew name;[8] but it may have been introduced into Egypt after the departure of the Israelites.

Field-Pea—*Pisum arvense*, Linnæus.

This pea is grown on a large scale for the seed, and also sometimes for fodder. Although its appearance and botanical characters allow of its being easily distinguished from the garden-pea, Greek and Roman authors confounded them, or are not explicit about them. Their writings do not prove that it was cultivated in their time. It has not been found in the lake-dwellings of

---

[1] Caruel, *Fl. Tosc.*, p. 136.
[2] Gussone, *Fl. Sic. Syn.*, ii. p. 267; Moris, *Fl. Sardoa*, i. p. 596.
[3] Boissier, *Fl. Orient.*, ii. p. 29.   [4] *Aufzählung*, etc., p. 257.
[5] Schweinfurth, *Plantæ Nilot. a Hartman Coll.*, p. 6.
[6] Unger, *Pflanzen d. Alt. Ægyp.*, p. 65.
[7] Wilkinson, *Manners and Customs of the Ancient Egyptians*, ii. p. 403.
[8] Rosenmüller, *Bibl. Alterth.*, vol. i.

Switzerland, France, and Italy.   Bobbio has a legend
(A.D. 930), in which it is said that the Italian peasants
called a certain seed *herbilia*, whence it has been supposed to be the modern *rubiglia* or the *Pisum sativum* of
botanists.[1]   The species is cultivated in the East, and as
far as the north of India.[2]   It is of recent cultivation in
the latter country, for there is no Sanskrit name, and
Piddington gives only one name in one of the modern
languages.

Whatever may be the date of the introduction of its
culture, the species is undoubtedly wild in Italy, not only
in hedges and near cultivated ground, but also in forests
and wild mountainous districts.[3]   I find no positive
indication in the floras that it grows in like manner
in Spain, Algeria, Greece, and the East.   The plant is
said to be indigenous in the south of Russia, but sometimes its wild character is doubtful, and sometimes the
species itself is not certain, from a confusion with *Pisum
sativum* and *P. elatius*.   Of all Anglo-Indian botanists,
only Royle admits it to be indigenous in the north of
India.

Garden-Pea—*Pisum sativum,* Linnæus.

The pea of our kitchen gardens is more delicate than
the field-pea, and suffers from frost and drought.   Its
natural area, previous to cultivation, was probably more
to the south and more restricted.   It has not hitherto
been found wild, either in Europe or in the west of Asia,
whence it is supposed to have come.   Bieberstein's indication of the species in the Crimea is not correct, according
to Steven, who was a resident in the country.[4]   Perhaps
botanists have overlooked its habitation; perhaps the
plant has disappeared from its original dwelling; perhaps
also it is a mere modification, effected by culture, of
Pisum arvense.   Alefeld held the latter opinion,[5] but he

---

[1] Muratori, *Antich. Ital.*, i. p. 347; *Diss.*, 24, quoted by Targioni, *Cenni Storici*, p. 31.

[2] Boissier, *Fl. Orient.*, ii. p. 623 ; Royle, *Ill. Himal.*, p. 200.

[3] Bertoloni, *Fl. Ital.*, vii. p. 419; Caruel, *Fl. Tosc.*, p. 184; Gussone, *Fl. Sic. Synopsis*, ii. p. 279 ; Moris, *Fl. Sardoa*, i. p. 577.

[4] Steven, *Verzeichniss*, p. 134.

[5] Alefeld, *Bot. Zeitung.*, 1860, p. 204.

has published too little on the subject for us to be able
to conclude anything from it. He only says that, having
cultivated a great number of varieties both of the field
and garden pea, he concludes that they belong to the
same species. Darwin [1] learnt through a third person
that Andrew Knight had crossed the field-pea with a
garden variety known as the Prussian pea, and that the
product was fertile. This would certainly be a proof
of specific unity, but further observation and experiment is required. In the mean time, in the search for
geographic origin, etc., I am obliged to consider the two
forms separately.

Botanists who distinguish many species in the genus
Pisum, admit eight, all European or Asiatic. *Pisum
sativum* was cultivated by the Greeks in the time of
Theophrastus.[2] They called it *pisos*, or *pison*. The
Albanians, descendants of the Pelasgians, call it *pizelle*.[3]
The Latins had *pisum*.[4] This uniformity of nomenclature seems to show that the Aryans knew the plant
when they arrived in Greece and Italy, and perhaps
brought it with them. Other Aryan languages have
several names for the generic sense of *pea ;* but it is
evident, from Adolphe Pictet's learned discussion on the
subject,[5] that none of these names can be applied to
Pisum sativum in particular. Even when one of the
modern languages, Slav or Breton, limits the sense to the
garden-pea, it is very probable that formerly the word
signified field-pea, lentil, or any other leguminous plant.

The garden-pea [6] has been found among the remains
in the lake-dwellings of the age of bronze, in Switzerland
and Savoy. The seed is spherical, wherein it differs from
Pisum arvense. It is smaller than our modern pea.
Heer says he found it also among relics of the stone age,

---

[1] Darwin, *Animals and Plants under Domestication*, p. 326.

[2] Theophrastus, *Hist.*, lib. viii. c. 3 and 5.

[3] Heldreich, *Nutzpflanzen Griechenlands*, p. 71.

[4] Pliny, *Hist.*, lib. xviii. c. 7 and 12. This is certainly *P. sativum*,
for the author says it cannot bear the cold.

[5] Ad. Pictet, *Origines Indo-Européennes*, edit. 2, vol. i. p. 359.

[6] Heer, *Pflanzen der Pfahlbaüten*, xxiii. fig. 48; Perrin, *Études Préhistoriques sur la Savoie*, p. 22.

at Moosseedorf; but he is less positive, and only gives figures of the less ancient pea of St. Peter's Island. If the species dates from the stone age in Switzerland, it would be anterior to the immigration of the Aryans.

There is no indication of the culture of *Pisum sativum* in ancient Egypt or in India. On the other hand, it has long been cultivated in the north of India, if it had, as Piddington says, a Sanskrit name, *harenso*, and if it has several names very different to this in modern Indian languages.[1] It has been introduced into China from Western Asia. The *Pent-sao*, drawn up at the end of the sixteenth century, calls it the Mahometan pea.[2] In conclusion: the species seems to have existed in Western Asia, perhaps from the south of the Caucasus to Persia, before it was cultivated. The Aryans introduced it into Europe, but it perhaps existed in Northern India before the arrival of the eastern Aryans. It no longer exists in a wild state, and when it occurs in fields, half-wild, it is not said to have a modified form so as to approach some other species.

Soy—*Dolichos soja*, Linnæus ; *Glycine soja*, Bentham.

This leguminous annual has been cultivated in China and Japan from remote antiquity. This might be gathered from the many uses of the soy bean and from the immense number of varieties. But it is also supposed to be one of the farinaceous substances called *shu* in Chinese writings of Confucius' time, though the modern name of the plant is *ta-tou*.[3] The bean is nourishing, and contains a large proportion of oil, and preparations similar to butter, oil, and cheese are extracted from it and used in Chinese and Japanese cooking.[4] Soy is also grown in the Malay Archipelago, but at the end of the eighteenth century it was still rare in Amboyna,[5] and Forster did not see it in the Pacific Isles at the time of Cook's voyages. It is of modern introduction in India,

---

[1] Piddington, *Index*. Roxburgh does not give a Sanskrit name.
[2] Bretschneider, *Study and Value*, etc., p. 16.
[3] *Ibid.*, p. 9.
[4] See Pailleux, in *Bull. de la Soc. d'Acclim.*, Sept. and Oct., 1880.
[5] Rumphius, *Amb.*, vol. v. p. 388.

for Roxburgh had only seen the plant in the botanical gardens at Calcutta, where it was brought from the Moluccas.[1] There are no common Indian names.[2] Besides, if its cultivation had been ancient in India, it would have spread westward into Syria and Egypt, which is not the case.

Kæmpfer[3] formerly published an excellent illustration of the soy bean, and it had existed for a century in European botanical gardens, when more extensive information about China and Japan excited about ten years ago a lively desire to introduce it into our countries. In Austria, Hungary, and France especially, attempts have been made on a large scale, of which the results have been summed up in works worthy of consultation.[4] It is to be hoped these efforts may be successful; but we must not digress from the aim of our researches, the probable origin of the species.

Linnæus says, in his *Species*, "habitat in India," and refers to Kæmpfer, who speaks of the plant in Japan, and to his own flora of Ceylon, where he gives the plant as *cultivated*. Thwaites's modern flora of Ceylon makes no mention of it. We must evidently go further east to find the origin both of the species and of its cultivation. Loureiro says that it grows in Cochin-China and that it is often cultivated in China.[5] I find no proof that it is wild in the latter country, but it may perhaps be discovered, as its culture is so ancient. Russian botanists[6] have only found it cultivated in the north of China and in the basin of the river Amur. It is certainly wild in Japan.[7] Junghuhn[8] found it in Java on Mount Gunung-Gamping, and a plant sent also from Java by Zollinger is supposed to belong to this species, but it is not certain that the

[1] Roxburgh, *Fl. Ind.*, iii. p. 314.   [2] Piddington, *Index*.

[3] Kaempfer, *Amer. Exot.*, p. 837, pl. 838.

[4] Haberlandt, *Die Sojabohne*, in 8vo, Vienna, 1878, quoted by Pailleux, *ubi supra*.

[5] Loureiro, *Fl. Cochin.*, ii. p. 538.

[6] Bunge, *Enum. Plant. Chin.*, 118; Maximowicz, *Primit. Fl. Amur.*, p. 87.

[7] Miquel, *Prolusio*, in *Ann. Mus. Lugd. Bat.*, iii. p. 52; Franchet and Savatier, *Enum. Plant. Jap.*, i. p. 108.

[8] Junghuhn, *Plantæ Jungh.*, p. 255.

specimen was wild.[1]   A Malay name, *kadelee*,[2] quite
different to the Japanese and Chinese common names, is
in favour of its indigenous character in Java.

Known facts and historical and philological probabilities
tend to show that the species was wild from Cochin-China
to the south of Japan and to Java when the ancient
inhabitants of this region began to cultivate it at a very
remote period, to use it for food in various ways, and to
obtain from it varieties of which the number is remarkable, especially in Japan.

Pigeon-Pea — *Cajanus indicus*, Sprengel; *Cytisus
Cajan*, Linnæus.

This leguminous plant, often grown in tropical countries, is a shrub, but it fruits in the first year, and in
some countries it is grown as an annual.   Its seed is an
important article of the food of the negroes and natives,
but the European colonists do not care for it unless
cooked green like our garden-pea.   The plant is easily
naturalized in poor soil round cultivated plots, even in
the West India Islands, where it is not indigenous.[3]

In Mauritius it is called *ambrevade ;* in the English
colonies, *doll, pigeon-pea*; and in the French Antilles,
pois d'Angola, pois de Congo, pois pigeon.

It is remarkable that, though the species is diffused in
three continents, the varieties are not numerous.   Two
are cited, based only upon the yellow or reddish colour
of the flower, which were formerly regarded as distinct
species; but a more attentive examination has resulted in
their being classed as one, in accordance with Linnæus'
opinion.[4]   The small number of variations obtained even
in the organ for which the species is cultivated is a sign
of no very ancient culture.   Its habitation previous to
culture is uncertain.   The best botanists have sometimes
supposed it to be a native of India, sometimes of tropical

---

[1] *Soja angustifolia*, Miquel; see Hooker, *Fl. Brit. Ind.*, ii. p. 184.

[2] Rumphius, *Amb.*, vol. v. p. 388.

[3] Tussac, *Flore des Antilles*, vol. iv. p. 94, pl. 32; Grisebach, *Fl. of Brit. W. Indies*, i. p. 191.

[4] See Wight and Arnott, *Prod. Fl. Penins. Ind.*, p. 256 ; Klotzsch, in Peters, *Reise nach Mozambique*, i. p. 36.   The yellow variety is figured in Tussac, that with the red flowers in the *Botanical Register*, 1845, pl. 31.

Africa. Bentham, who has made a careful study of the leguminous plants, believed in 1861 in the African origin; in 1865 he inclined rather to Asia.[1] The problem is, therefore, an interesting one. There is no question of an American origin. The cajan was introduced into the West Indies from the coast of Africa by the slave trade, as the common names quoted above show,[2] and the unanimous opinion of authors or American floras. It has also been taken to Brazil, Guiana, and into all the warm parts of the American continent.

The facility with which the species is naturalized would alone prevent attaching great importance to the statements of collectors, who have found it more or less wild in Asia or in Africa; and besides, these assertions are not precise, but are usually doubtful. Most writers on the flora of continental India have only seen the plant cultivated,[3] and none, to my knowledge, affirms that it exists wild. For the island of Ceylon Thwaites says,[4] "It is said not to be really wild, and the country names seem to confirm this." Sir Joseph Hooker, in his *Flora of British India*, says, "Wild (?) and cultivated to the height of six thousand feet in the Himalayas." Loureiro[5] gives it as cultivated and non-cultivated in China and Cochin-China. Chinese authors do not appear to have spoken of it, for the species is not named by Bretschneider in his work *On the Study*, etc. In the Sunda Isles it is mentioned as cultivated, and that rarely, at Amboyna at the end of the eighteenth century, according to Rumphius.[6] Forster had not seen it in the Pacific Isles at the time of Cook's voyages, but Seemann says that it has been recently introduced by missionaries into the Fiji Isles.[7] All this argues no very ancient extension of cultivation to the east and south of the continent of Asia. Besides the quotation from Loureiro, I find the species

---

[1] Bentham, *Flora Hongkongensis*, p. 89 ; *Flora Brasil.*, vol. xv. p. 199; Bentham and Hooker, i. p. 541.

[2] Tussac, *Flore des Antilles ;* Jacquin, *Obs.*, p. 1.

[3] Rheede, Roxburgh, Kurz, *Burm. Fl.*, etc.

[4] Thwaites, *Enum. Pl. Ceylan.*   [5] Loureiro, *Fl. Cochin.*, p. 565.

[6] Rumphius, *Amb.*, vol. v. t. 135.

[7] Seemann, *Fl. Vitiensis*, p. 74.

indicated on the mountain of Magelang, Java;[1] but, supposing this to be a true and ancient wild growth in both
cases, it would be very extraordinary not to find the
species in many other Asiatic localities.

The abundance of Indian and Malay names[2] shows
a somewhat ancient cultivation. Piddington even gives
a Sanskrit name, *arhuku*, which was not known to Roxburgh, but he gives no proof in support of his assertion.
The name may have been merely supposed from the
Hindu and Bengali names *urur* and *orol*. No Semitic
name is known.

In Africa the cajan is often found from Zanzibar to
the coast of Guinea.[3] Authors say it is cultivated, or
else make no statement on this head, which would seem
to show that the specimens are sometimes wild. In
Egypt this cultivation is quite modern, of the nineteenth
century.[4]

Briefly, then, I doubt that the species is really wild
in Asia, and that it has been grown there for more than
three thousand years. If more ancient peoples had known
it, it would have come to the knowledge of the Arabs and
Egyptians before our time. In tropical Africa, on the
contrary, it is possible that it has existed wild or cultivated for a very long time, and that it was introduced
into Asia by ancient travellers trading between Zanzibar
and India or Ceylon.

The genus Cajanus has only one species, so that no
analogy of geographical distribution leads us to believe it
to be rather of Asiatic than African origin, or *vice versâ*.

Carob Tree[5]—*Ceratonia siliqua*, Linnæus.

The seeds and pods of the carob are highly prized in
the hotter parts of the Mediterranean basin, as food for
animals and even for man. De Gasparin[6] has given in-

---

[1] Junghuhn, *Plantœ Jungh.*, fasc. i. p. 241.
[2] Piddington, *Index ;* Rheede, *Malab.*, vi. p. 23, etc.
[3] Pickering, *Chron. Arrang. of Plants*, p. 442; Peters, *Reise*, p. 33;
R. Brown, *Bot. of Congo*, p. 53; Oliver, *Fl. of Trop. Afr.*, ii. p. 216.
[4] *Bulletin de la Société d'Acclimation*, 1871, p. 663.
[5] The species is given here in order not to separate it from the other
leguminous plants cultivated for the seeds alone.
[6] De Gasparin, *Cours. d'Agric.*, iv. p. 328.

teresting details about the raising, uses, and habitation of
the species as a cultivated tree.   He notes that it does
not pass the northern limit beyond which the orange
cannot be grown without shelter.   This fine evergreen
tree does not thrive either in very hot countries, especially
where there is much humidity.   It likes the neighbourhood of the sea and rocky places.   Its original country,
according to Gasparin, is "probably the centre of Africa.
Denham and Clapperton found it in Burnou."   This
proof seems to me insufficient, for in all the Nile Valley
and in Abyssinia the carob is not wild nor even cultivated.[1]   R. Brown does not mention it in his account of
Denham and Clapperton's journey.   Travellers have seen
it in the forests of Cyrenaica between the high-lands
and the littoral ; but the able botanists who have drawn
up the catalogue of the plants of this country are careful
to say,[2] "perhaps indigenous."   Most botanists merely
mention the species in the centre and south of the Mediterranean basin, from Spain and Marocco to Syria and
Anatolia, without inquiring closely whether it is indigenous or cultivated, and without entering upon the
question of its true country previous to cultivation.
Usually they indicate the carob tree, as "cultivated and
subspontaneous, or nearly wild."   However, it is stated to
be wild in Greece by Heldreich, in Sicily by Gussone and
Bianca, in Algeria by Munby ;[3] and these authors have
each lived long enough in the country for which each is
quoted to form an enlightened opinion.

Bianca remarks, however, that the carob tree is not
always healthy and productive in those restricted localities
where it exists in Sicily, in the small adjacent islands,
and on the coast of Italy.   He puts forward the opinion,
moreover, based upon the similarity of the Italian name
carrubo with the Arabic word, that the species was

[1] Schweinfurth and Ascherson, *Aufzählung,* p. 255 ; Richard, *Tentamen
Fl. Abyss.*

[2] Ascherson, etc., in Rohls, *Kufra,* 1 vol. in 8vo, 1881, p. 519.

[3] Heldreich, *Nutzpflanzen Griechenlands,* p. 73 ; *Die Pflanzen der
Attischen Ebene,* p. 477 ; Gussone, *Syn. Fl. Sic.,* p. 646 ; Bianca, *Il Carrubo,*
in the *Giornale d'Agricoltura Italiana,* 1881 ; Munby, *Catal. Pl. in Alg.
Spont.,* p. 13.

anciently introduced into the south of Europe, the species being of Syrian or north African origin. He maintains as probable the theory of Hœfer and Bonné,[1] that the lotus of the lotophagi was the carob tree, of which the flower is sweet and the fruit has a taste of honey, which agrees with the expressions of Homer. The lotus-eaters dwelt in Cyrenaica, so that the carob must have been abundant in their country. If we admit this hypothesis we must suppose that Pliny and Herodotus did not know Homer's plant, for the one describes the lotos as bearing a fruit like a mastic berry (*Pistacia lentiscus*), the other as a deciduous tree.[2]

An hypothesis regarding a doubtful plant formerly mentioned by a poet can hardly serve as the basis of an argument upon facts of natural history. After all, Homer's lotus plant perhaps existed only in the fabled garden of Hesperides. I return to more serious arguments, on which Bianca has said a few words.

The carob has two names in ancient languages—the one Greek, *keraunia* or *kerateia*;[3] the other Arabic, *chirnub* or *charûb*. The first alludes to the form of the pod, which is like a slightly curved horn; the other means merely pod, for we find in Ebn Baithar's[4] work that four other leguminous plants bear the same name, with a qualifying epithet. The Latins had no special name; they used the Greek word, or the expression *siliqua, siliqua græca* (Greek pod).[5] This dearth of names is the sign of a once restricted area, and of a culture which probably does not date from prehistoric time. The Greek name is still retained in Greece. The Arab name persists among the Kabyles, who call the fruit *kharroub*, the tree *takharrout*,[6] and the Spaniards *algarrobo*. Curiously enough,

[1] Hœfer, *Hist. Bot. Minér. et Géol.*, 1 vol. in 12mo, p. 20; Bonné, *Le Caroubier, ou l'Arbre des Lotophages*, Algiers, 1869 (quoted by Hœfer). See above, the article on the jujube tree.

[2] Pliny, *Hist.*, lib. i. cap. 30.

[3] Theophrastus, *Hist. Plant.*, lib. i. cap. 11; Dioscorides, lib. i. cap. 155; Fraas, *Syn. Fl. Class.*, p. 65.

[4] Ebn Baithar, German trans., i. p. 354; Forskal, *Fl. Ægypt.*, p. 77.

[5] Columna, quoted by Lenz, *Bot. der Alten*, p. 73; Pliny, *Hist.*, lib. xiii. cap. 8.

[6] *Dict. Franç.-Berbère*, at the word *Caroube.*

the Italians also took the Arab name *currabo, carubio,*
whence the French *caroubier.* It seems that it must
have been introduced after the Roman epoch by the
Arabs of the Middle Ages, when there was another name
for it. These details are all in favour of Bianca's
theory of a more southern origin than Sicily. Pliny
says the species belonged to Syria, Ionia, Cnidos, and
Rhodes, but he does not say whether it was wild or
cultivated in these places. Pliny also says that the
carob tree did not exist in Egypt. Yet it has been
recognized in monuments belonging to a much earlier
epoch than that of Pliny, and Egyptologists even
attribute two Egyptian names to it, *kontrates* or *jiri.*[1]
Lepsius gives a drawing of a pod which appears to
him to be certainly a carob, and the botanist Kotschy
made certain by microscopic investigation that a stick
taken from a sarcophagus was made from the wood of
the carob tree.[2] There is no known Hebrew name for
the species, which is not mentioned in the Old Testament.
The New Testament speaks of it by the Greek name in
the parable of the prodigal son. It is a tradition of the
Christians in the East that St. John Baptist fed upon
the fruit of the carob in the desert, and hence came
the names given to it in the Middle Ages—*bread of
St. John,* and *Johannis brodbaum.*

Evidently this tree became important at the beginning
of the Christian era, and it spread, especially through
the agency of the Arabs, towards the West. If it had
previously existed in Algeria, among the Berbers, and in
Spain, older names would have persisted, and the species
would probably have been introduced into the Canaries
by the Phœnicians.

The information gained on the subject may be
summed up as follows :—

The carob grew wild in the Levant, probably on the
southern coast of Anatolia and in Syria, perhaps also in

---

[1] *Lexicon Oxon.,* quoted by Pickering, *Chron. Hist. of Plants,* p. 141.
[2] The drawing is reproduced in Unger's *Pflanzen des Alten Ægyptens,*
fig. 22. The observation which he quotes from Kotschy needs confirmation by a special anatomist.

Cyrenaica. Its cultivation began within historic time. The Greeks diffused it in Greece and Italy; but it was afterwards more highly esteemed by the Arabs, who propagated it as far as Marocco and Spain. In all these countries the tree has become naturalized here and there in a less productive form, which it is needful to graft to obtain good fruit.

The carob has not been found in the tufa and quater-nary deposits of Southern Europe. It is the only one of its kind in the genus *Ceratonia*, which is somewhat exceptional among the *Leguminosœ*, especially in Europe. Nothing shows that it existed in the ancient tertiary or quaternary flora of the south-west of Europe.

Common Haricot Kidney Bean—*Phaseolus vulgaris,* Savi.

When, in 1855, I wished to investigate the origin of the genera *Phaseolus* and *Dolichos*,[1] the distinction of species was so little defined, and the floras of tropical countries so rare, that I was obliged to leave several questions on one side. Now, thanks to the works of Bentham and Georg von Martens,[2] completing the previous labours of Savi,[3] the *Leguminœ* of hot countries are better known; lastly, the seeds discovered quite recently in the Peruvian tombs of Ancon, examined by Wittmack, have completely modified the question of origin.

I will speak first of the common haricot bean, afterwards of some other species, without, however, enumerating all those which are cultivated, for several of these are still ill defined.

Botanists held for a long time that the common haricot was of Indian origin. No one had found it wild, nor has it yet been found, but it was supposed to be of Indian origin, although the species was also cultivated in Africa and America, in temperate and hot regions, at least in those where the heat and humidity are not excessive. I called attention to the fact that there is

---

[1] A. de Candolle, *Géogr. Bot. Rais.*, p. 961.

[2] Bentham, in *Ann. Wiener Museum*, vol. ii.; Martens, *Die Gartenbohnen*, in 4to, Stuttgart, 1860, edit. 2, 1869.

[3] Savi, *Osserv. sopra Phaseolus e Dolichos*, 1, 2, 3.

no Sanskrit name, and that sixteenth-century gardeners often called the species *Turkish bean*. Convinced, moreover, that the Greeks cultivated this plant under the names *fasiolos* and *dolichos*, I suggested that it came originally from Western Asia, and not from India. Georg von Martens adopted this hypothesis.

However, the meaning of the words *dolichos* of Theophrastus, *fasiolos* of Dioscorides, *faseolus* and *phaseolus* of the Romans,[1] is far from being sufficiently defined to allow them to be attributed with certainty to *Phaseolus vulgaris*. Several cultivated *Leguminosæ* are supported by the trellises mentioned by authors, and have pods and seeds of a similar kind. The best argument for translating these names by *Phaseolus vulgaris* is that the modern Greeks and Italians have names derived from *fasiolus* for the common haricot. In modern Greek it is *fasoulia*, in Albanian (Pelasgic?) *fasulé*, in Italian *fagiolo*. It is possible, however, that the name has been transferred from a species of pea or vetch, or from a haricot formerly cultivated, to our modern haricot. It is rather bold to determine a species of *Phaseolus* from one or two epithets in an ancient author, when we see how difficult is the distinction of species to modern botanists with the plants under their eyes. Nevertheless, the *dolichos* of Theophrastus has been definitely referred to the *scarlet runner*, and the *fasiolos* to the dwarf haricot of our gardens, which are the two principal modern varieties of the common haricot, with an immense number of sub-varieties in the form of the pods and seed. I can only say it may be so.

If the common haricot was formerly known in Greece, it was not one of the earliest introductions, for the *faseolos* did not exist at Rome in Cato's time, and it is only at the beginning of the empire that Latin authors speak of it. Virchow brought from the excavations at Troy the seeds of several leguminæ, which Wittmack[2]

---

[1] Theophrastus, *Hist.*, lib. viii. cap. 3; Dioscorides, lib. ii. cap. 130; Pliny, *Hist.*, lib. xviii. cap. 7, 12, interpreted by Fraas, *Syn. Fl. Class.*, p. 52; Lenz, *Bot. der Alten*, p. 731; Martens, *Die Gartenbohnen*, p. 1.

[2] Wittmack, *Bot. Vereins Brandenburg*, Dec. 19, 1879.

has ascertained to belong to the following species: broad bean (*Faba vulgaris*), garden-pea (*Pisum sativum*), ervilla (*Ervum ervilia*), and perhaps the flat-podded vetchling (*Lathyrus Cicera*), but no haricot. Nor has the species been found in the lake-dwellings of Switzerland, Savoy, Austria, and Italy.

There are no proofs or signs of its existence in ancient Egypt. No Hebrew name is known answering to the *Phaseolus* or *Dolichos* of botanists. A less ancient name, for it is Arabic, *loubia*, exists in Egypt for *Dolichos lubia*, and in Hindustani as *loba* for *Phaseolus vulgaris.*[1] As regards the latter species, Piddington only gives two names in modern languages, and those both Hindustani, *loba* and *bakla*. This, together with the absence of a Sanskrit name, points to a recent introduction into Southern Asia. Chinese authors do not mention *P. vulgaris,*[2] which is a further indication of a recent introduction into India, and also into Bactriana, whence the Chinese have imported plants from the second century of our era.

All these circumstances incline me to doubt whether the species was known in Asia before the Christian era. The argument based upon the modern Greek and Italian names for the haricot, derived from *fasiolos*, needs some support. It may be said in its favour that it was used in the Middle Ages, probably for the common haricot. In the list of vegetables which Charlemagne commanded to be sown in his farms, we find *fasiolum*,[3] without explanation. Albertus Magnus describes under the name *faseolus* a leguminous plant which appears to be our dwarf haricot.[4] I notice, on the other hand, that writers

---

[1] Delile, *Plantes Cultivées en Égypte*, p. 14 ; Piddington, *Index*.

[2] Bretschneider does not mention any, either in his pamphlet *On the Study and Value of Chinese Botanical Works*, or in his private letters to me.

[3] E. Meyer, *Geschichte der Botanique*, iii. p. 404.

[4] " *Faseolus est species leguminis et grani, quod est in quantitate parum minus quam Faba, et in figura est columnare sicut faba, herbaque ejus minor est aliquantulum quam herba Fabæ. Et sunt faseoli multorum colorum, sed quodlibet granorum habet maculam nigram in loco cotyledonis*" (Jessen, *Alberti Magni, De Vegetabilibus*, edit. critica, p. 515).

in the fifteenth century, such as Pierre Crescenzio [1] and Macer Floridus,[2] mention no *faseolus* or similar name. On the other hand, after the discovery of America, from the sixteenth century all authors publish descriptions and drawings of *Phaseolus vulgaris*, with a number of varieties.

It is doubtful that its cultivation is ancient in tropical Africa. It is indicated there less often than that of other species of the Dolichos and Phaseolus genera.

It had not occurred to any one to seek the origin of the haricot in America till, quite recently, some remarkable discoveries of fruits and seeds were made in Peruvian tombs at Ancon, near Lima. Rochebrune [3] published a list of the species of different families from the collection made by Cossac and Savatier. Among the number are three kinds of haricot, none of which, says the author, is *Phaseolus vulgaris*; but Wittmack,[4] who studied the leguminæ brought from these same tombs by Reiss and Stubel, says he made out several varieties of the common haricot among other seeds belonging to *Phaseolus lunatus*, Linnæus. He had identified them with the varieties of *P. vulgaris* called by botanists *Oblongus purpureus* (Martens), *Ellipticus præcox* (Alefeld), and *Ellipticus atrofuscus* (Alefeld), which belong to the category of dwarf or branchless haricots.

It is not certain that the tombs in question are all anterior to the advent of the Spaniards. The work of Reiss and Stubel, now in the press, will perhaps give some information on this head; but Wittmack admits, on their authority, that some of the tombs are not ancient. I notice a fact, however, which has passed without observation. The fifty species of Rochebrune are all American. There is not one which can be suspected to be of European origin. Evidently these plants and seeds

---

[1] P. Crescens, French trans., 1539.

[2] Macer Floridus, edit. 1485, and Choulant's commentary, 1832.

[3] De Rochebrune, *Actes de la Soc. Linn. de Bordeaux*, vol. xxxiii. Jan., 1880, of which I saw an analysis in *Botanisches Centralblatt*, 1880, p. 1633.

[4] Wittmack, *Sitzungsbericht des Bot. Vereins Brandenburg*, Dec. 19, 1879, and a private letter.

were either deposited before the conquest, or, in certain tombs which perhaps belong to a subsequent epoch, the inhabitants took care not to put species of foreign origin. This was natural enough according to their ideas, for the custom of depositing plants in the tombs was not a result of the Catholic religion, but was an inheritance from the customs and opinions of the natives. The presence of the common haricot among exclusively American plants seems to me important, whatever the date of the tombs.

It may be objected that the seeds are insufficient ground for determining the species of a *phaseolus*, and that several species of this genus which are not yet well known were cultivated in South America before the arrival of the Spaniards. Molina [1] speaks of thirteen or fourteen species (or varieties ?) cultivated formerly in Chili alone.

Wittmack insists upon the general and ancient use of the haricot in several parts of South America. This proves at least that several species were indigenous and cultivated. He quotes the testimony of Joseph Acosta, one of the first writers after the conquest, who says that "the Peruvians cultivated vegetables which they called *frisoles* and *palares*, and which they used as the Spaniards use *garbanzos* (chick-pea), beans and lentils. I have not found," he adds, "that these or other European vegetables were found here before the coming of the Europeans." *Frisole, fajol, fasoler*, are Spanish names for the common haricot, corruptions of the Latin *faselus, fasolus, faseolus. Paller* is American.

I may take this opportunity of explaining the origin of the French name haricot. I sought for it formerly in vain; [2] but I noticed that Tournefort [3] (*Instit.*, p. 415) was the first to use it. I called attention also to the existence of the word *arachos* (ἀραχος) in Theophrastus, probably for a kind of vetch, and of the Sanskrit word

---

[1] Molina (*Essai sur l'Hist. Nat. du Chili*, French trans., p. 101) mentions *Phaseoli*, which he calls *pallar* and *asellus*, and Cl. Gay's *Fl. du Chili* adds, without much explanation, *Ph. Cumingii*, Bentham.

[2] A. de Candolle, *Géog. Bot. Rais.*, p. 691.

[3] Tournefort, *Eléments* (1694), i. p. 328; *Instit.*, p. 415.

harenso for the common pea. I rejected as improbable
the notion that the name of a vegetable could come from
the dish called haricot or laricot of mutton, as suggested
by an English author, and criticized Bescherelle, who
derived the word from Keltic, while the Breton words are
totally different, and signify small bean (*fa-munno*) or
kind of pea (*pis-ram*). Lettré, in his dictionary, also seeks
the etymology of the word. Without any acquaintance
with my article, he inclines to the theory that *haricot,* the
plant, comes from the ragout, seeing that the latter is
older in the language, and that a certain resemblance
may be traced between the haricot bean and the morsels
of meat in the ragout, or else that this bean was suitable
to the making of the dish. It is certain that this
vegetable was called in French *faséole* or *fazéole,* from the
Latin name, until nearly the end of the seventeenth
century; but chance has led me to discover the real
origin of the word haricot. An Italian name, *araco,*
found in Durante and Matthioli, in Latin *Aracus niger,*[1]
was given to a leguminous plant which modern botanists
attribute to *Lathyrus ochrus.* It is not surprising that
an Italian seventeenth-century name should be transported by French cultivators of the following century to
another leguminous plant, and that *ara* should have been
ari. It is the sort of mistake which is common now.
Besides, *aracos* or *arachos* has been attributed by commentators to several *Leguminosæ* of the genera *Lathyrus,*
Vicia, etc. Durante gives the Greek *arachos* as the
synonym for his *araco,* whereby we see the etymology.
Père Feuillée [2] wrote in French *aricot;* before him Tournefort spelt it *haricot,* in the belief, perhaps, that the
Greek word was written with an aspirate, which is not
the case, at least in the best authors.

I may sum up as follows :—(1) *Phaseolus vulgaris* has
not been long cultivated in India, the south-west of Asia,
and Egypt ; (2) it is not certain that it was known in
Europe before the discovery of America; (3) at this epoch

---

[1] Durante, *Herbario Nuovo,* 1585, p. 39 ; Matthioli ed Valgris, p. 322 ;
Targioni, *Dizion. Bot. Ital.,* i. p. 13.
[2] Feuillée, *Hist. des Plan. Medic. du Pérou,* etc., in 4to, 1725, p. 54.

the number of varieties suddenly increased in European gardens, and all authors commenced to mention them; (4) the majority of the species of the genus exist in South America; (5) seeds apparently belonging to the species have been discovered in Peruvian tombs of an uncertain date, intermixed with many species, all American.

I do not examine whether *Phaseolus vulgaris* existed in both hemispheres previous to cultivation, because examples of this nature are exceedingly rare among non-aquatic phanerogamous plants of tropical countries. Perhaps there is not one in a thousand, and even then human agency may be suspected.[1] To open this question in the case of *Ph. vulgaris*, it should at least be found wild in both old and new worlds, which has not happened. If it had occupied so vast an area, we should see signs of it in individuals really wild in widely separate regions on the same continent, as is the case with the following species, *Ph. lunatus*.

Scimetar-podded Kidney Bean, or Sugar Bean.—*Phaseolus lunatus*, Linnæus; *Phaseolus lunatus macrocarpus*; Bentham, *Ph. inamœnus*, Linnæus.

This haricot, as well as that called *Lima*, is so widely diffused in tropical countries, that it has been described under different names.[2] All these forms can be classed in two groups, of which Linnæus made different species. The commonest in our gardens is that which has been called since the beginning of the century the *Lima haricot*. It may be distinguished by its height, by the size of its pods and beans. It lasts several years in countries which are favourable to it.

Linnæus believed that his *Ph. lunatus* came from Bengal and the other from Africa, but he gives no proof. For a century his assertions were repeated. Now, Bentham,[3] who is careful about origins, believes the species and its variety to be certainly American; he only doubts about its presence as a wild plant both in Africa

---

[1] A. de Candolle, *Géogr. Bot. Rais.*, chapter on *disjunctive* species.

[2] *Ph. bipunctatus*, Jacquin; *Ph. inamœnus*, Linnæus; *Ph. puberulus*, Kunth; *Ph. saccharatus*, MacFadyen; etc., etc.

[3] Bentham, in *Fl. Brasil.*, vol. xv. p. 181.

and Asia. I see no indication whatever of ancient existence in Asia. The plant has never been found wild, and
it has no name in the modern languages of India or
in Sanskrit.[1] It is not mentioned in Chinese works.
Anglo-Indians call it French bean,[2] like the common
haricot, which shows how modern is its cultivation.

It is cultivated in nearly all tropical Africa. However, Schweinfurth and Ascherson[3] do not mention it
for Abyssinia, Nubia, or Egypt. Oliver[4] quotes a number
of specimens found in Guinea and the interior of Africa,
without saying whether they were wild or cultivated.
If we suppose the species of African origin or of very early
introduction, it would have spread to Egypt and thence
to India.

The facts are quite different for South America.
Bentham mentions wild specimens from the Amazon
basin and Central Brazil. They belong especially to the
large variety (*macrocarpus*), which abounds also in the
Peruvian tombs of Ancon, according to Wittmack.[5] It is
evidently a Brazilian species, diffused by cultivation, and
perhaps long since naturalized here and there in tropical
America. I am inclined to believe it was introduced into
Guinea by the slave trade, and that it spread thence
into the interior and the coast of Mozambique.

Moth, or **Aconite-leaved Kidney Bean** — *Phaseolus
aconitifolius*, Willdenow.

An annual species grown in India as fodder, and of
which the seeds are eatable, though but little valued.
The Hindustani name is *mout*, among the Sikhs *moth*. It
is somewhat like *Ph. trilobus*, which is cultivated for the
seed. *Ph. aconitifolius* is wild in British India from
Ceylon to the Himalayas.[6] The absence of a Sanskrit
name, and of different names in modern Indian languages,
points to a recent cultivation.

Three-lobed Kidney Bean — *Phaseolus trilobus*, Willdenow.

---

[1] Roxburgh, Piddington, etc.   [2] Royle, *Ill. Himalaya*, p. 190.
[3] *Aufäzhlung*, etc., p. 257.   [4] Oliver, *Fl. of Trop. Afr.*, p. 192.
[5] Wittmack, *Sitz. Bot. Vereins Branden.*, Dec. 19, 1879.
[6] Roxburgh, *Fl. Ind.* edit. 1832, vol. iii. p. 299; Aitchison, *Catal. of
Punjab*, p. 48; Sir J. Hooker, *Fl. of Brit. Ind.*, ii. p. 202.

One of the most commonly cultivated species in India;[1] at least in the last few years, for Roxburgh,[2] at the end of the eighteenth century, had only seen it wild. All authors agree in considering it as wild from the foot of the Himalayas to Ceylon. It also exists in Nubia, Abyssinia, and Zambesi;[3] it is not said whether wild or cultivated. Piddington gives a Sanskrit name, and several names in modern Indian languages, which shows that the species has been cultivated, or at least known for three thousand years.

Green Gram, or **Múng**—*Phaseolus mungo*, Linnæus.

A species commonly cultivated in India and in the Nile Valley. The considerable number of varieties, and the existence of three different names in the modern languages of India, point to a cultivation of one or two thousand years, but there is no Sanskrit name.[4] In Africa it is probably recent. Anglo-Indian botanists agree that it is wild in India.

Lablab, or **Wall**—*Dolichos Lablab*, Linnæus.

This species is much cultivated in India and tropical Africa. Roxburgh counts as many as seven varieties with Indian names. Piddington quotes in his *Index* a Sanskrit name, *schimbi*, which recurs in modern languages. Its culture dates perhaps from three thousand years. Yet the species was not anciently diffused in China, or in Western Asia and Egypt; at least, I can find no trace of it. The little extension of these edible *Leguminosæ* beyond India in ancient times is a singular fact. It is possible that their cultivation is not of ancient date.

The lablab is undoubtedly wild in India, and also, it is said, in Java.[5] It has become naturalized from cultivation in the Seychelles.[6] The indications of authors are not positive enough to say whether it is wild in Africa.[7]

---

[1] Sir J. Hooker, *Fl. of Brit. Ind.*, ii. p. 201.   [2] Roxburgh, *Fl. Ind.*, p. 299.
[3] Schweinfurth, *Beitr. z. Fl. Ethiop.*, p. 15; *Aufzählung*, p. 257; Oliver, *Fl. Trop. Afr.*, p. 194.
[4] See authors quoted for *P. tribolus.*
[5] Sir J. Hooker, *Fl. Brit. Ind.*, ii. p. 209; Junghuhn, *Plantæ Jungh.,* fasc ii. p. 240.
[6] Baker, *Fl. of Mauritius*, p. 83.
[7] Oliver, *Fl. of Trop. Africa*, ii. p. 210.

Lubia—*Dolichos Lubia,* Forskal.

This species, cultivated in Europe under the name of *lubia, loubya, loubyé,* according to Forskal and Delile,[1] is little known to botanists. According to the latter author it exists also in Syria, Persia, and India; but I do not find this in any way confirmed in modern works on these two countries. Schweinfurth and Ascherson[2] admit it as a distinct species, cultivated in the Nile Valley. Hitherto no one has found it wild. No *Dolichos* or *Phaseolus* is known in the monuments of ancient Egypt. We shall see from the evidence of the common names that these plants were probably introduced into Egyptian agriculture after the time of the Pharaohs.

The name *lubia* is used by the Berbers, unchanged, and by the Spaniards as *alubia* for the common haricot, *Phaseolus vulgaris.* Although *Phaseolus* and *Dolichos* are very similar, this is an example of the little value of common names as a proof of species. *Loba* is, as we have seen, one of the Hindustani names for *Phaseolus vulgaris,*[3] and *lobia* that of *Dolichos sinensis* in the same language.[4] Orientalists should tell us whether *lubia* is an old word in Semitic languages. I do not find a similar name in Hebrew, and it is possible that the Armenians or the Arabs took *lubia* from the Greek *lobos* (λοβος), which means any projection, like the lobe of the ear, a fruit of the nature of a pod, and more particularly, according to Galen, *Ph. vulgaris. Lobion* (λοβιον) in Dioscorides is the fruit of *Ph. vulgaris,* at least in the opinion of commentators.[5] It remains as *loubion* in modern Greek, with the same meaning.[6]

Bambarra Ground Nut—*Glycine subterranea,* Linnæus, junr.; *Voandzeia subterranea,* Petit Thouars.

---

[1] Forskal, *Descript.,* p. 133; Delile, *Plant. Cult. en Égypte,* p. 14.

[2] Schweinfurth and Ascherson, *Aufzählung,* p. 256.

[3] *Dict. Franç.-Berbère,* at the word *haricot;* Willkomm and Lange, *Prod. Fl. Hisp.,* iii. p. 324. The common haricot has no less than five different names in the Iberian peninsula.

[4] Piddington, *Index.*

[5] Lenz, *Bot. der Alt. Gr. und Röm.,* p. 732.

[6] Langkavel, *Bot. der Späteren Griechen,* p. 4; Heldreich, *Nutzpfl. Griechenl.,* p. 72.

The earliest travellers in Madagascar remarked this leguminous annual, cultivated by the natives for the pod or seed, dressed like peas, French beans, etc. It resembles the earth, particularly in that the flower-stem curves downwards, and plunges the young fruit or pod into the earth. Its cultivation is common in the gardens of tropical Africa, and it is found, but less frequently, in those of Southern Asia.[1] It seems that it is not much grown in America,[2] except in Brazil, where it is called *mandubi di Angola*.[3]

Early writers on Asia do not mention it; its origin must, therefore, be sought in Africa. Loureiro[4] had seen it on the eastern coast of this continent, and Petit Thouars in Madagascar, but they do not say that it was wild. The authors of the flora of Senegambia[5] described it as " cultivated and probably wild " in Galam. Lastly, Schweinfurth and Ascherson[6] found it wild on the banks of the Nile from Khartoum to Gondokoro. In spite of the possibility of naturalization from cultivation, it is extremely probable that the plant is wild in tropical Africa.

Buckwheat—*Polygonum fagopyrum*, Linnæus; *Fagopyrum esculentum*, Mœnch.

The history of this species has been completely cleared up in the last few years. It grows wild in Mantschuria, on the banks of the river Amur,[7] in Dahuria, and near Lake Baikal.[8] It is also indicated in China and in the mountains of the north of India,[9] but I do not find that in these regions its wild character is certain. Roxburgh

---

[1] Sir J. Hooker, *Flora of Brit. Ind.*, ii. p. 205; Miquel, *Fl. Indo-Batava*, i. p. 175.

[2] Linnæus, junr., *Decad.*, ii. pl. 19, seems to have confounded this plant with *Arachis*, and he gives, perhaps because of this error, *Voandzeia* as cultivated at his time in Surinam. Modern writers on America either have not seen it or have omitted to mention it.

[3] *Gardener's Chronicle*, Sept. 4, 1880.

[4] Loureiro, *Fl. Cochin.*, ii. p. 523.

[5] Guillemin, Perottet, Richard, *Fl. Senegambia Tentamen*, p. 254.

[6] *Aufzählung*, p. 259.

[7] Maximowicz, *Primitiæ Fl. Amur.*, p. 236.

[8] Ledebour, *Fl. Ross.*, iii. 517.

[9] Meissner, in De Candolle, *Prodr.*, xiv. p. 143.

has only seen it in a cultivated state in the north of India, and Bretschneider[1] thinks it doubtful that it is indigenous in China. Its cultivation is not ancient, for the first Chinese author who mentions it lived in the tenth or eleventh century of the Christian era.

Buckwheat is cultivated in the Himalayas under the names *ogal* or *ogla* and *kouton*.[2] As there is no Sanskrit name for this species nor for the two following, I doubt the antiquity of their cultivation in the mountains of Central Asia. It was certainly unknown to the Greeks and Romans. The name *fagopyrum* is an invention of modern botanists from the similarity in the shape of the seed to a beech-nut, whence also the German *buch-weitzen*[3] (corrupted in English into buckwheat) and the Italian *faggina*.

The names of this plant in European languages of Aryan origin have not a common root. Thus the western Aryans did not know the species any more than the Sanskrit-speaking Orientals, a further sign of the non-existence of the plant in the mountains of Central Asia. Even at the present day it is probably unknown in the north of Persia and in Turkey, since floras do not mention it.[4] Bosc states, in the *Dictionnaire d'Agriculture*, that Olivier had seen it wild in Persia, but I do not find this in this naturalist's published account of his travels.

The species came into Europe in the Middle Ages, through Tartary and Russia. The first mention of its cultivation in Germany occurs in a Mecklenburg register of 1436.[5] In the sixteenth century it spread towards the centre of Europe, and in poor soil, as in Brittany, it became important. Reynier, who, as a rule, is very accurate, imagined that the French name *sarrasin* was Keltic;[6] but M. le Gall wrote to me formerly that the Breton names simply mean black wheat or black corn, *ed-du*

---

[1] Bretschneider, *On Study*, etc., p. 9.
[2] Madden, *Trans. Edinburgh Bot. Soc.*, **v.** p. 118.
[3] The English name *buckwheat* and the French name of some localities, *buscail*, come from the German.
[4] Boissier, *Fl. Orient.*; Buhse and Boissier, *Pflanzen Transcaucasien.*
[5] Pritzel, *Sitzungsbericht Naturforsch. freunde zu Berlin*, May 15, 1866.
[6] Reynier, *Économie des Celtes*, p. 425.

and *gwinis-du*. There is no original name in Keltic languages, which seems natural now that we know the origin of the species.[1]

When the plant was introduced into Belgium and into France, and even when it became known in Italy, that is to say in the sixteenth century, the name *blé sarrasin* (Saracen wheat) or *sarrasin* was commonly adopted. Common names are often so absurd, and so unthinkingly bestowed, that we cannot tell in this particular case whether the name refers to the colour of the grain which was that attributed to the Saracens, or to the supposed introduction from the country of the Arabs or Moors. It was not then known that the species did not exist in the countries south of the Mediterranean, nor even in Syria and Persia. It is also possible that the idea of a southern origin was taken from the name *sarrasin*, which was given from the colour. This origin was admitted until the end of the last and even in the present century.[2] Reynier was, fifty years ago, the first to oppose it.

Buckwheat sometimes escapes from cultivation and becomes quasi-wild. The nearer we approach its original country the more often this occurs, whence it results that it is hard to define the limit of the wild plant on the confines of Europe and Asia, in the Himalayas, and in China. In Japan these semi-naturalizations are not rare.[3]

Tartary Buckwheat—*Polygonum tataricum*, Linnæus; *Fagopyrum tataricum*, Gærtner.

Less sensitive to cold than the common buckwheat, but yielding a poorer kind of seed, this species is sometimes cultivated in Europe and Asia—in the Himalayas,[4] for instance; but its culture is recent. Authors of the sixteenth and seventeenth centuries do not mention it, and Linnæus was one of the first to speak of it as of Tartar

---

[1] I have given the vernacular names at greater length in *Géogr. Bot. Rais.*, p. 953.

[2] Nemnich, *Polyglott. Lexicon*, p. 1030; Bosc, *Dict. d'Agric.*, xi. p. 379.

[3] Franchet and Savatier, *Enum. Pl. Japon.*, i. p. 403.

[4] Royle, *Ill. Himal.*, p. 317.

origin. Roxburgh and Hamilton had not seen it in Northern India in the beginning of this century, and I find no indication of it in China and Japan.

It is undoubtedly wild in Tartary and Siberia, as far as Dauria;[1] but Russian botanists have not found it further east, in the basin of the river Amur.[2]

As this plant came from Tartary into Eastern Europe later than the common buckwheat, it is the latter which bears in several Slav languages the names *tatrika, tatarka,* or *tattar,* which would better suit the Tartary buckwheat.

It seems that the Aryan peoples must have known the species, and yet no name is mentioned in the ancient Indo-European languages. No trace of it has hitherto been found in the lake-dwellings of Switzerland or of Savoy.

Notch-seeded Buckwheat—*Polygonum emarginatum,* Roth; *Fagopyrum emarginatum,* Meissner.

This third species of buckwheat is grown in the highlands of the north-east of India, under the name *phaphra* or *phaphar,*[3] and in China.[4] I find no positive proof that it has been found wild. Roth only says that it "inhabits China," and that the grain is used for food. Don,[5] who was the first of Anglo-Indian botanists to mention it, says that it is hardly considered wild. It is not mentioned in floras of the Amur valley, nor of Japan. Judging from the countries where it is cultivated, it is probably wild in the Eastern Himalayas and the north-west of China.

The genus *Fagopyrum* has eight species, all of temperate Asia.

Quinoa—*Chenopodium quinoa,* Willdenow.

The quinoa was a staple food of the natives of New Granada, Peru, and Chili, in the high and temperate parts at the time of the conquest. Its cultivation has

---

[1] Gmelin, *Flora Sibirica,* iii. p. 64; Ledebour, *Fl. Rossica,* iii. p. 576.

[2] Maximowicz, *Primitiæ;* Regel, *Opit. Flori,* etc.; Schmidt, *Reisen in Amur,* do not mention it.

[3] Royle, *Ill. Himal.,* p. 317; Madden, *Trans. Bot. Soc. Edin.,* v. p. 118.

[4] Roth, *Catalecta Botanica,* i. p. 48.

[5] Don, *Prodr. Fl. Nepal.,* p. 74.

persisted in these countries from custom, and on account of the abundance of the product.

From all time the distinction has existed between the quinoa with coloured leaves, and the quinoa with green leaves and white seed.[1] The latter was regarded by Moquin[2] as a variety of a little known species, believed to be Asiatic; but I believe that I showed conclusively that the two American quinoas are two varieties, probably very ancient, of a single species.[3] The less coloured, which is also the most farinaceous, is probably derived from the other.

The white quinoa yields a grain which is much esteemed at Lima, according to information furnished by the *Botanical Magazine*, where a good drawing may be seen (pl. 3641). The leaves may be dressed in the same manner as spinach.[4]

No botanist has mentioned the quinoa as wild or semi-wild. The most recent and complete work on one of the countries where the species is cultivated, the *Flora of Chili*, by Cl. Gay, speaks of it only as a cultivated plant. Père Feuillée and Humboldt said the same for Peru and New Granada. It is perhaps due to the insignificance of the plant and its aspect of a garden weed that collectors have neglected to bring back wild specimens.

Kiery—*Amarantus frumentaceus*, Roxburgh.

This annual is cultivated in the Indian peninsula for its small farinaceous grain, which is in some localities the principal food of the natives.[5] Fields of this species, of a red or golden colour, produce a beautiful effect.[6] From Roxburgh's account, Dr. Buchanan " discovered it on the hills of Mysore and Coimbatore," which seems to indicate a wild condition. *Amarantus speciosus*, cultivated in gardens and figured on pl. 2227 of the *Botanical Maga-*

---

[1] Molina, *Hist. Nat. du Chili*, p. 101.

[2] Moquin, in De Candolle, *Prodromus*, xiii. part 1, p. 67.

[3] A. de Candolle, *Géogr. Bot. Rais.*, p. 952.

[4] *Bon Jardinier*, 1880, p. 562.

[5] Roxburgh, *Fl. Ind.*, edit. 2, vol. iii. p. 609; Wight, *Icones*, pl. 720; Aitchison, *Catalogue of Punjab Plants*, p. 130.

[6] Madden, *Trans. Edin. Bot. Soc.*, v. p. 118.

zine, appears to be the same species. Hamilton found it in Nepal.[1] A variety or allied species, *Amarantus anardana,* Wallich,[2] is grown on the slopes of the Himalayas, but has been hitherto ill defined by botanists. Other species are used as vegetables (see p. 100, *Amarantus gangeticus*).

Chestnut—*Castanea vulgaris,* Lamarck.

The chestnut, belonging to the order *Cupuliferæ,* has an extended but disjunctive natural area. It forms forests and woods in mountainous parts of the temperate zone from the Caspian Sea to Portugal. It has also been found in the mountains of Edough in Algeria, and more recently towards the frontier of Tunis (Letourneux). If we take into account the varieties *japonica* and *americana,* it exists also in Japan and in the temperate region of North America.[3] It has been sown or planted in several parts of the south and west of Europe, and it is now difficult to know if it is wild or cultivated. However, cultivation consists chiefly in the operation of grafting good varieties on the trees which yield indifferent fruit. For this purpose the variety which produces but one large kernel is preferred to those which bear two or three, separated by a membrane, which is the natural state of the species.

The Romans in Pliny's time [4] already distinguished eight varieties, but we cannot discover from the text of this author whether they possessed the variety with a single kernel (Fr. *marron*). The best chestnuts came from Sardis in Asia Minor, and from the neighbourhood of Naples. Olivier de Serres,[5] in the sixteenth century, praises the chestnuts *Sardonne* and *Tuscane,* which produced the single-kernelled fruit called the *Lyons marron.*[6]

[1] Don, *Prodr. Fl. Nepal,* p. 76.

[2] Wallich, *List,* No. 6903; Moquin, in D. C., *Prodr.,* xiii. sect. 2, p. 256.

[3] For further details, see my article in *Prodromus,* vol. xvi. part 2, p. 114; and Boissier, *Flora Orientalis,* iv. p. 1175.

[4] Pliny, *Hist. Nat.,* lib. xix. c. 23.

[5] Olivier de Serres, *Théâtre de l'Agric.,* p. 114.

[6] Lyons *marrons* now come chiefly from Dauphiné and Vivarais. Some are also obtained from Luc in the department of Var (Gasparin, *Traité d'Agric.,* iv. p. 744).

He considered that these varieties came from Italy, and
Targioni [1] tells us that the name *marrone* or *marone* was
employed in that country in the Middle Ages (1170).

Wheat and Kindred Species.—The innumerable varieties of wheat, properly so called, of which the ripened
grain detaches itself naturally from the husk, have been
classed into four groups by Vilmorin,[2] which form distinct species, or modifications of the common wheat
according to different authors. I am obliged to distinguish them in order to study their history, but this, as
will be seen, supports the opinion of a single species.[3]

1. **Common Wheat**—*Triticum vulgare*, Villars; *Triticum hybernum* and *T. æstivum*, Linnæus.

According to the experiments of the Abbé Rozier, and
later of Tessier, the distinction between autumn and
spring wheat has no importance. "All wheats," says the
latter,[4] "are either spring or autumn sown, according to
the country. They all pass with time from the one state
to the other, as I have ascertained. They only need to
be gradually accustomed to the change, by sowing the
autumn wheat a little later, spring wheat a little earlier,
year by year." The fact is that among the immense
number of varieties there are some which feel the cold of
the winter more than others, and it has become the custom to sow them in the spring.[5] We need take no note
of this distinction in studying the question of origin,
especially as the greater number of the varieties thus
obtained date from a remote period.

The cultivation of wheat is prehistoric in the old
world. Very ancient Egyptian monuments, older than
the invasion of the shepherds, and the Hebrew Scriptures
show this cultivation already established, and when the

---

[1] Targioni, *Cenni Storici*, p. 180.
[2] Vilmorin, *Essai d'un Catalogue Méthodique et Synonymique des Froments*, Paris, 1850.
[3] The best drawings of the different kinds of wheat may be found in
Metzger's *Europæische Cerealien*, in folio, Heidelberg, 1824; and in Host,
Graminæ, in folio, vol. iii.
[4] Tessier, *Dict. d'Agric.*, vi. p. 198.
[5] Loiseleur Deslongchamps, *Consid. sur les Céréales*, 1 vol. in 8vo,
p. 219.

Egyptians or Greeks speak of its origin, they attribute it to mythical personages, Isis, Ceres, Triptolemus.[1] The earliest lake-dwellings of Western Switzerland cultivated a small-grained wheat, which Heer[2] has carefully described and figured under the name *Triticum vulgare antiquorum*. From various facts, taken collectively, we gather that the first lake-dwellers of Robenhausen were at least contemporary with the Trojan war, and perhaps earlier. The cultivation of their wheat persisted in Switzerland until the Roman conquest, as we see from specimens found at Buchs. Regazzoni also found it in the rubbish-heaps of the lake-dwellers of Varese, and Sordelli in those of Lagozza in Lombardy.[3] Unger found the same form in a brick of the pyramid of Dashur, Egypt, to which he assigns a date, 3359 B.C. (Unger, *Bot. Streifzüge*, vii.; *Ein Ziegel*, etc., p. 9). Another variety (*Triticum vulgare compactum muticum*, Heer) was less common in Switzerland in the earliest stone age, but it has been more often found among the less ancient lake-dwellers of Western Switzerland and of Italy.[4] A third intermediate variety has been discovered at Aggtelek in Hungary, cultivated in the stone age.[5] None of these is identical with the wheat now cultivated, as more profitable varieties have taken their place.

The Chinese, who grew wheat 2700 B.C., considered it a gift direct from heaven.[6] In the annual ceremony of sowing five kinds of seed, instituted by the Emperor Shen-nung or Chin-nong, wheat is one species, the others being rice, sorghum, *Setaria italica*, and soy.

The existence of different names for wheat in the most ancient languages confirms the belief in a great antiquity

[1] These questions have been discussed with learning and judgment by four authors: Link, *Ueber die ältere Geschichte der Getreide Arten*, in *Abhandl. der Berlin Akad.*, 1816, vol. xvii. p. 122; 1826, p. 67; and in *Die Urwelt und das Alterthum*, 2nd edit., Berlin, 1834, p. 399; Reynier, *Économie des Celtes et des Germains*, 1818, p. 417; Dureau de la Malle, *Ann. des Sciences Nat.*, vol. ix. 1826; and Loiseleur Deslongchamps, *Consid. sur les Céréales*, 1812, part i. p. 52.

[2] Heer, *Pflanzen der Pfahlbauten*, p. 13, pl. 1, figs. 14–18.

[3] Sordelli, *Sulle piante della torbiera di Lagozza*, p. 31.

[4] Heer, *ibid.*; Sordelli, *ibid.*   [5] Nyari, quoted by Sordelli, *ibid.*

[6] Bretschneider, *Study and Value*, etc., pp. 7 and 8.

of cultivation. The Chinese name is *mai*, the Sanskrit *sumana* and *gôdhûma*, the Hebrew *chittah*, Egyptian *br*, Guancho *yrichen*, without mentioning several names in languages derived from the primitive Sanskrit, nor a Basque name, *ogaia* or *okhaya*, which dates perhaps from the Iberians,[1] and several Finn, Tartar, and Turkish names, etc.,[2] which are probably Turanian. This great diversity might be explained by a wide natural area in the case of a very common wild plant, but this is far from being the case of wheat. On the contrary, it is difficult to prove its existence in a wild state in a few places in Western Asia, as we shall see. If it had been widely diffused before cultivation, descendants would have remained here and there in remote countries. The manifold names of ancient languages must, therefore, be attributed to the extreme antiquity of its culture in the temperate parts of Europe, Asia, and Africa—an antiquity greater than that of the most ancient languages. We have two methods of discovering the home of the species previous to cultivation in the immense zone stretching from China to the Canaries : first, the opinion of ancient authors ; second, the existence, more or less proved, of wheat in a wild state in a given country.

According to the earliest of all historians, Berosus, a Chaldean priest, fragments of whose writings have been preserved by Herodotus, wild wheat (*Frumentum agreste*[3]) might be seen growing in Mesopotamia. The texts of the Bible alluding to the abundance of wheat in Canaan prove no more than that the plant was cultivated there, and that it was very productive. Strabo,[4] born 50 B.C., says that, according to Aristobulus, a grain very similar to wheat grew wild upon the banks of the Indus on the 25th parallel of latitude. He also says[5] that in Hircania

[1] Bretschneider, *Study and Value*, etc. ; Ad. Pictet, *Les Origines Indo-Euro.*, edit. 2, vol. i. p. 328 ; Rosenmüller, *Bibl. Naturgesch.*, i. p. 77 ; Pickering, *Chronol. Arrang.*, p. 78 ; Webb and Berthelot, *Canaries*, *Ethnogr.*, p. 187 ; D'Abadie, *Notes MSS. sur les Noms Basques*; De Charencey, *Recherches sur les Noms Basques*, in *Actes Soc. Philolog.*, March, 1869.

[2] Nemnich, *Lexicon*, p. 1492.

[3] G. Syncelli, *Chronogr.*, fol. 1652, p. 28.

[4] Strabo, edit. 1707, vol. ii. p. 1017.   [5] *Ibid.*, vol. i. p. 124 ; ii. p. 776.

(the modern Mazanderan) the grains of wheat which fell from the ear sowed themselves. This may be observed to some degree at the present day in all countries, and the author says nothing upon the important question whether this accidental sowing reproduced itself in the same place from generation to generation. According to the *Odyssey*,[1] wheat grew in Sicily without the help of man. But it is impossible to attach great importance to the words of a poet, and of a poet whose very existence is contested. Diodorus Siculus at the beginning of the Christian era says the same thing, and deserves greater confidence, since he is a Sicilian. Yet he may easily have been mistaken as to the wild character, as wheat was then generally cultivated in Sicily. Another passage in Diodorus [2] mentions the tradition that Osiris found wheat and barley growing promiscuously with other plants at Nisa, and Dureau de la Malle has proved that this town was in Palestine. Among all this evidence, that of Berosus and that of Strabo for Mesopotamia and Western India alone appear to me of any value.

The five species of seed of the ceremony instituted by Chin-nong are considered by Chinese scholars to be natives of their country,[3] and Bretschneider adds that communication between China and Western Asia dates only from the embassy of Chang-kien in the second century before Christ. A more positive assertion is needed, however, before we can believe wheat to be indigenous in China; for a plant cultivated in western Asia two or three thousand years before the epoch of Chin-nong, and of which the seeds are so easily transported, may have been introduced into the north of China by isolated and unknown travellers, as the stones of peaches and apricots were probably carried from China into Persia in prehistoric time.

Botanists have ascertained that wheat is not wild in Sicily at the present day.[4] It sometimes escapes from

[1] Lib. ix. v. 109.

[2] Diodorus, Terasson's trans., ii. pp. 186, 190.

[3] Bretschneider, *ibid.*, p. 15.

[4] Parlatore, *Fl. Ital.*, i. pp. 46, 568. His assertion is the more worthy of attention that he was a Sicilian.

cultivation, but it does not persist indefinitely.[1] The plant which the inhabitants call wild wheat, *Frumentu sarvaggiu*, which covers uncultivated ground, is *Ægilops ovata*, according to Inzenga.[2]

A zealous collector, Balansa, believed that he had found wheat growing on Mount Sipylus, in Asia Minor, under circumstances in which it was impossible not to believe it wild;[3] but the plant he brought back is a spelt, *Triticum monococcum*, according to a very careful botanist, to whom it was submitted for examination.[4] Olivier,[5] before him, when he was on the right bank of the Euphrates, to the north-west of Anah, a country unfit for cultivation, "found in a kind of ravine, wheat, barley, and spelt, which," he adds, " we have already seen several times in Mesopotamia."

Linnæus says,[6] that Heintzelmann found wheat in the country of the Baschkirs, but no one has confirmed this statement, and no modern botanist has seen the species really wild in the neighbourhood of the Caucasus or the north of Persia. Bunge,[7] whose attention was drawn to this point, declares that he has seen no indication which leads him to believe that cereals are indigenous in that country It does not even appear that wheat has a tendency in these regions to spring up accidentally outside cultivated ground. I have not discovered any mention of it as a wild plant in the north of India, in China, or Mongolia.

It is remarkable that wheat has been twice asserted to be indigenous in Mesopotamia, at an interval of twenty-three centuries, once by Berosus, and once by Olivier in our own day. The Euphrates valley lying nearly in the middle of the belt of cultivation which formerly extended from China to the Canaries, it is infinitely probable that it was the principal habitation of the species in very early

[1] Strobl, in *Flora*, 1880, p. 348.   [2] Inzenga, *Annali Agric. Sicil.*
[3] *Bull. de la Soc. Bot. de France*, 1854, p. 108.
[4] J. Gay, *Bull. Soc. Bot. de France*, 1860, p. 30.
[5] Olivier, *Voy. dans l'Emp. Othoman* (1807), vol. iii. p. 460.
[6] Linnæus, *Sp. Plant.*, edit. 2, vol. i. p. 127.
[7] Bunge, *Bull. Soc. Bot. France*, 1860, p. 29.

prehistoric times. The area may have extended towards Syria, as the climate is very similar, but to the east and west of Western Asia wheat has probably never existed but as a cultivated plant; anterior, it is true, to all known civilization.

2. **Turgid, and Egyptian Wheat**—*Triticum turgidum* and *T. compositum*, Linnæus.

Among the numerous common names of the varieties which come under this head, we find that of Egyptian wheat. It appears that it is now much cultivated in that country and in the whole of the Nile valley. A. P. de Candolle says[1] that he recognized this wheat amongst seeds taken from the sarcophagi of ancient mummies, but he had not seen the ears. Unger[2] thinks it was cultivated by the ancient Egyptians, yet he gives no proof founded on drawings or specimens. The fact that no Hebrew or Armenian name[3] can be attributed to the species seems to me important. It proves at least that the remarkable forms with branching ears, commonly called *wheat of miracle, wheat of abundance,* did not exist in antiquity, for they would not have escaped the knowledge of the Israelites. No Sanskrit name is known, nor even any modern Indian names, and I cannot discover any Persian name. The Arab names which Delile[4] attributes to the species belong perhaps to other varieties of wheat. There is no Berber name.[5] From all this it results, I think, that the plants united under the name of *Triticum turgidum,* and especially the varieties with branching ears, are not ancient in the north of Africa or in the west of Asia.

Oswald Heer,[6] in his curious paper upon the plants of the lake-dwellers of the stone age in Switzerland, attributes to *T. turgidum* two non-branched ears, the one bearded, the other almost without beard, of which he gives drawings. Later, in an exploration of the lake-

---

[1] De Candolle, *Physiologie Botanique,* ii. p. 696.
[2] Unger, *Die Pflanzen des Alten Ægyptens,* p. 31.
[3] See Rosenmüller, *Bibl. Naturgesch.;* and Löw, *Aramaische Pflanzen Namen,* 1881.
[4] Delile, *Pl. Cult. en Égypte,* p. 3 ; *Fl. Ægypt. Illus.,* p. 5.
[5] *Dict. Fr.-Berb.,* published by the Government.
[6] Heer, *Pflanzen der Pfahlbauten.* p. 5, fig. 4; p. 52, fig. 20.

dwellings of Robenhausen, Messicommer did not find it, although there was abundant store of grain.[1] Strœbel and Pigorini said they found wheat with *grano grosso duro* (*T. turgidum*), in the lake-dwellings of Parmesan.[2] For the rest, Heer[3] considers this to be a variety or race of the common wheat, and Sordelli inclines to the same opinion.

Fraas thinks that the *krithanias* of Theophrastus was *T. turgidum*, but this is absolutely uncertain. According to Heldreich,[4] the great wheat is of modern introduction into Greece. Pliny[5] spoke briefly of a wheat with branching ears, yielding one hundred grains, which was most likely our *miraculous wheat*.

Thus history and philology alike lead us to consider the varieties of *Triticum turgidum* as modifications of the common wheat obtained by cultivation. The form with branching ears is not perhaps earlier than Pliny's time.

These deductions would be overthrown by the discovery of the *T. turgidum* in a wild state, which has not hitherto been made with certainty. In spite of C. Koch,[6] no one admits that it grows, outside cultivation, at Constantinople and in Asia Minor. Boissier's herbarium, so rich in Eastern plants, has no specimen of it. It is given as wild in Egypt by Schweinfurth and Ascherson, but this is the result of a misprint.[7]

3. Hard Wheat—*Triticum durum*, Desfontaines.

Long cultivated in Barbary, in the south of Switzerland and elsewhere, it has never been found wild. In the different provinces of Spain it has no less than fifteen names,[8] and none are derived from the Arab name *quemah* used in Algeria[9] and Egypt.[10] The

---

[1] Messicommer, in *Flora*, 1869, p. 320.
[2] Quoted from Sordelli, *Notizie sull. Lagozza*, p. 32.
[3] Heer, *ubi supra*, p. 50.
[4] Heldreich, *Die Nutzpflanzen Griechenlands*, p. 5.
[5] Pliny, *Hist.*, lib. xviii. cap. 10.   [6] Koch, *Linnœa*, **xxi.** p. 427.
[7] Letter from Ascherson, 1881.   [8] *Dict. MS. of Vernacular Names.*
[9] Debeaux, *Catal. des Plan. de Boghar*, p. 110.
[10] Delile says (*ubi supra*) that wheat is called *qamh*, and a red variety *qamh-ahmar*.

absence of names in several other countries, especially of original names, is very striking. This is a further indication of a derivation from the common wheat obtained in Spain and the north of Africa at an unknown epoch, perhaps within the Christian era.

4. **Polish Wheat**—*Triticum polonicum*, Linnæus.

This other hard wheat, with yet longer grain, cultivated chiefly in the east of Europe, has not been found wild. It has an original name in German, *Gäner, Gommer, Gümmer,*[1] and in other languages names which are connected only with persons or with countries whence the seed was obtained. It cannot be doubted that it is a form obtained by cultivation, probably in the east of Europe, at an unknown, perhaps recent epoch.

Conclusion as to the Specific Unity of the Principal Races of Wheat.

We have just shown that the history and the vernacular names of the great races of wheat are in favour of a derivation contemporary with man, probably not very ancient, from the common kind of wheat, perhaps from the small-grained wheat formerly cultivated by the Egyptians, and by the lake-dwellers of Switzerland and Italy. Alefeld[2] arrived at the specific unity of *T. vulgare, T. turgidum,* and *T. durum,* by means of an attentive observation of the three cultivated together, under the same conditions. The experiments of Henri Vilmorin[3] on the artificial fertilization of these wheats lead to the same result. Although the author has not yet seen the product of several generations, he has ascertained that the most distinct principal forms can be crossed with ease and produce fertile hybrids. If fertilization be taken as a measure of the intimate degree of affinity which leads to the grouping of individuals into the same species, we cannot hesitate in the case in question, especially with the support of the historical considerations which I have given.

[1] Nemnich, *Lexicon,* p. 1488. [2] Alefeld, *Bot. Zeitung,* 1865, p. 9.
[3] H. Vilmorin, *Bull. Soc. Bot. de France,* 1881, p. 356.

On the supposed Mummy Wheat.

Before concluding this article, I think it pertinent to say that no grain taken from an ancient Egyptian sarcophagus and sown by horticulturists has ever been known to germinate. It is not that the thing is impossible, for grains are all the better preserved that they are protected from the air and from variations of temperature or humidity, and certainly these conditions are fulfilled by Egyptian monuments; but, as a matter of fact, the attempts at raising wheat from these ancient seeds have not been successful. The experiment which has been most talked of is that of the Count of Sternberg, at Prague.[1] He had received the grains from a trustworthy traveller, who assured him they were taken from a sarcophagus. Two of these seeds germinated, it is said; but I have ascertained that in Germany well-informed persons believe there is some imposture, either on the part of the Arabs, who sometimes slip modern seeds into the tombs (even maize, an American plant), or on that of the *employés* of the Count of Sternberg. The grain known in commerce as mummy wheat has never had any proof of antiquity of origin.

Spelt and Allied Varieties or Species.[2]

Louis Vilmorin,[3] in imitation of Seringe's excellent work on cereals,[4] has grouped together those wheats whose seeds when ripe are closely contained in their envelope or husk, necessitating a special operation to free them from it, a character rather agricultural than botanical. He then enumerates the forms of these wheats under three names, which correspond to as many species of most botanists.

1. **Spelt**—*Triticum spelta*, Linnæus.

Spelt is now hardly cultivated out of south Germany and German-Switzerland. This was not the case formerly. The descriptions of cereals by Greek authors are so brief

---

[1] Journal, *Flora*, 1835, p. 4.
[2] See the plates of Metzger and Host, in the works previously quoted.
[3] *Essai d'un Catal. Méthod. des Froments*, Paris, 1850.
[4] Seringe, *Monogr. des Céré. de la Suisse*, in 8vo, Berne, 1818.

and insignificant that there is always room for hesitation as to the sense of the words they use. Yet, judging from the customs of which they speak, scholars think [1] that the Greeks first called spelt *olyra*, afterwards *zeia*, names which we find in Herodotus and Homer. Dioscorides [2] distinguishes two sorts of *zeia*, which apparently answer to *Triticum spelta* and *T. monococcum*. It is believed that spelt was the *semen* (corn, *par excellence*) and the *far* of Pliny, which he said was used as food by the Latins for 360 years before they knew how to make bread.[3] As spelt has not been found among the lake-dwellers of Switzerland and Italy, and as the former cultivated the allied varieties called *T. dicoccum* and *T. monococcum*,[4] it is possible that the *far* of the Latins was rather one of these.

The existence of the true spelt in ancient Egypt and the neighbouring countries seems to me yet more doubtful. The *olyra* of the Egyptians, of which Herodotus speaks, was not the *olyra* of the Greeks; some authors have supposed it to be rice, *oryza*.[5] As to spelt, it is a plant which is not grown in such hot countries. Modern travellers from Rauwolf onwards have not seen it in Egyptian cultivation,[6] nor has it been found in the ancient monuments. This is what led me to suppose [7] that the Hebrew word *kussemeth*, which occurs three times in the Bible,[8] ought not to be attributed to spelt, as it is by Hebrew scholars.[9] I imagined it was perhaps the allied form, *T. monococcum*, but neither is this grown in Egypt.

[1] Fraas, *Syn. Fl. Class.*, p. 307 ; Lenz, *Bot. der Alten*, p. 257.

[2] Dioscorides, *Mat. Med.*, ii., 111–115.

[3] Pliny, *Hist.*, lib. xviii. cap. 7 ; Targioni, *Cenni Storici*, p. 6.

[4] Heer, *Pflanzen der Pfahlbauten*, p. 6 ; Unger, *Pflanzen des Alten Ægyptens*, p. 32.

[5] Delile, *Pl. Cult. en Égypte*, p. 5.

[6] Reynier, *Écon. des Égyptiens*, p. 337 ; Dureau de la Malle, *Ann. Sc. Nat.*, ix. p. 72 ; Schweinfurth and Ascherson, *Aufzäh. Tr. spelta* of Forskal is not admitted by any subsequent author.

[7] *Géogr. Bot. Rais.*, p. 933.

[8] Exod. ix. 32 ; Isa. xxviii. 25 ; Ezek. iv. 9.

[9] Rosenmüller, *Bibl. Alterth.*, iv. p. 83 ; Second, *Trans. of Old Test.*, 1874.

Spelt has no name in Sanskrit, nor in any modern Indian languages, nor in Persian,[1] and therefore, of course, none in Chinese. European names, on the contrary, are numerous, and bear witness to an ancient cultivation, especially in the east of Europe. *Spelta* in Saxon, whence the English name, and the French, *épeautre;* *Dinkel* in modern German, *orkiss* in Polish, *pobla* in Russian,[2] are names which seem to come from very different roots. In the south of Europe the names are rarer. There is a Spanish one, however, of Asturia, *escandia,*[3] but I know of none in Basque.

History, and especially philology, point to an origin in eastern temperate Europe and the neighbouring countries of Asia. We have to discover whether the plant has been found wild.

Olivier,[4] in a passage already quoted, says that he several times found it in Mesopotamia, in particular upon the right bank of the Euphrates, north of Anah, in places unfit for cultivation. Another botanist, André Michaux, saw it in 1783, near Hamadan, a town in the temperate region of Persia. Dureau de la Malle says that he sent some grains of it to Bosc, who sowed them at Paris and obtained the common spelt; but this seems to me doubtful, for Lamarck, in 1786,[5] and Bosc himself, in the *Dictionnaire d'Agriculture,* article *Épeautre* (spelt), published in 1809, says not a word of this. The herbariums of the Paris Museum contain no specimens of the cereals mentioned by Olivier.

There is, as we have seen, much uncertainty as to the origin of the species as a wild plant. This leads me to attribute more importance to the hypothesis that spelt is derived by cultivation from the common wheat, or from an intermediate form at some not very early prehistoric time. The experiments of H. Vilmorin[6] support this theory, for cross fertilizations of the spelt

---

[1] Ad. Pictet, *Orig. Indo-Europ.*, edit. 2, vol. i. p. 348.
[2] Ad. Pictet, *ibid.;* Nemnich, *Lexicon.*
[3] Willkomm and Lange, *Prodr. Fl. Hisp.,* i. p. 107.
[4] Olivier, *Voyage,* 1807, vol. iii. p. 460.
[5] Lamarck, *Dict. Encycl.,* ii. p. 560.
[6] H. Vilmorin, *Bull. Soc. Bot. de France,* 1881, p. 858.

by the downy white wheat, and *vice versâ,* yield "hybrids whose fertility is complete, with a mixture of the characters of both parents, those of the spelt preponderating."

2. **Starch Wheat**—*Triticum dicoccum,* Schrank; *Triticum amyleum,* Seringe.

This form (*Emmer,* or *Aemer* in German), cultivated for starch chiefly in Switzerland, resists a hard winter. It contains two grains in each little ear, like the true spelt.

Heer[1] attributes to a variety of *T. dicoccum* an ear found in a bad state of preservation in the lake-dwellings of Wangen, Switzerland. Messicommer has since found some at Robenhausen.

It has never been found wild; and the rarity of common names is remarkable. These two circumstances, and the slight value of the botanical characters which serve to distinguish it from *Tr. spelta,* lead to the conclusion that it is an ancient cultivated variety of the latter.

3. **One-grained Wheat**—*Triticum monococcum,* Linnæus.

The one-grained wheat, or little spelt, *Einkorn* in German, is distinguished from the two preceding by a single seed in the little ear, and by other characters which lead the majority of botanists to consider it as a really distinct species. The experiments of H. Vilmorin confirm this opinion so far, for he has not yet succeeded in crossing *T. monococcum* with other spelts or wheats. This may be due, as he says himself, to some detail in the manner of operating. He intends to renew his attempts, and may perhaps succeed. [In the *Bulletin de la Sociéte Botanique de France,* 1883, p. 62, Mr. Vilmorin says that he has not met with better success in the third and fourth years in his attempts at crossing *T. monococcum* with other species. He intends to make the experiment with *T. bœoticum,* Boissier, wild in Servia, of which I sent him some seeds gathered by Pancic. As this species is supposed to be the original stock of *T. monococcum,* the experiment is an interesting one.—AUTHOR'S NOTE,

---

[1] Heer, *Pflanz. der. Pfahlb.,* p. 5, fig. 23, and p. 15.

1884.] In the mean time let us see whether this form of spelt has been long in cultivation, and if it has anywhere been found growing wild.

The one-grained wheat thrives in the poorest and most stony soil. It is not very productive, but yields excellent meal. It is sown especially in mountainous districts, in Spain, France, and the east of Europe, but I do not find it mentioned in Barbary, Egypt, the East, or in India or China.

From some expressions it has been believed to be the *tiphai* of Theophrastus.[1] It is easier to invoke Dioscorides,[2] for he distinguishes two kinds of *zeia*, one with two seeds, another with only one. The latter would be the one-grained wheat. Nothing proves that it was commonly cultivated by the Greeks and Romans. Their modern descendants do not sow it.[3] There are no Sanskrit, Persian, or Arabic names. I suggested formerly that the Hebrew word *kussemeth* might apply to this species, but this hypothesis now seems to me difficult to maintain.

Marschall Bieberstein[4] mentions *Triticum monococcum,* or a variety of it, growing wild in the Crimea and the eastern Caucasus, but no botanist has confirmed this assertion. Steven,[5] who lived in the Crimea, declares that he never saw the species except cultivated by the Tartars. On the other hand, the plant which Balansa gathered in a wild state near Mount Sipylus, in Anatolia, is *T. monococcum,* according to J. Gay,[6] who takes with this form *Triticum bœoticum,* Boissier, which grows wild in the plains of Bœotia [7] and in Servia.[8]

---

[1] Fraas, *Syn. Fl. Class.*, p. 307.
[2] Dioscorides, *Mat. Med.*, 2, c. iii. 155.
[3] Heldreich, *Nutz. Griech.*
[4] Bieberstein, *Fl. Tauro-Caucasaica*, vol. i. p. 85.
[5] Steven, *Verzeichniss Taur. Halbins. Pflan.*, p. 354.
[6] *Bull. Soc. Bot. Fran.*, 1860, p. 30.
[7] Boissier, *Diagnoses*, 1st series, vol. ii. fasc. 13, p. 69.
[8] Balansa, 1854, No. 137 in Boissier's Herbarium, in which there is also a specimen found in the fields in Servia, and a variety with brown beards sent by Pancic, growing in Servian meadows. The same botanist (of Belgrade) has just sent me wild specimens from Servia, which I cannot distinguish from *T. monococcum*, which he assures me is not cultivated in Servia. Bentham writes to me that *T. bœoticum,*

Admitting these facts, *T. monococcum* is a native of Servia, Greece, and Asia Minor, and as the attempts to cross it with other spelts or wheats have not been successful, it is rightly termed a species in the Linnæan sense.

The separation of wheat with free grains from spelt must have taken place before all history, perhaps before the beginning of agriculture. Wheat must have appeared first in Asia, and then spelt, probably in Eastern Europe and Anatolia. Lastly, among spelts *T. monococcum* seems to be the most ancient form, from which the others have gradually developed in several thousand years of cultivation and selection.

Two-rowed Barley—*Hordeum distichon*, Linnæus.

Barley is among the most ancient of cultivated plants. As all its forms resemble each other in nature and uses, we must not expect to find in ancient authors and in common names that precision which would enable us to recognize the species admitted by botanists. In many cases the name barley has been taken in a vague or generic sense. This is a difficulty which we must take into account. For instance, the expression of the Old Testament, of Berosus, of Moses of Chorene, Pausanias, Marco Polo, and more recently of Olivier, indicating "wild and cultivated barley" in a given country, prove nothing, because we do not know to which species they refer. There is the same obscurity in China. Dr. Bretschneider says [1] that, according to a work published in the year A.D. 100, the Chinese cultivated barley, but he does not specify the kind. At the extreme west of the old world the Guanchos also cultivated a barley, of which we know the name but not the species.

The common variety of the two-rowed barley, in which the husk remains attached to the ripened grain, has been found wild in Western Asia, in Arabia Petrea,[2]

of which he saw several specimens, is, he thinks, the same as *T. monococcum.*

[1] Bretschneider, *On the Study*, etc., p. 8.
[2] A specimen determined by Reuter in Boissier's Herbarium.

near Mount Sinai,[1] in the ruins of Persepolis,[2] near the Caspian Sea,[3] between Lenkoran and Baku, in the desert of Chirvan and Awhasia, to the south of the Caucasus,[4] and in Turcomania.[5]   No author mentions it in Greece, Egypt, or to the east of Persia.   Willdenow [6] indicates it at Samara, in the south-east of Russia; but more recent authors do not confirm this.   Its modern area is, therefore, from the Red Sea to the Caucasus and the Caspian Sea.

Hence this barley should be one of the forms cultivated by Semitic and Turanian peoples.   Yet it has not been found in Egyptian monuments.   It seems that the Aryans must have known it, but I find no proof in vernacular names or in history.

Theophrastus [7] speaks of the two-rowed barley.   The lake-dwellers of Eastern Switzerland cultivated it before they possessed metals,[8] but the six-rowed barley was more common among them.

The variety in which the grain is bare at maturity (*H. distichon nudum*, Linnæus), which in France has all sorts of absurd names, *orge à café, orge du Pérou* (coffee barley, Peruvian barley), has never been found wild.

The fan-shaped barley (*Hordeum Zeocriton*, Linnæus) seems to me to be a cultivated form of the two-rowed barley.   It is not known in a wild state, nor has it been found in Egyptian monuments, nor the lake-dwellings of Switzerland, Savoy, and Italy.

Common Barley—*Hordeum vulgare*, Linnæus.

The common barley with four rows of grain is mentioned by Theophrastus,[9] but it seems to have been

---

[1] Figari and de Notaris, *Agrostologiæ Ægypt. Fragm.*, p. 18.

[2] A very starved plant gathered by Kotschy, No. 290, of which I possess a specimen.   Boissier terms it *H. distichon, varietas*.

[3] C. A. Meyer, *Verzeichniss*, p. 26, from specimens seen also by Ledebour, *Fl. Ross.*, iv. p. 327.

[4] Ledebour, *ibid.*

[5] Regel, *Descr. Plant.*, Nov., 1881, fasc. 8, p. 37.

[6] Willdenow, *Sp. Plant.*, i. p. 473.

[7] Theophrastus, *Hist. Plant.*, lib. viii. cap. 4.

[8] Heer, *Pflanzen der Pfahlbauten*, p. 13 ; Messicommer, *Flora Bot. Zeitung*, 1869, p. 320.

[9] Theophrastus, *Hist.*, lib. viii. cap. 4.

less cultivated in antiquity than that with two rows, and considerably less than that with six rows. It has not been found in Egyptian monuments, nor in the lake-dwellings of Switzerland, Savoy, and Italy.

Willdenow[1] says that it grows in Sicily and in the south-east of Russia, at Samara, but the modern floras of these two countries do not confirm this. We do not know what species of barley it was that Olivier saw growing wild in Mesopotamia; consequently the common barley has not yet been found certainly wild.

The multitude of common names which are attributed to it prove nothing as to its origin, for in most cases it is impossible to know if they are names of barley in general, or of a particular kind of barley cultivated in a given country.

Six-rowed Barley—*Hordeum hexastichon*, Linnæus.

This was the species most commonly cultivated in antiquity. Not only is it mentioned by Greek authors, but it has also been found in the earliest Egyptian monuments,[2] and in the remains of the lake-dwellings of Switzerland (age of stone), of Italy, and of Savoy (age of bronze).[3] Heer has even distinguished two varieties of the species formerly cultivated in Switzerland. One of them answers to the six-rowed barley represented on the medals of Metapontis, a town in the south of Italy, six centuries before Christ.

According to Roxburgh,[4] it was the only kind of barley grown in India at the end of the last century. He attributes to it the Sanskrit name *yuva*, which has become *juba* in Bengali. Adolphe Pictet[5] has carefully studied the names in Sanskrit and other Indo-European languages which answer to the generic name

---

[1] Willdenow, *Species Plant.*, i. p. 472.

[2] Unger, *Pflanzen des Alten Egyptens*, p. 33; *Ein Ziegel der Dashur Pyramide*, p. 109.

[3] Heer, *Pflanzen der Pfahlbauten*, p. 5, figs. 2 and 3; p. 13, fig. 9; *Flora Bot. Zeitung*, 1869, p. 320; de Mortillet, according to Perrin, *Études préhistoriques sur la Savoie*, p. 23; Sordelli, *Sulle piante della torbiera di Lagozza*, p. 33.

[4] Roxburgh, *Fl. Ind.*, edit. 1832, vol. i. p. 358.

[5] Ad. Pictet, *Origines Indo-Europ.*, edit. 2, vol. i. p. 333.

barley, but he has not been able to go into the details of each species.

The six-rowed barley has not been seen in the conditions of a wild plant, of which the species has been determined by a botanist. I have not found it in Boissier's herbarium, which is so rich in Eastern plants. It is possible that the wild barleys mentioned by ancient authors and by Olivier were *Hordeum hexastichon*, but there is no proof of this.

On Barleys in general.

We have seen that the only form which is now found wild is the simplest, the least productive, *Hordeum distichon*, which was, like *H. hexastichon*, cultivated in prehistoric time. Perhaps *H. vulgare* has not been so long in cultivation as the two others.

Two hypotheses may be drawn from these facts: 1. That the barleys with four and six rows were, in prehistoric agriculture anterior to that of the ancient Egyptians who built the monuments, derived from *H. distichon.* 2. The barleys with six and four ranks were species formerly wild, extinct since the historical epoch. It would be strange in this case that no trace of them has remained in the floras of the vast region comprised between India, the Black Sea, and Abyssinia, where we are nearly sure of their cultivation, at least of that of the six-ranked barley.

Rye—*Secale cereale*, Linnæus.

Rye has not been very long in cultivation, unless, perhaps, in Russia and Thrace. It has not been found in Egyptian monuments, and has no name in Semitic languages, even in the modern ones, nor in Sanskrit and the modern Indian languages derived from Sanskrit. These facts agree with the circumstance that rye thrives better in northern than in southern countries, where it is not usually cultivated in modern times. Dr. Bretschneider[1] thinks it is unknown to Chinese agriculture. He doubts the contrary assertion of a modern writer,

---

[1] Bretschneider, *On Study and Value*, etc., pp. 18, 44.

and remarks that the name of a cereal mentioned in the memoirs of the Emperor Kanghi, which may be supposed to be this species, signifies Russian wheat. Now rye, he says, is much cultivated in Siberia. There is no mention of it in Japanese floras.

The ancient Greeks did not know it. The first author who mentions it in the Roman empire is Pliny,[1] who speaks of the *secale* cultivated at Turin at the foot of the Alps, under the name of *Asia*. Galen,[2] born in A.D. 131, had seen it cultivated in Thrace and Macedonia under the name *briza*. Its cultivation does not seem ancient, at least in Italy, for no trace of rye has been found in the remains of the lake-dwellings of the north of that country, or of Switzerland and Savoy, even of the age of bronze. Jetteles found remains of rye near Olmutz, together with instruments of bronze, and Heer,[3] who saw the specimens, mentions others of the Roman epoch in Switzerland.

Failing archæological proofs, European languages show an early knowledge of rye in German, Keltic, and Slavonic countries. The principal names, according to Adolphe Pictet,[4] belong to the peoples of the north of Europe : Anglo-Saxon, *ryge, rig ;* Scandinavian, *rûgr ;* Old High German, *roggo ;* Ancient Slav, *ruji, roji ;* Polish, *rez ;* Illyrian, *raz,* etc. The origin of this name must date, he says, from an epoch previous to the separation of the Teutons from the Lithuano-Slavs. The word *secale* of the Latins recurs in a similar form among the Bretons, *segal,* and the Basques, *cekela, zekhalea ;* but it is not known whether the Latins borrowed it from the Gauls and Iberians, or whether, conversely, the latter took the name from the Romans. This second hypothesis appears to be the more probable of the two, since the Cisalpine Gauls of Pliny's time had quite a different name. I also find mentioned a Tartar name, *aresch,*[5] and an Ossete name, *syl, sil,*[6] which points to an ancient cultivation to the east of Europe.

[1] Pliny, *Hist.,* lib. xviii. c. 16.
[2] Galen, *De Alimentis,* lib. xiii., quoted by Lenz, *Bot. de Alten,* p. 259.
[3] Heer, *Die Pflanzen der Pfahlbauten,* p. 16.
[4] Ad. Pictet, *Origines Indo-Europ.,* edit. 2, vol. i. p. 344.
[5] Nemnich, *Lexicon Naturgesch.*   [6] Ad. Pictet, *ubi supra.*

Thus historical and philological data show that the species probably had its origin in the countries north of the Danube, and that its cultivation is hardly earlier than the Christian era in the Roman empire, but perhaps more ancient in Russia and Tartary.

The indication of wild rye given by several authors should scarcely ever be accepted, for it has often happened that *Secale cereale* has been confounded with perennial species, or with others of which the ear is easily broken, which modern botanists have rightly distinguished.[1] Many mistakes which thus arose have been cleared up by an examination of original specimens. Others may be suspected. Thus I do not know what to think of the assertions of L. Ross, who said he had found rye growing wild in several parts of Anatolia,[2] and of the Russian traveller Ssaewerzoff, who said he saw it in Turkestan.[3] The latter fact is probable enough, but it is not said that any botanist verified the species. Kunth[4] had previously mentioned it in "the desert between the Black Sea and the Caspian," but he does not say on what authority of traveller or of specimens. Boissier's herbarium has shown me no wild *Secale cereale*, but it has persuaded me that another species of rye might easily be mistaken for this one, and that assertions require to be carefully verified.

Failing satisfactory proofs of wild plants, I formerly urged, in my *Géographie Botanique Raisonnée*, an argument of some value. *Secale cereale* sows itself from cultivation, and becomes almost wild in parts of the Austrian empire,[5] which is seldom seen elsewhere.[6] Thus

---

[1] *Secale fragile*, Bieberstein; *S. anatolicum*, Boissier; *S. montanum*, Gussone; *S. villosum*, Linnæus. I explained in my *Géogr. Botanique*, p. 936, the errors which result from this confusion, when rye was said to be wild in Sicily, Crete, and sometimes in Russia.

[2] *Flora, Bot. Zeitung*, 1856, p. 520.

[3] *Flora, Bot. Zeitung*, 1869, p. 93.   [4] Kunth, *Enum.*, i. p. 449.

[5] Sadler, *Fl. Pesth.*, i. p. 80; Host, *Fl. Austr.*, i. p. 177; Baumgarten, *Fl. Transylv.*, p. 225; Neilreich, *Fl. Wien.*, p. 58; Viviani, *Fl. Dalmat.*, i. p. 97; Farkas, *Fl. Croat.*, p. 1288.

[6] Strobl saw it, however, in the woods on the slopes of Etna, a result of its introduction into cultivation in the eighteenth century (*Œster. Bot. Zeit.*, 1881, p. 159).

in the east of Europe, where history points to an ancient cultivation, rye finds at the present day the most favourable conditions for living without the aid of man.   It can hardly be doubted, from these facts, that its original area was in the region comprised between the Austrian Alps and the north of the Caspian Sea.   This seems the more probable that the five or six known species of the genus *Secale* inhabit western temperate Asia or the south-east of Europe.

Admitting this origin, the Aryan natives would not have known the species, as philology already shows us; but in their migrations westward they must have met with it under different names, which they transported here and there.

Common Oats and **Eastern Oats**—*Avena sativa*, Linnæus; *Avena orientalis*, Schreber.

The ancient Egyptians and the Hebrews did not cultivate oats, but they are now grown in Egypt.[1]   There is no Sanskrit name, nor any in modern Indian languages. They are only now and then planted by the English in India for their horses.[2]   The earliest mention of oats in China is in an historical work on the period 618 to 907 A.D.;   it refers to the variety known to botanists as *Avena sativa nuda*.[3]   The ancient Greeks knew the genus very well;   they called it *bromos*,[4] as the Latins called it *avena;* but these names were commonly applied to species which are not cultivated, and which are weeds mixed with cereals.   There is no proof that they cultivated the common oats.   Pliny's remark[5] that the Germans lived on oatmeal, implies that the species was not cultivated by the Romans.

The cultivation of oats was, therefore, practised anciently to the north of Italy and of Greece.   It was diffused later and partially in the south of the Roman empire.   It is possible that it was more ancient in Asia Minor, for Galen[6] says that oats were abundant in

---

[1] Schweinfurth and Ascherson, *Beitrage zur Fl. Æthiop.*, p. 298.
[2] Royle, *Ill.*, p. 419.
[3] Bretschneider, *On Study and Value*, etc., pp. 18, 44.
[4] Fraas, *Syn. Fl. Class.*, p. 303; Lenz. *Bot. der Alten*, p. 243.
[5] Pliny, *Hist.*, lib. xviii. cap. 17.   [6] Galen, *De Alimentis*, lib. i. cap. 12.

Mysia, above Pergamus; that they were given to horses, and that men used them for food in years of scarcity. A colony of Gauls had formerly penetrated into Asia Minor. Oats have been found among the remains of the Swiss lake-dwellings of the age of bronze,[1] and in Germany, near Wittenburg, in several tombs of the first centuries of the Christian era, or a little earlier.[2] Hitherto none have been found in the lake-dwellings of the north of Italy, which confirms the belief that oats were not cultivated in Italy in the time of the Roman republic.

The vernacular names also prove an ancient existence north and west of the Alps, and on the borders of Europe towards Tartary and the Caucasus. The most widely diffused of these names is indicated by the Latin *avena*, Ancient Slav *ovisu, ovesu, ovsa*, Russian *ovesu*, Lithuanian *awiza*, Lettonian *ausas*, Ostias *abis*.[3] The English word *oats* comes, according to A. Pictet, from the Anglo-Saxon *ata* or *ate*. The Basque name, *olba* or *oloa*,[4] argues a very ancient Iberian cultivation.

The Keltic names are quite different:[5] Irish *coirce, cuirce, corca*, Armorican *kerch*. Tartar *sulu*, Georgian *kari*, Hungarian *zab*, Croat *zob*, Esthonian *kaer*, and others are mentioned by Nemnich[6] as applying to the generic name oats, but it is not likely that names so varied do not belong to a cultivated species. It is strange that there should be an independent Berber name *zekkoum*,[7] as there is nothing to show that the species was anciently cultivated in Africa.

All these facts show how erroneous is the opinion which reigned in the last century,[8] that oats were brought originally from the island of Juan Fernandez, a belief which came apparently from an assertion of the navigator Anson.[9] It is evidently not in the Austral

---

[1] Heer, *Pflanzen der Pfahlbauten*, p. 6, fig. 24.
[2] Lenz, *Bot. der Alten*, p. 245.
[3] Ad. Pictet, *Orig. Indo.-Europ.*, edit. 2, vol. i. p. 350.
[4] Notes communicated by M. Clos.   [5] Ad. Pictet, *ubi supra*.
[6] Nemnich, *Polyglott. Lexicon*, p. 548.
[7] *Dict. Fr.-Berbère*, published by the French Government.
[8] Linnæus, *Species*, p. 118; Lamarck, *Dict. Enc.*, i. p. 431.
[9] Phillips, *Cult. Veget.*, ii. p. 4.

hemisphere that we must seek for the home of the species, but in those countries of the northern hemisphere where it was anciently cultivated.

Oats sow themselves on rubbish-heaps, by the wayside, and near cultivated ground more easily than other cereals, and sometimes persist in such a way as to appear wild. This has been observed in widely separate places, as Algeria and Japan, Paris and the north of China.[1] Instances of this nature render us sceptical as to the wild nature of the oats which Bové said he found in the desert of Sinai. It has also been said [2] that the traveller Olivier saw oats wild in Persia, but he does not mention the fact in his work. Besides, several annual species nearly resembling oats may deceive the traveller. I cannot discover either in books or herbaria the existence of really wild oats either in Europe or Asia, and Bentham has assured me that there are no such specimens in the herbarium at Kew; but certainly the half-wild or naturalized condition is more frequent in the Austrian states from Dalmatia to Transylvania [3] than elsewhere. This is an indication of origin which may be added to the historical and philological arguments in favour of eastern temperate Europe.

Avena strigosa, Schreber, appears to be a variety of the common oats, judging from the experiments in cultivation mentioned by Bentham, who adds, it is true, that these need confirmation.[4] There is a good drawing of the variety in Host, *Icones Graminum Austriacorum,* ii. pl. 56, which may be compared with *A. sativa,* pl. 59. For the rest, *Avena strigosa* has not been found wild. It exists in Europe in deserted fields, which confirms the hypothesis that it is a form derived by cultivation.

Avena orientalis, Schreber, of which the spikelets

[1] Munby, *Catal. Alger.,* edit. 2, p. 36; Franchet and Savatier, *Enum. Pl. Jap.,* ii. p. 175; Cosson, *Fl. Paris,* ii. p. 637; Bunge, *Enum. Chin.,* p. 71, for the variety *nuda.*

[2] Lamarck, *Dict. Encycl.,* i. p. 331.

[3] Viviani, *Fl. Dalmat.,* i. p. 69; Host, *Fl. Austr.,* i. p. 138; Neilreich, *Fl. Wien.,* p. 85; Baumgarten, *Enum. Transylv.,* iii. p. 259; Farkas, *Fl. Croatica,* p. 1277.

[4] Bentham, *Handbook of British Flora,* edit. 4, p. 544.

lean all to one side, has also been grown in Europe from the end of the eighteenth century. It is not known in a wild state. Often mixed with common oats, it is not to be distinguished from them at a glance. The names it bears in Germany, Turkish or Hungarian oats, points to a modern introduction from the East. Host gives a good drawing of it (*Gram. Austr.*, i. pl. 44).

As all the varieties of oats are cultivated, and none have been discovered in a truly wild state, it is very probable that they are all derived from a single pre-historic form, a native of eastern temperate Europe and of Tartary.

Common Millet—*Panicum miliaceum*, Linnæus.

The cultivation of this plant is prehistoric in the south of Europe, in Egypt, and in Asia. The Greeks knew it by the name *kegchros*, and the Latins by that of *milium*.[1] The Swiss lake-dwellers of the age of stone made great use of millet,[2] and it has also been found in the remains of the lake-dwellings of Varese in Italy.[3] As we do not elsewhere find specimens of these early times, it is impossible to know what was the *panicum* or the *sorghum* mentioned by Latin authors which was used as food by the inhabitants of Gaul, Panonia, and other countries. Unger[4] counts *P. miliaceum* among the species of ancient Egypt, but it does not appear that he had positive proof of this, for he has mentioned no monument, drawing, or seed found in the tombs. Nor is there any material proof of ancient cultivation in Mesopotamia India, and China. For the last-named country it is a question whether the *shu*, one of the five cereals sown by the emperors in the great yearly ceremony, is *Panicum miliaceum*, an allied species, or sorghum; but it appears that the sense of the word *shu* has changed, and that formerly it was perhaps sorghum which was sown.[5]

---

[1] The passages from Theophrastus, Cato, and others, are translated in Lenz, *Botanik der Alten*, p. 232.

[2] Heer, *Pflanzen der Pfahlbauten*, p. 17.

[3] Regazzoni, *Riv. Arch. Prov. di Como*, 1880, fasc. 7.

[4] Unger, *Pflanzen des Alten Ægyptens*, p. 34.

[5] Bretschneider, *Study and Value of Chinese Botanical Works*, pp. 7, 8, 45.

Anglo-Indian botanists[1] attribute two Sanskrit names to the modern species, *ûnû* and *vreehib-heda*, although the modern Hindu and Bengali name *cheena* and the Telinga name *worga* are quite different. If the Sanskrit names are genuine, they indicate an ancient cultivation in India. No Hebrew nor Berber name is known;[2] but there are Arab names, *dokhn*, used in Egypt, and *kosjœjb* in Arabia.[3] There are various European names. Besides the Greek and Latin words, there is an ancient Slav name, *proso*,[4] retained in Russia and Poland, an old German word *hirsi*, and a Lithuanian name *sora*.[5] The absence of Keltic names is remarkable. It appears that the species was cultivated especially in Eastern Europe, and spread westward towards the end of the Gallic dominion.

With regard to its wild existence, Linnæus says[6] that it inhabits India, and most authors repeat this; but Anglo-Indian botanists[7] always give it as cultivated. It is not found in Japanese floras. In the north of China de Bunge only saw it cultivated,[8] and Maximowicz near the Ussuri, on the borders of fields and in places near Chinese dwellings.[9] Ledebour says[10] it is nearly wild in Altaic Siberia and Central Russia, and wild south of the Caucasus and in the country of Talysch. He quotes Hohenacker for the last-named locality, who, however, says only "nearly wild."[11] In the Crimea, where it furnishes bread for the Tartars, it is found here and there nearly wild,[12] which is also the case in the south of France, in Italy, and in Austria.[13] It is not wild in

[1] Roxburgh, *Fl. Ind.*, edit. 1832, p. 310 ; Piddington, *Index*.
[2] Rosenmüller, *Bibl. Alterth.* ; *Dict. Franç.-Berbère*.
[3] Delile, *Fl. Ægypt.*, p. 3 ; Forskal, *Fl. Arab.*, civ.
[4] Ad. Pictet, *Origines Indo-Européennes*, edit. 2, vol. i. p. 351.
[5] *Ibid.*   [6] Linnæus, *Spec. Plant.*, i. p. 86.
[7] Roxburgh, *Fl. Ind.*, edit. 1832, p. 310; Aitchison, *Cat. of Punjab Pl.*, p. 159.
[8] Bunge, *Enum.*, No. 400.   [9] Maximowicz, *Primitiœ Amur.*, p. 330.
[10] Ledebour, *Fl. Ross.*, iv. p. 469.
[11] Hohenacker, *Plant. Talysch.*, p. 13.
[12] Steven, *Verzeich. Halb. Taur.*, p. 371.
[13] Mutel, *Fl. Franç.*, iv. p. 20 ; Parlatore, *Fl. Ital.*, i. p. 122 ; Viviani, *Fl. Damat.*, i. p. 60 ; Neilreich, *Fl. Nied. Œsterr.*, p. 32.

Greece,[1] and no one has found it in Persia or in Syria. Forskal and Delile indicated it in Egypt, but Ascherson does not admit this;[2] and Forskal gives it in Arabia.[3] The species may have become naturalized in these regions, as the result of frequent cultivation from the time of the ancient Egyptians. However, its wild nature is so doubtful elsewhere, that its Egypto-Arabian origin is very probable.

Italian Millet—*Panicum Italicum*, Linnæus; *Setaria Italica*, Beauvois.

The cultivation of this species was very common in the temperate parts of the old world in prehistoric times. Its seeds served as food for man, though now they are chiefly given to birds.

In China it is one of the five plants which the emperor sows each year in a public ceremony, according to the command issued by Chin-nong 2700 B.C.[4] The common name is *siao mi* (little seed), the more ancient name being *ku;* but the latter seems to be applied also to a very different species.[5] Pickering says he recognized it in two ancient Egyptian drawings, and that it is now cultivated in Egypt[6] under the name *dokhn;* but that is the name of *Panicum miliaceum*. It is, therefore, very doubtful that the ancient Egyptians cultivated it. It has been found among the remains of the Swiss lake-dwellings of the stone epoch, and therefore *à fortiori* among the lake-dwellers of the subsequent epoch in Savoy.[7]

The ancient Greeks and Latins did not mention it, or at least it has not been possible to certify it from what they say of several panicums and millets. In our own day the species is rarely cultivated in the south of Europe, not at all in Greece,[8] for instance, and I do not

---

[1] Heldreich, *Nutz. Griechenl.*, p. 3 ; *Pflanz. Attisch. Ebene.*, p. 516.

[2] M. Ascherson informs me in a letter that in his *Aufzählung* the word *cult.* has been omitted by mistake after *Panicum miliaceum*.

[3] Forskal, *Fl. Arab.*, p. civ.

[4] Bretschneider, *Study and Value*, etc., pp. 7, 8

[5] Bretschneider, *ibid.*

[6] According to Unger, *Pflanz. d. Alt. Ægypt.*, p. 34.

[7] Heer, *Pflanzen d. Pfahlbaut.*, p. 5, fig. 7 ; p. 17, figs. 28, 29 ; Perrin, *Études Préhistoriques sur la Savoie*, p. 22.

[8] Heldreich, *Nutzpfl. Griech.*

find it indicated in Egypt, but it is common in Southern Asia.[1]

The Sanskrit names *kungû* and *priyungû*, of which the first is retained in Bengali,[2] are attributed to this species. Piddington mentions several other names in Indian languages in his *Index*. Ainslie [3] gives a Persian name, *arzun*, and an Arabic name ; but the latter is commonly attributed to *Panicum miliaceum*. There is no Hebrew name, and the plant is not mentioned in botanical works upon Egypt and Arabia. The European names have no historical value. They are not original, and commonly refer to the transmission of the species or to its cultivation in a given country. The specific name, *italicum*, is an absurd example, the plant being rarely cultivated and never wild in Italy.

Rumphius says it is wild in the Sunda Isles, but not very positively.[4] Linnæus probably started from this basis to exaggerate and even promulgate an error, saying, "inhabits the Indies." [5] It certainly does not come from the West Indies ; and further, Roxburgh asserts that he never saw it wild in India. The Graminæ have not yet appeared in Sir Joseph Hooker's flora ; but Aitchison [6] gives the species as only cultivated in the northwest of India. The Australian plant which Robert Brown said belonged to this species belongs to another.[7] *P. italicum* appears to be wild in Japan, at least in the form called *germanica* by different authors,[8] and the Chinese consider the five cereals of the annual ceremony to be natives of their country. Yet Bunge, in the north of China, and Maximowicz in the basin of the river Amur, only saw the species cultivated on a large scale, in the form of the *germanica* variety.[9] In

---

[1] Roxburgh, *Fl. Ind.*, edit. 1832, vol. i. p. 302; Rumphius, *Amboin.*, v. p. 202, t. 75.

[2] Roxburgh, *ibid.*   [3] Ainslie, *Mat. Med. Ind.*, i. p. 226.

[4] "Obcurrit in Baleya," etc. (*Rumphius*, v. p. 202).

[5] "Habitat in Indiis " (Linnæus, *Species*, i. p. 83).

[6] Aitchison, *Catal. of Punjab Pl.*, p. 162.

[7] Bentham, *Flora Austral.*, vii. p. 493.

[8] Franchet and Savatier, *Enum. Japon.*, ii. p. 262.

[9] Bunge, *Enum.*, No. 399; Maximowicz, *Primitiæ Amur.*, p. 330.

Persia,[1] the Caucasus Mountains, and Europe, I only find in floras the plant indicated as cultivated, or escaped sometimes from cultivation on rubbish-heaps, waysides, waste ground, etc.[2]

The sum of the historical, philological, and botanical data make me think that the species existed before all cultivation, thousands of years ago in China, Japan, and in the Indian Archipelago. Its cultivation must have early spread towards the West, since we know of Sanskrit names, but it does not seem to have been known in Syria, Arabia, and Greece, and it is probably through Russia and Austria that it early arrived among the lake-dwellers of the stone age in Switzerland.

Common Sorghum—*Holcus sorghum*, Linnæus; *Andropogon sorghum*, Brotero; *Sorghum vulgare*, Persoon.

Botanists are not agreed as to the distinction of several of the species of sorghum, and even as to the genera into which this group of the Graminæ should be divided. A good monograph on the sorghums is needed, as in the case of the panicums. In the mean time I will give some information on the principal species, because of their immense importance as food for man, rearing of poultry, and as fodder for cattle.

We may take as a typical species the sorghum cultivated in Europe, as it is figured by Host in his *Graminæ Austriacæ* (iv. pl. 2). It is one of the plants most commonly cultivated by the modern Egyptians, under the name of *dourra*, and also in equatorial Africa, India, and China.[3] It is so productive in hot countries that it is a staple food of immense populations in the old world.

Linnæus and all authors, even our contemporaries, say that it is of Indian origin; but in the first edition of Roxburgh's flora, published in 1820, this botanist, who should have been consulted, asserts that he had only seen it cultivated. He makes the same remark for the allied forms (*bicolor*, *saccharatus*, etc.), which are often regarded

---

[1] Buhse, *Aufzählung*, p. 232.
[2] See Parlatore, *Fl. Ital.*, i. p. 113 ; Mutel, *Fl. Franç.*, iv. p. 20, etc.
[3] Delile, *Plantes Cult. en Égypte*, p. 7 ; Roxburgh, *Fl. Ind.*, edit. 1832, vol. i. p. 269; Aitchison, *Catal. of Punjab Pl.*, p. 175; Bretschneider, *Study and Value*, etc., p. 9.

as mere varieties.   Aitchison also had only seen the sorghum cultivated.   The absence of a Sanskrit name also
renders the Indian origin very doubtful.   Bretschneider,
on the other hand, says the sorghum is indigenous in
China, although he says that ancient Chinese authors
have not spoken of it.   It is true that he quotes a name,
common at Pekin, *kao-liang* (tall millet), which also
applies to *Holcus saccharatus,* and to which it is better
suited.

The sorghum has not been found among the remains
of the lake-dwellings of Switzerland and Italy.   The
Greeks never spoke of it.   Pliny's phrase [1] about a *milium*
introduced into Italy from India in his time has been
supposed to refer to the sorghum; but it was a taller plant,
perhaps *Holcus saccharatus.*   The sorghum has not been
found in a natural state in the tombs of ancient Egypt.
Dr. Hannerd thought he recognized it in some crushed
seeds brought by Rosellini from Thebes ; [2] but Mr. Birch,
the keeper of Egyptian antiquities in the British Museum,
has more recently declared that the species has not been
found in the ancient tombs.[3]   Pickering says he recognized its leaves mixed with those of the papyrus.   He
says he also saw paintings of it ; and Leipsius has copies
of drawings which he, as well as Unger and Wilkinson,
takes to be the *dourra* of modern cultivation.[4]   The height
and the form of the ear are undoubtedly those of the
sorghum.   It is possible that this species is the *dochan,*
once mentioned in the Old Testament [5] as a cereal from
which bread was made ; yet the modern Arabic word
dokhn refers to the sweet sorghum.

Common names tell us nothing, either from their lack
of meaning, or because in many cases the same name
has been applied to the different kinds of panicum and
sorghum.   I can find none which is certain in the
ancient languages of India or Western Asia, which

---

[1] Pliny, *Hist.*, lib. xviii. c. 7.
[2] Quoted by Unger, *Die Pflanzen des Alten Egyptens,* p. 34.
[3] S. Birch, in Wilkinson, *Man. and Cust. of Anc. Egyptians,* 1878, vol. ii.
p. 427.
[4] Lepsius' drawings are reproduced by Unger and by Wilkinson.
[5] Ezek. iv. 9.

argues an introduction of but few centuries before the
Christian era.

No botanist mentions the *dourra* as wild in Egypt
or in Arabia. An analogous form is wild in equatorial
Africa, but R. Brown has not been able to identify it,[1]
and the flora of tropical Africa in course of publication at
Kew has not yet reached the order Graminæ. There
remains, therefore, the single assertion of Dr. Bretschneider, that the tall sorghum is indigenous in China.
If it is really the species in question, it spread westward
very late. But it was known to the ancient Egyptians,
and how could they have received it from China while
it remained unknown to the intermediate peoples? It
is easier to understand that it is indigenous in tropical
Africa, and was introduced into Egypt in prehistoric
time, afterwards into India, and finally into China, where
its cultivation does not seem to be very ancient, for the
first work which mentions it belongs to the fourth century of our era.

In support of the theory of African origin, I may quote
the observation of Schmidt,[2] that the species abounds in
the island of San Antonio, in the Cape Verde group, in
rocky places. He believes it to be "completely naturalized," which perhaps conceals a true origin.

Sweet Sorghum—*Holcus saccharatus*, Linnæus; *Andropogon saccharatus*, Roxburgh; *Sorghum saccharatum*, Persoon.

This species, taller than the common sorghum and
with a loose panicle,[3] is cultivated in tropical countries
for the seed—which, however, is not so good as that of
the common sorghum—and in less hot countries as fodder,
or even for the sugar which the stem contains in considerable quantities. The Chinese extract a spirit from
it, but not sugar.

The opinion of botanists and of the public in general
is that it comes from India; but Roxburgh says that it
is only cultivated in that country. It is the same in

[1] Brown, *Bot. of Congo*, p. 544.
[2] Schmidt, *Beiträge zur Flora Capverdischen Inseln*, p. 158.
[3] See Host, *Graminæ Austriacæ*, vol. iv. pl. 4.

the Sunda Isles, where the *battari* is certainly this species. It is the *kao-liang*, or great millet of the Chinese. It is not said to be indigenous in China, nor is it mentioned by Chinese authors who lived before the Christian era.[1] From these facts, and the absence of any Sanskrit name, the Asiatic origin seems to me a delusion.

The plant is now cultivated in Egypt less than the common sorghum, and in Arabia under the name *dokhna* or *dokhn*.[2] No botanist has seen it wild in these countries. There is no proof that the ancient Egyptians cultivated it. Herodotus[3] spoke of a "tree-millet" in the plains of Assyria. It might be the species in question, but it is not possible to prove it.

The Greeks and Romans were not acquainted with it, not at least before the Roman empire, but it is possible that this was the millet, seven feet high, which Pliny mentions[4] as having been introduced from India in his lifetime.

We must probably seek its origin in tropical Africa, where the species is generally cultivated. Sir William Hooker[5] mentions specimens from the banks of the river Nun, which were perhaps wild. The approaching publication of the Graminæ in the flora of tropical Africa will probably throw some light on this question. The spread of its cultivation from the interior of Africa to Egypt after the Pharaohs, to Arabia, the Indian Archipelago, and, after the epoch of Sanskrit, to India, lastly to China, towards the beginning of our era, tallies with historical data, and is not difficult to admit. The inverse hypothesis of a transmission from east to west presents a number of objections.

Several varieties of sorghum are cultivated in Asia and in Africa; for instance, *cernuus* with drooping

---

[1] Roxburgh, *Fl. Ind.*, edit. 2, vol. i. p. 271; Rumphius, *Amboin.*, v. p. 194, pl. 75, fig. 1; Miquel, *Fl. Indo-Batava*, iii. p. 503; Bretschneider, *Study and Value*, etc., pp. 9, 46; Loureiro, *Fl. Cochin.*, ii. p. 792.

[2] Forskal, Delile, Schweinfurth, and Ascherson, *ubi supra*.

[3] Herodotus, lib. i. cap. 193.

[4] Pliny, *Hist.*, lib. xviii. cap. 7. This may also be the variety or species known as *bicolor*.

[5] W. Hooker, *Niger Flora*.

panicles, mentioned by Roxburgh, and which Prosper Alpin had seen in Egypt; *bicolor,* which in height resembles the *saccharatus;* and *niger* and *rubens,* which also seem to be varieties of cultivation. None of these has been found wild, and it is probable that a monograph would connect them with one or other of the above-mentioned species.

Coracan—*Eleusine coracana,* Gærtner

This annual grass, which resembles the millets, is cultivated especially in India and the Malay Archipelago. It is also grown in Egypt[1] and in Abyssinia;[2] but the silence of many botanists, who have mentioned the plants of the interior and west of Africa, shows that its cultivation is not widely spread on that continent. In Japan[3] it sometimes escapes from cultivation. The seeds will ripen in the south of Europe, but the plant is valueless there except as fodder[4]

No author mentions having found it in a wild state in Asia or in Africa. Roxburgh,[5] who is attentive to such matters, after speaking of its cultivation, adds, " I never saw it wild." He distinguishes under the name *Eleusine stricta* a form even more commonly cultivated in India, which appears to be simply a variety of *E. coracana,* and which also he has not found uncultivated.

We shall discover its country by other means.

In the first place, the species of the genus *Eleusine* are more numerous in the south of Asia than in other tropical regions. Besides the cultivated plant, Royle[6] mentions other species, of which the poorer natives of India gather the seeds in the plains. According to Piddington's *Index,* there is a Sanskrit name, *rajika,* and several other names in the modern languages of India. That of *coracana* comes from an old name used in Ceylon, *kourakhan.*[7] In the Malay Archipelago the names appear less numerous and less original.

---

[1] Schweinfurth and Ascherson, *Aufzählung,* p. 299.
[2] *Bon Jardinier,* 1880, p. 585.
[3] Franchet and Savatier, *Enum. Plant. Japon.,* ii. p. 172.
[4] *Bon Jardinier, ibid.*   [5] Roxburgh, *Fl. Indica,* edit. 2, vol. i. p. 343.
[6] Royle, *Ill. Him. Plants.*   [7] Thwaites, *Enum. Pl. Zeylan.,* p. 371.

In Egypt the cultivation of this species is perhaps not very ancient. The monuments of antiquity bear no trace of it. Græco-Roman authors who knew the country did not speak of it, nor later Prosper Alpin, Forskal, and Delile. We must refer to a modern work, that of Schweinfurth and Ascherson, to find mention of the species, and I cannot even discover an Arab name.[1] Thus botany, history, and philology point to an Indian origin. The flora of British India, in which the Graminæ have not yet appeared, will perhaps tell us the plant has been found wild in recent explorations.

A nearly allied species is grown in Abyssinia, *Eleusine Tocussa*, Fresenius,[2] a plant very little known, which is perhaps a native of Africa.

Rice—*Oryza sativa*, Linnæus.

In the ceremony instituted by the Chinese Emperor Chin-nong, 2800 years B.C., rice plays the principal part. The reigning emperor must himself sow it, whereas the four other species are or may be sown by the princes of his family [3] The five species are considered by the Chinese as indigenous, and it must be admitted that this is probably the case with rice, which is in general use, and has been so for a long time, in a country intersected by canals and rivers, and hence peculiarly favourable to aquatic plants. Botanists have not sufficiently studied Chinese plants for us to know whether rice is often found outside cultivated ground; but Loureiro[4] had seen it in marshes in Cochin-China.

Rumphius and modern writers upon the Malay Archipelago give it only as a cultivated plant. The multitude of names and varieties points to a very ancient cultivation. In British India it dates at least from the Aryan invasion, for rice has Sanskrit names, *vrihi*,

[1] Several synonyms and the Arabic name in Linnæus, Delile, etc., apply to *Dactyloctenium ægyptiacum*, Willdenow, or *Eleusine ægyptiaca* of some authors, which is not cultivated.

[2] Fresenius, *Catal. Sem. Horti. Francof.*, 1834, *Beitr. z. Fl. Abyss.*, p. 141.

[3] Stanislas Julien, in Loiseleur, *Consid. sur les Céréales*, part i. p. 29; Bretschneider, *Study and Value of Chinese Botanical Works*, pp. 8 and 9.

[4] Loureiro, *Fl. Cochin.*, i. p. 267.

arunya,[1] whence come, probably, several names in modern Indian languages, and *oruza* or *oruzon* of the ancient Greeks, *rouz* or *arous* of the Arabs. Theophrastus[2] mentioned rice as cultivated in India. The Greeks became acquainted with it through Alexander's expedition. "According to Aristobulus," says Strabo,[3] "rice grows in Bactriana, Babylonia, Susida;" and he adds, "we may also add in Lower Syria." Further on he notes that the Indians use it for food, and extract a spirit from it. These assertions, doubtful perhaps for Bactriana, show that this cultivation was firmly established, at least, from the time of Alexander (400 B.C.), in the Euphrates valley, and from the beginning of our era in the hot and irrigated districts of Syria. The Old Testament does not mention rice, but a careful and judicious writer, Reynier,[4] has remarked several passages in the Talmud which relate to its cultivation. These facts lead us to suppose that the Indians employed rice after the Chinese, and that it spread still later towards the Euphrates—earlier, however, than the Aryan invasion into India. A thousand years elapsed between the existence of this cultivation in Babylonia and its transportation into Syria, whence its introduction into Egypt after an interval of probably two or three centuries. There is no trace of rice among the grains or paintings of ancient Egypt.[5] Strabo, who had visited this country as well as Syria, does not say that rice was cultivated in Egypt in his time, but that the Garamantes[6] grew it, and this people is believed to have inhabited an oasis to the south of Carthage. It is possible that they received it from Syria. At all events, Egypt could not long fail

---

[1] Piddington, *Index*; Hehn, *Culturpflanzen*, edit. 3, p. 437.

[2] Theophrastus, *Hist.*, lib. iv. cap. 4, 10.

[3] Strabo, *Géographie*, Tardieu's translation, lib. xv. cap. 1, § 18; lib. xv. cap. 1, § 53.

[4] Reynier, *Économie des Arabes et des Juifs* (1820), p. 450; *Économie Publique et Rurale des Égyptiens et des Carthaginois* (1823), p. 324.

[5] Unger mentions none; Birch, in 1878, furnishes a note to Wilkinson's *Manners and Customs of the Ancient Egyptians*, ii. p. 402, "There is no proof of the cultivation of rice, of which no grains have been found."

[6] Reynier, *ibid.*

to possess a crop so well suited to its peculiar conditions of irrigation. The Arabs introduced the species into Spain, as we see from the Spanish name *arroz*. Rice was first cultivated in Italy in 1468, near Pisa.[1] It is of recent introduction into Louisiana.

When I said that the cultivation of rice in India was probably more recent than in China, I did not mean that the plant was not wild there. It belongs to a family of which the species cover wide areas, and, besides, aquatic plants have commonly more extensive habitations than others. Rice existed, perhaps, before all cultivation in Southern Asia from China to Bengal, as is shown by the variety of names in the monosyllabic languages of the races between India and China.[2] It has been found outside cultivation in several Indian localities, according to Roxburgh.[3] He says that wild rice, called *newaree* by the Telingas, grows in abundance on the shores of lakes in the country of the Circars. Its grain is prized by rich Hindus, but it is not planted because it is not very productive. Roxburgh has no doubt that this is the original plant. Thomson [4] found wild rice at Moradabad, in the province of Delhi. Historical reasons support the idea that these specimens are indigenous. Otherwise they might be supposed to be the result of the habitual cultivation of the species, all the more that there are examples of the facility with which rice sows itself and becomes naturalized in warm, damp climates.[5] In any case historical evidence and botanical probability tend to the belief that rice existed in India before cultivation.[6]

Maize—*Zea mays*, Linnæus.

" Maize is of American origin, and has only been introduced into the old world since the discovery of the new.

---

[1] Targioni, *Cenni Storici.*

[2] Crawfurd, in *Journal of Botany*, 1866, p. 324.

[3] Roxburgh, *Fl. Ind.*, edit. 1832, vol. ii. p. 200.

[4] Aitchinson, *Catal. Punjab.*, p. 157.

[5] Nees, in Martius, *Fl. Brasil.*, in 8vo, ii. p. 518; Baker, *Fl. of Mauritius*, p. 458.

[6] Von Mueller writes to me that rice is certainly wild in tropical Australia. It may have been accidentally sown, and have become naturalized.—AUTHOR'S NOTE, 1884.

I consider these two assertions as positive, in spite of the contrary opinion of some authors, and the doubts of the celebrated agriculturist Bonafous, to whom we are indebted for the most complete treatise upon maize."[1] I used these words in 1855, after having already contested the opinion of Bonafous at the time of the publication of his work.[2] The proofs of an American origin have been since reinforced. Yet attempts have been made to prove the contrary, and as the French name, *blé de Turquie*, gives currency to an error, it is as well to resume the discussion with new data.

No one denies that maize was unknown in Europe at the time of the Roman empire, but it has been said that it was brought from the East in the Middle Ages. The principal argument is based upon a charter of the thirteenth century, published by Molinari,[3] according to which two crusaders, companions in arms of Boniface III., Marquis of Monferrat, gave in 1204 to the town of Incisa a piece of the true cross . . . and a purse containing a kind of seed of a golden colour and partly white, unknown in the country and brought from Anatolia, where it was called *meliga*, etc. The historian of the crusades, Michaux, and later Daru and Sismondi, said a great deal about this charter; but the botanist Delile, as well as Targioni-tozzetti and Bonafous himself, thought that the seed in question might belong to some sorghum and not to maize. These old discussions have been rendered absurd by the Comte de Riant's discovery [4] that the charter of Incisa is the fabrication of a modern impostor. I quote this instance to show how scholars who are not naturalists may make mistakes in the interpretation of the names of plants, and also how dangerous it is to rely upon an isolated proof in historical questions.

The names *blé de Turquie,* Turkish wheat (Indian

---

[1] Bonafous, *Hist. Nat. Agric. et Économique du Maïs*, 1 vol. in folio, Paris and Turin, 1836.

[2] A. de Candolle, *Bibliothèque Universelle de Genève*, Aug. 1836, *Géogr. Bot. Rais.*, p. 942.

[3] Molinari, *Storia d'Incisa*, Asti, 1810.

[4] Riant, *La Charte d'Incisa*, 8vo pamphlet, 1877, reprinted from the *Revue des Questions Historiques*.

corn), given to maize in almost all modern European languages no more prove an Eastern origin than the charter
of Incisa. These names are as erroneous as that of *coq
d'Inde,* in English *turkey,* given to an American bird.
Maize is called in Lorraine and in the Vosges Roman corn;
in Tuscany, Sicilian corn; in Sicily, Indian corn; in the
Pyrenees, Spanish corn; in Provence, Barbary or Guinea
corn. The Turks call it Egyptian corn, and the Egyptians, Syrian *dourra.* This last case proves at least that
it is neither Egyptian nor Syrian. The widespread
name of Turkish wheat dates from the sixteenth century.
It sprang from an error as to the origin of the plant,
which was fostered perhaps by the tufts which terminate
the ears of maize, which were compared to the beard of
the Turks, or by the vigour of the plant, which may have
given rise to an expression similar to the French *fort
comme un turc.* The first botanist who uses the name,
Turkish wheat, is Ruellius, in 1536.[1] Bock or Tragus,[2] in
1552, after giving a drawing of the species which he calls
Frumentum turcicum, Welschkorn, in Germany, having
learnt by merchants that it came from India, conceived
the unfortunate idea that it was a certain *typha* of Bactriana, to which ancient authors alluded in vague terms.
Dodoens in 1583, Camerarius in 1588, and Matthiole[3] rec-
tified these errors, and positively asserted the American
origin. They adopted the name *mays,* which they knew
to be American. We have seen (p. 363) that the zea of
the Greeks was a spelt. Certainly the ancients did not
know maize. The first travellers[4] who described the
productions of the new world were surprised at it, a clear
proof that they had not known it in Europe. Hernandez,[5]
who left Europe in 1571, according to some authorities,
in 1593 according to others,[6] did not know that from the

---

[1] Ruellius, *De Natura Stirpium,* p. 428, "Hanc quoniam nostrorum
ætate e Græcia vel Asia venerit *Turcicum frumentum* nominant." Fuch-
sius, p. 824, repeats this phrase in 1543.

[2] Tragus, *Stirpium,* etc., edit. 1552, p. 650.

[3] Dodoens, *Pemptades,* p. 509; Camerarius, *Hort.,* p. 94; Matthiole,
deit. 1570, p. 305.

[4] P. Martyr, Ercilla, Jean de Lery, etc., 1516–1578.

[5] Hernandez, *Thes. Mexic.,* p. 242.   [6] Lasègue, *Musée Delessert,* p. 467.

year 1500 maize had been sent to Seville for cultivation. This fact, attested by Fée, who has seen the municipal records,[1] clearly shows the American origin, which caused Hernandez to think the name of Turkish wheat a very bad one.

It may perhaps be urged that maize, new to Europe in the sixteenth century, existed in some parts of Asia or Africa before the discovery of America. Let us see what truth there may be in this.

The famous orientalist D'Herbelot [2] had accumulated several errors pointed out by Bonafous and by me, on the subject of a passage in the Persian historian Mirkoud of the fifteenth century, about a cereal which Rous, son of Japhet, sowed upon the shores of the Caspian Sea, and which he takes to be the Indian corn of our day. It is hardly worth considering these assertions of a scholar to whom it had never occurred to consult the works of the botanists of his own day, or earlier. What is more important is the total silence on the subject of maize of the travellers who visited Asia and Africa before the discovery of America; also the absence of Hebrew and Sanskrit names for this plant; and lastly, that Egyptian monuments present no specimen or drawing of it.[3] Rifaud, it is true, found an ear of maize in a sarcophagus at Thebes, but it is believed to have been the trick of an Arab impostor. If maize had existed in ancient Egypt, it would be seen in all monuments, and would have been connected with religious ideas like all other remarkable plants. A species so easy of cultivation would have spread into all neighbouring countries. Its cultivation would not have been abandoned; and we find, on the contrary, that Prosper Alpin, visiting Egypt in 1592, does not speak of it, and that Forskal,[4] at the end of the eighteenth century, mentioned maize as still but little grown in Egypt, where it had no name distinct from the sorghums. Ebn Baithar,

---

[1] Fée, *Souvenirs de la Guerre d'Espagne*, p. 128.

[2] *Bibliothèque Orientale*, Paris, 1697, at the word *Rous*.

[3] Kunth, *Ann. Sc. Nat.*, sér. 1, vol. viii. p. 418; Raspail, *ibid.*; Unger, *Pflanzen des Alten Ægyptens*; A. Braun, *Pflanzenreste Ægypt. Mus. in Berlin*; Wilkinson, *Manners and Customs of Ancient Egyptians*.

[4] Forskal, p. liii.

an Arab physician of the thirteenth century, who had travelled through the countries lying between Spain and Persia, indicates no plant which can be supposed to be maize.

J. Crawfurd,[1] having seen maize generally cultivated in the Malay Archipelago under a name *jarung*, which appears to be indigenous, believed that the species was a native of these islands. But then how is it Rumphius makes no mention of it. The silence of this author points to an introduction later than the seventeenth century. Maize was so little diffused on the continent of India in the last century, that Roxburgh[2] wrote in his flora, which was published long after it was drawn up, " Cultivated in different parts of India in gardens, and only as an ornament, but nowhere on the continent of India as an object of cultivation on a large scale." We have seen that there is no Sanskrit name.

Maize is frequently cultivated in China in modern times, and particularly round Pekin for several generations,[3] although most travellers of the last century make no mention of it. Dr. Bretschneider, in his work published in 1870, does not hesitate to say that maize is not indigenous in China; but some words in his letter of 1881 make me think that he now attributes some importance to an ancient Chinese author, of whom Bonafous and afterwards Hance and Mayers have said a great deal. This is a work by Li-chi-tchin, entitled *Phen-thsao-kang-mou*, or *Pên-tsao-kung-mu*, a species of treatise on natural history, which Bretschneider[4] says was written at the end of the sixteenth century. Bonafous says it was concluded in 1578, and the edition which he had seen in the Huzard library was of 1637. It contains a drawing of maize with the Chinese character. This plate is copied in Bonafous' work, at the beginning of the chapter on the original country of the maize. It is clear that it repre-

[1] Crawfurd, *History of the Indian Archipelago*, Edinburgh, 1820, vol. i.; *Journal of Botany*, 1866, p. 326.

[2] Roxburgh, *Flora Indica*, edit. 1832, vol. iii. p. 563.

[3] Bretschneider, *Study and Value*, etc., pp. 7, 18.

[4] *Ibid.*

sents the plant. Dr. Hance[1] appears to have based his
arguments upon the researches of Mayers, who says that
early Chinese authors assert that maize was imported
from Sifan (Lower Mongolia, to the west of China) long
before the end of the fifteenth century, at an unknown
date. The article contains a copy of the drawing in the
Pên-tsao-kung-mu, to which he assigns the date 1597.

The importation through Mongolia is improbable to
such a degree that it is hardly worth speaking of it, and
as for the principal assertion of the Chinese author, the
dates are uncertain and late. The work was finished in
1578 according to Bonafous, in 1597 according to Mayers.
If this be true, and especially if the second of these dates
is the true one, it may be admitted that maize was brought
to China after the discovery of America. The Portuguese
came to Java in 1496,[2] that is to say four years after the
discovery of America, and to China in 1516.[3] Magellan's
voyage from South America to the Philippine Islands took
place in 1520. During the fifty-eight or seventy-seven
years between 1516 and the dates assigned to the Chinese
work, seeds of maize may have been taken to China by
navigators from America or from Europe. Dr. Bretschneider wrote to me recently that the Chinese did not
know the new world earlier than the Europeans, and that
the lands to the east of their country, to which there are
some allusions in their ancient writings, are the islands of
Japan. He had already quoted the opinion of a Chinese
savant, that the introduction of maize in the neighbourhood
of Pekin dates from the last years of the Ming dynasty,
which ended in 1644. This date agrees with the other
facts. The introduction into Japan was probably of later
date, since Kæmpfer makes no mention of the species.[4]

From all these facts, we conclude that maize is not a
native of the old world. It became rapidly diffused in it

---

[1] The article is in the *Pharmaceutical Journal* of 1870; I only know
it from a short extract in Seemann's *Journal of Botany*, 1871, p. 62.

[2] Rumphius, *Amboin.*, vol. v. p. 525.

[3] Malte-Brun, *Géographie*, i. p. 493.

[4] A plant engraved on an ancient weapon which Siebold had taken
for maize is a sorghum, according to Rein, quoted by Wittmack, *Ueber
Antiken Maïs.*

after the discovery of America, and this very rapidity completes the proof that, had it existed anywhere in Asia or Africa, it would have played an important part in agriculture for thousands of years.

We shall see that the facts are quite contrary to these in America.

At the time of the discovery of the new continent, maize was one of the staples of its agriculture, from the La Plata valley to the United States. It had names in all the languages.[1] The natives planted it round their temporary dwellings where they did not form a fixed population. The burial-mounds of the natives of North America who preceded those of our day, the tombs of the Incas, the catacombs of Peru, contain ears or grains of maize, just as the monuments of ancient Egypt contain grains of barley and wheat and millet-seed. In Mexico, a goddess who bore a name derived from that of maize (*Cinteutl*, from *Cintli*) answered to the Ceres of the Greeks, for the first-fruits of the maize harvest were offered to her, as the first-fruits of our cereals to the Greek goddess. At Cusco the virgins of the sun offered sacrifices of bread made from Indian corn. Nothing is better calculated to show the antiquity and generality of the cultivation of a plant than this intimate connection with the religious rites of the ancient inhabitants. We must not, however, attribute to these indications the same importance in America as in the old world. The civilization of the Peruvians under the Incas, and that of the Toltecs and Aztecs in Mexico, has not the extraordinary antiquity of the civilizations of China, Chaldea, and Egypt. It dates at earliest from the beginning of the Christian era; but the cultivation of maize is more ancient than the monuments, to judge from the numerous varieties of the species found in them, and their dispersal into remote regions.

A yet more remarkable proof of antiquity has been discovered by Darwin. He found ears of Indian corn, and eighteen species of shells of our epoch, buried in the soil of the shore in Peru, now at least eighty-five feet

[1] See Martius, *Beiträge zur Ethnographie Amerikas*, p. 127.

above the level of the sea.[1] This maize was perhaps not
cultivated, but in this case it would be yet more
interesting, as an indication of the origin of the species.

Although America has been explored by a great
number of botanists, none have found maize in the
conditions of a wild plant.

Auguste de Saint-Hilaire[2] thought he recognized the
wild type in a singular variety, of which each grain is
enclosed within its sheath or bract. It is known at
Buenos-Ayres under the name *pinsigallo*. It is *Zea Mays
tunicata* of Saint-Hilaire, of which Bonafous gives an
illustration, pl. 5, *bis*, under the name *Zea cryptosperma*.
Lindley[3] also gives a description and a drawing from
seeds brought, it is said, from the Rocky Mountains, but
this is not confirmed by recent Californian floras. A
young Guarany, born in Paraguay on its frontiers, had
recognized this maize, and told Saint-Hilaire that it grew
in the damp forests of his country. This is very insufficient proof that it is indigenous. No traveller to my
knowledge has seen this plant wild in Paraguay or
Brazil. But it is an interesting fact that it has been
cultivated in Europe, and that it often passes into the
ordinary state of maize. Lindley observed it when it
had been only two or three years in cultivation, and
Professor Radic obtained from one sowing 225 ears of the
form *tunicata*, and 105 of the common form with naked
grains.[4] Evidently this form, which might be believed a
true species, but whose country is, however, doubtful, is
hardly even a race. It is one of the innumerable varieties,
more or less hereditary, of which botanists who are considered authorities make only a single species, because of
their want of stability and the transitions which they
frequently present.

On the condition of *Zea Mays,* and its habitation in
America before it was cultivated, we have nothing but con-

[1] Darwin, *Var. of Plants and Anim. under Domest.,* i. p. 320.

[2] A. de Saint-Hilaire, *Ann. Sc. Nat.,* xvi. p. 143.

[3] Lindley, *Journ. of the Hortic. Soc.,* i. p. 114.

[4] I quote these facts from Wittmack, *Ueber Antiken Maïs aus Nord
und Sud Amerika,* p. 87, in *Berlin Anthropol. Ges.,* Nov. 10, 1879.

jectural knowledge. I will state what I take to be the sum of this, because it leads to certain probable indications.

I remark first that maize is a plant singularly unprovided with means of dispersion and protection. The grains are hard to detach from the ear, which is itself enveloped. They have no tuft or wing to catch the wind, and when the ear is not gathered by man the grains fall still fixed in the receptacle, and then rodents and other animals must destroy them in quantities, and all the more that they are not sufficiently hard to pass intact through the digestive organs. Probably so unprotected a species was becoming more and more rare in some limited region, and was on the point of becoming extinct, when a wandering tribe of savages, having perceived its nutritious qualities, saved it from destruction by cultivating it. I am the more disposed to believe that its natural area was small that the species is unique; that is to say, that it constitutes what is called a single-typed genus. The genera which contain few species, and especially the monotypes, have as a rule more restricted areas than others. Palæontology will perhaps one day show whether there ever existed in America several species of *Zea*, or similar Graminæ, of which maize is the last survivor. Now, the genus Zea is not only a monotype, but stands almost alone in its family. A single genus, *Euchlœna* of Schrader, may be compared with it, of which there is one species in Mexico and another in Guatemala; but it is a quite distinct genus, and there are no intermediate forms between it and *Zea*.

Wittmack has made some curious researches in order to discover which variety of maize probably represents the form belonging to the epoch anterior to cultivation. For this purpose he has compared ears and grains taken from the mounds of North America with those from Peru. If these monuments offered only one form of maize, the result would be important, but several different varieties have been found in the mounds and in Peru. This is not very surprising; these monuments are not very ancient. The cemetery of Ancon in Peru, whence Wittmack obtained his best specimens, is nearly contemporary with the discovery of America.[1] Now, at that epoch the number of varieties was already considerable, which proves a much more ancient cultivation.

Experiments in sowing varieties of maize in uncultivated ground several years in succession would perhaps show a reversion to some common form which might then be considered as the original stock, but nothing of this kind has been attempted. The varieties have only been observed to lack stability in spite of their great diversity.

As to the habitation of the unknown primitive form, the following considerations may enable us to guess it. Settled populations can only have been formed where nutritious species existed naturally in soil easy of cultivation. The potato, the sweet potato, and maize doubtless fulfilled these conditions in America, and as the great populations of this part of the world existed first in the high grounds of Chili and Mexico, it is there probably that wild maize existed. We must not look for it in the low-lying regions such as Paraguay and the banks of the Amazon, or the hot districts of Guiana, Panama, and Mexico, since their inhabitants were formerly less numerous. Besides, forests are unfavourable to annuals, and maize does not thrive in the warm damp climates where manioc is grown.[2] On the other hand, its transmission from one tribe to another is easier to comprehend if we suppose the point of departure in the centre, than if we place it at one of the limits of the area over which the species was cultivated at the time of the Incas and the Toltecs, or rather of the Mayas, Nahuas, and Chibchas, who preceded these. The migrations of peoples have not always followed a fixed course from north to south, or from south to north: They have taken different directions according to the epoch and the country.[3] The

[1] Rochebrune, *Recherches Ethnographiques sur les Sépultures Péruviennes d'Ancon*, from an extract by Wittmack in Uhlworm, *Bot. Central-Blatt.*, 1880, p. 1633, where it may be seen that the burial-ground was used before and after the discovery of America.

[2] Sagot, *Cult. des Céréales de la Guyane Franç.* (*Journ. de la Soc. Centr. d'Hortic. de France*, 1872, p. 94).

[3] De Naidaillac, in his work entitled *Les Premiers Hommes et les*

ancient Peruvians scarcely knew the Mexicans, and *vice versâ*, as the total difference of their beliefs and customs shows. As they both early cultivated maize, we must suppose an intermediate point of departure. New Granada seems to me to fulfil these conditions. The nation called Chibcha which occupied the table-land of Bogota at the time of the Spanish conquest, and considered itself aboriginal, was an agricultural people. It enjoyed a certain degree of civilization, as the monuments recently investigated show. Perhaps this tribe first possessed and cultivated maize. It marched with Peru, then but little civilized, on the one hand, and with the Mayas on the other, who occupied Central America and Yucatan. These were often at war with the Nahuas, predecessors of the Toltecs and the Aztecs in Mexico. There is a tradition that Nahualt, chief of the Nahuas, taught the cultivation of maize.[1]

I dare not hope that maize will be found wild, although its habitation before it was cultivated was probably so small that botanists have perhaps not yet come across it. The species is so distinct from all others, and so striking, that natives or unscientific colonists would have noticed and spoken of it. The certainty as to its origin will probably come rather from archæological discoveries. If a great number of monuments in all parts of America are studied, if the hieroglyphical inscriptions of some of these are deciphered, and if dates of migrations and economical events are discovered, our hypothesis will be justified, modified, or rejected.

Article II.—Seeds used for Different Purposes.

Poppy—*Papaver somniferum*, Linnæus.

The poppy is usually cultivated for the oil contained in the seed, and sometimes, especially in Asia, for the sap,

Temps Préhistoriques, gives briefly the sum of our knowledge of these migrations of the ancient peoples of America in general. See especially vol. ii. chap. 9.

[1] De Naidaillac, ii. p. 69, who quotes Bancroft, *The Native Races of the Pacific States.*

extracted by making incisions in the capsules, and from which opium is obtained.

The variety which has been cultivated for centuries escapes readily from cultivation, or becomes almost naturalized in certain localities of the south of Europe.[1] It cannot be said to exist in a really wild state, but botanists are agreed in regarding it as a modification of the poppy called *Papaver setigerum,* which is wild on the shores of the Mediterranean, notably in Spain, Algeria, Corsica, Sicily, Greece, and the island of Cyprus. It has not been met with in Eastern Asia,[2] consequently this is really the original of the cultivated form. Its cultivation must have begun in Europe or in the north of Africa. In support of this theory we find that the Swiss lake-dwellers of the stone age cultivated a poppy which is nearer to *P. setigerum* than to *P. somniferum.* Heer[3] has not been able to find any of the leaves, but the capsule is surmounted by eight stigmas, as in *P. setigerum,* and not by ten or twelve, as in the cultivated poppy. This latter form, unknown in nature, seems therefore to have been developed within historic times. *P. setigerum* is still cultivated in the north of France, together with *P. somniferum,* for the sake of its oil.[4]

The ancient Greeks were well acquainted with the cultivated poppy. Homer, Theophrastus, and Dioscorides mention it. They were aware of the somniferous properties of the sap, and Dioscorides[5] mentions the variety with white seeds. The Romans cultivated the poppy before the republic, as we see by the anecdote of Tarquin and the poppy-heads. They mixed its seeds with their flour in making bread.

The Egyptians of Pliny's time[6] used the juice of the poppy as a medicament, but we have no proof that this

---

[1] Willkomm and Lange, *Prodr. Fl. Hisp.,* iii. p. 872.

[2] Boissier, *Fl. Orient.;* Tchihatcheff, *Asie Mineure;* Ledebour, *Fl. Ross.,* and others.

[3] Heer, *Pflanzen der Pfahlbauten,* p. 32, figs. 65, 66.

[4] De Lanessan, in his translation from Flückiger and Hanbury, *Histoire des Drogues d'Origine Végétale,* i. p. 129.

[5] Dioscorides, *Hist. Plant.,* lib. iv. c. 65.

[6] Pliny, *Hist. Plant.,* lib. xx. c. 18.

plant was cultivated in Egypt in more ancient times.[1] In the Middle Ages[2] and in our own day it is one of the principal objects of cultivation in that country, especially for the manufacture of opium. Hebrew writings do not mention the species. On the other hand, there are one or two Sanskrit names. Piddington gives *chosa*, and Adolphe Pictet *khaskhasa*, which recurs, he says, in the Persian *chashchâsh*, the Armenian *chashchash*,[3] and in Arabic. Another Persian name is *kouknar*.[4] These names, and others I could quote, very different from the *maikôn* (Μήκων) of the Greeks, are an indication of an ancient cultivation in Europe and Western Asia. If the species was first cultivated in prehistoric time in Greece, as appears probable, it may have spread eastward before the Aryan invasion of India, but it is strange that there should be no proof of its extension into Palestine and Egypt before the Roman epoch. It is also possible that in Europe the variety called *Papaver setigerum*, employed by the Swiss lake-dwellers, was first cultivated, and that the variety now grown came from Asia Minor, where the species has been cultivated for at least three thousand years. This theory is supported by the existence of the Greek name *maikôn*, in Dorian *makon*, in several Slav languages, and in those of the peoples to the south of the Caucasus, under the form *mack*.[5]

The cultivation of the poppy in India has been recently extended, because of the importation of opium into China; but the Chinese will soon cease to vex the English by buying this poison of them, for they are beginning eagerly to produce it themselves. The poppy is now grown over more than half of their territory.[6] The species is never wild in the east of Asia, and even as regards China its cultivation is recent.[7]

---

[1] Unger, *Die Pflanze als Errerungs und Betaübungsmittel*, p. 47 ; *Die Pflanzen des Alten Ægyptens*, i. p. 50.

[2] Ebn Baithar, German trans., i. p. 64.

[3] Ad. Pictet, *Origines Indo-Européennes*, edit. 3, vol. i. p. 3(

[4] Ainslie, *Mat. Med. Indica*, i. p. 326.

[5] Nemnich, *Polygl. Lexicon*, p. 848.

[6] Martin, in *Bull. Soc. d'Acclimatation*, 1872, p. 200.

[7] Sir J. Hooker, *Flora of Brit. Ind.*, i. p. 117 ; Bretschneider, *Study and Value*, etc., 47.

The name opium given to the drug extracted from the juice of the capsule is derived from the Greek. Dioscorides wrote *opos* (Οπος). The Arabs converted it into *afiun*,[1] and spread it eastwards even to China.

Flückiger and Hanbury [2] give a detailed and interesting account of the extraction, trade, and use of opium in all countries, particularly in China. Yet I imagine my readers may like to read the following extracts from Dr. Bretschneider's letters, dated from Pekin, Aug. 23, 1881, Jan. 28, and June 18, 1882. They give the most certain information which can be derived from accurately translated Chinese works.

" The author of the *Pent-sao-kang-mou*, who wrote in 1552 and 1578, gives some details concerning the *a-fou-yong* (that is *afioun, opiun*), a foreign drug produced by a species of *ying-sou* with red flowers in the country of Tien-fang (Arabia), and recently used as a medicament in China. In the time of the preceding dynasty there had been much talk of the *a-fou-yong*. The Chinese author gives some details relative to the extraction of opium in his native country, but he does not say that it is also produced in China, nor does he allude to the practice of smoking it. In the *Descriptive Dictionary of the Indian Islands*, by Crawfurd, p. 312, I find the following passage : ' The earliest account we have of the use of opium, not only from the Archipelago, but also from India and China, is by the faithful, intelligent Barbosa.[3] He rates it among the articles brought by the Moorish and Gentile merchants of Western India, to exchange for the cargoes of Chinese junks.' "

"It is difficult to fix the exact date at which the Chinese began to smoke opium and to cultivate the poppy which produces it. As I have said, there is much confusion on this head, and not only European authors, but also the modern Chinese, apply the name *ying-sou* to *P. somniferum* as well as to *P. rhœas*. *P somniferum* is now extensively cultivated in all the provinces

[1] Ebn Baithar, i. p. 64.
[2] Flückiger and Hanbury, *Pharmacographia*, p. 40.
[3] Barbosa's work was published in 1516.

of the Chinese empire, and also in Mantchuria and Mongolia. Williamson (*Journeys in North China, Mantchuria, Mongolia,* 1868, ii. p. 55) saw it cultivated everywhere in Mantchuria. He was told that the cultivation of the poppy was twice as profitable as that of cereals. Potanin, a Russian traveller, who visited Northern Mongolia in 1876, saw immense plantations of the poppy in the valley of Kiran (between lat. 47° and 48°) This alarms the Chinese government, and still more the English, who dread the competition of native opium."

"You are probably aware that opium is eaten, not smoked, in India and Persia. The practice of smoking this drug appears to be a Chinese invention, and modern. Nothing proves that the Chinese smoked opium before the middle of the last century. The Jesuit missionaries to China in the seventeenth and eighteenth centuries do not mention it; Father d'Incarville alone says in 1750 that the sale of opium is forbidden because it was used by suicides. Two edicts forbidding the smoking of opium date from before 1730, and another in 1796 speaks of the progress made by the vice in question. Don Sinibaldo di Mas, who in 1858 published a very good book on China, where he had lived many years as Spanish ambassador, says that the Chinese took the practice from the people of Assam, where the custom had long existed."

So bad a habit, like the use of tobacco or absinth, is sure to spread. It is becoming gradually introduced into the countries which have frequent relations with China. It is to be hoped that it will not attack so large a proportion of the peoples of other countries as in Amoy, where the proportion of opium-smokers are as fifteen to twenty of the adult population.[1]

Arnotto, or **Anatto**—*Bisca orellana,* Linnæus.

The dye, called *rocou* in French, *arnotto* in English, is extracted from the pulp which encases the seed. The inhabitants of the West India Islands, of the Isthmus of Darien, and of Brazil, used it at the time of the discovery of America to stain their bodies red, and the Mexicans.

---

[1] Hughes, *Trade Report,* quoted by Flückiger and Hanbury.

in painting.[1]   The arnotto, a small tree of the order
Bixaceæ, grows wild in the West Indies,[2] and over a
great part of the continent of America between the
tropics. Herbaria and floras abound in indications of
locality, but do not generally specify whether the species
is cultivated, wild, or naturalized. I note, however, that
it is said to be indigenous by Seemann on the north-
west coast of Mexico and Panama, by Triana in New
Granada, by Meyer in Dutch Guiana, and by Piso and
Claussen in Brazil.[3] With such a vast area, it is not
surprising that the species has many names in American
languages; that of the Brazilians, *urucu,* is the origin of
rocou.

It was not very necessary to plant this tree in order
to obtain its product; nevertheless Piso relates that the
Brazilians, in the sixteenth century, were not content
with the wild plant, and in Jamaica, in the seventeenth
century, the plantations of Bixa were common. It was
one of the first species transported from America to the
south of Asia and to Africa. It has become so entirely
naturalized, that Roxburgh[4] believed it to be indigenous
in India.

   Cotton—*Gossypium herbaceum,* Linnæus.

When, in 1855, I sought the origin of the cultivated
cottons,[5] there was still great uncertainty as to the distinction of the species. Since then two excellent works
have appeared in Italy, upon which we can rely ; one by
Parlatore,[6] formerly director of the botanical gardens at
Florence, the other by Todaro,[7] of Palermo. These two

---

   [1] Sloane, *Jamaica,* ii. p. 53.
   [2] Sloane, *ibid.* ; Clos, *Ann. Sc. Nat.,* 4th series, vol. viii. p. 260 ;
Grisebach, *Fl. of Brit. W. Ind. Is.,* p. 20.
   [3] Seemann, *Bot. of Herald.,* pp. 79, 268 ; Triana and Planchon, *Prodr.
Fl. Novo-Granat.,* p. 94; Meyer, *Essequebo,* p. 202 ; Piso, *Hist. Nat.
Brasil,* edit. 1648, p. 65 ; Claussen, in Clos, *ubi supra.*
   [4] Roxburgh, *Fl. Ind.,* ii. p. 581; Oliver, *Fl. Trop. Africa,* i. p. 114.
   [5] *Géogr. Bot. Rais.,* p. 971.
   [6] Parlatore, *Le Specie dei Cotoni,* text in 4to, plates in folio, Florence,
1866.
   [7] Todaro, *Relazione della Coltura dei Cotoni in Italia, segnita da una
Monographia del Genere Gossypium,* text large 8vo, plates in folio, Rome
and Palermo, 1877–78 ; a work preceded by several others of less importance, which were known to Parlatore.

works are illustrated with magnificent coloured plates.
Nothing better can be desired for the cultivated cottons.
On the other hand, our knowledge of the true species,
I mean of those which exist naturally in a wild state,
has not increased as much as it might.   However, the
definition of species seems fairly accurate in the works
of Dr. Masters,[1] whom I shall therefore follow.   This
author agrees with Parlatore in admitting seven well-
known species and two doubtful, while Todaro counts
fifty-four, of which only two are doubtful, reckoning as
species forms with some distinguishing character, but
which originated and are preserved by cultivation.

The common names of the cottons give no assistance ;
they are even calculated to lead us completely astray as
to the origin of the species.   A cotton called Siamese
comes from America; another is called Brazilian or Ava
cotton, according to the fancy or the error of cultivators.

We will first consider *Gossypium herbaceum*, an
ancient species in Asiatic plantations, and now the commonest in Europe and in the United States.   In the
hot countries whence it came, its stem lasts several years,
but out of the tropics it becomes annual from the effect
of the winter's cold.   The flower is generally yellow, with
a red centre; the cotton yellow or white, according to
the variety.   Parlatore examined in herbaria several
wild specimens, and cultivated others derived from wild
plants of the Indian Peninsula.   He also admits it to be
indigenous in Burmah and in the Indian Archipelago,
from the specimens of collectors, who have not perhaps
been sufficiently careful to verify its wild character.

Masters regards as undoubtedly wild in Sindh a form
which he calls *Gossypium Stocksii*, which he says is
probably the wild condition of *Gossypium herbaceum*,
and of other cottons cultivated in India for a long time.
Todaro, who is not given to uniting many forms in a
single species, nevertheless admits the identity of this
variety with the common *G. herbaceum*.   The yellow
colour of the cotton is then the natural condition of the

[1] Masters, in Oliver, *Fl. Trop. Afr.*, i. p. 210; and in Sir J. Hooker,
Fl. Brit. Ind., i. p. 346.

species. The seed has not the short down which exists
between the longer hairs in the cultivated *G. herbaceum*.

Cultivation has probably extended the area of the
species beyond the limits of the primitive habitation.
This is, I imagine, the case in the Sunda Islands and the
Malay Peninsula, where certain individuals appear more
or less wild. Kurz,[1] in his Burmese flora, mentions
G. herbaceum, with yellow or white cotton, as cultivated
and also as wild in desert places and waste ground.

The herbaceous cotton is called *kapase* in Bengali,
kapas in Hindustani, which shows that the Sanskrit
word *karpassi* undoubtedly refers to this species.[2] It
was early cultivated in Bactriana, where the Greeks had
noticed it at the time of the expedition of Alexander.
Theophrastus speaks of it [3] in such a manner as to leave
no doubt. The tree-cotton of the Isle of Tylos, in the
Persian Gulf, of which he makes mention further on,[4]
was probably also *G. herbaceum ;* for Tylos is not far
from India, and in such a hot climate the herbaceous
cotton becomes a shrub. The introduction of a cotton
plant into China took place only in the ninth or tenth
century of our era, which shows that probably the area
of *G. herbaceum* was originally limited to the south and
east of India. The knowledge and perhaps the cultivation of the Asiatic cotton was propagated in the Græco-
Roman world after the expedition of Alexander, but
before the first centuries of the Christian era.[5] If the
byssos of the Greeks was the cotton plant, as most
scholars think, it was cultivated at Elis, according to
Pausanias and Pliny ;[6] but Curtius and C. Ritter[7] consider the word *byssos* as a general term for threads,
and that it was probably applied in this case to fine
linen. It is evident that the cotton was never, or very
rarely, cultivated by the ancients. It is so useful that
it would have become common if it had been introduced

[1] Kurz, *Forest Flora of British Burmah*, i. p. 129.
[2] Piddington, *Index.*   [3] Theophrastus, *Hist. Plant.*, lib. iv. cap. 5.
[4] *Ibid.*, lib. iv. cap. 9.   [5] Bretschneider, *Study and Value*, etc., p. 7.
[6] Pausanias, lib. v., cap. 5 ; lib. vi. cap. 26 ; Pliny, lib. xix. cap. 1.
See Brandes, *Baumwolle*, p. 96.
[7] C. Ritter, *Die Geographische Verbreitung der Baumwolle*, p. 25.

into a single locality—in Greece, for instance. It was afterwards propagated on the shores of the Mediterranean by the Arabs, as we see from the name *qutn* or *kutn*,[1] which has passed into the modern languages of the south of Europe as *cotone, coton, algodon.* Eben el Awan, of Seville, who lived in the twelfth century, describes its cultivation as it was practised in his time in Sicily, Spain, and the East.[2]

Gossypium herbaceum is the species most cultivated in the United States.[3] It was probably introduced there from Europe. It was a new cultivation a hundred years ago, for a bale of North American cotton was confiscated at Liverpool in 1774, on the plea that the cotton-plant did not grow there.[4] The silky cotton (*sea island*) is another species, American, of which I shall presently speak.

Tree-Cotton—*Gossypium arboreum*, Linnæus.

This species is taller and of longer duration than the herbaceous cotton; the lobes of the leaf are narrower, the bracts less divided or entire. The flower is usually pink, with a red centre. The cotton is always white.

According to Anglo-Indian botanists, this is not, as it was supposed, an Indian species, and is even rarely cultivated in India. It is a native of tropical Africa. It has been seen wild in Upper Guinea, in Abyssinia, Sennaar, and Upper Egypt.[5] So great a number of collectors have brought it from these countries, that there is no room for doubt; but cultivation has so diffused and mixed this species with others that it has been described under several names in works on Southern Asia.

---

[1] It is impossible not to remark the resemblance between this name and that of flax in Arabic, *kattan* or *kittan;* it is an example of the confusion which takes place in names where there is an analogy between the products.

[2] De Lasteyrie, *Du Cotonnier,* p. 290.

[3] Torrey and Asa Gray, *Flora of North America,* i. p. 230; Darlington, *Agricultural Botany,* p. 16.

[4] Schouw, *Naturschilderungen,* p. 152.

[5] Masters, in Oliver, *Fl. Trop. Afr.,* i. p. 211; Hooker, *Fl. of Brit. Ind.,* i. p. 347; Schweinfurth and Ascherson, *Aufzählung,* p. 265 (under the name *Gossypium nigrum*); Parlatore, *Specie dei Cotoni,* p. 25.

Parlatore attributed to *G. arboreum* some Asiatic specimens of *G. herbaceum,* and a plant but little known which Forskal found in Arabia. He suspected from this that the ancients had known *G. arboreum* as well as *G. herbaceum.* Now that the two species are better distinguished, and that the origin of both is known, this does not seem probable. They knew the herbaceous cotton through India and Persia, while the tree-cotton can only have come to them through Egypt. Parlatore himself has given a most interesting proof of this. Until his work appeared in 1866, it was not certain to what species belonged some seeds of the cotton plant which Rosellini found in a vase among the monuments of ancient Thebes.[1] These seeds are in the Florence museum. Parlatore examined them carefully, and declares them to belong to *Gossypium arboreum.*[2] Rosellini is certain he was not imposed upon, as he was the first to open both the tomb and the vase. No archæologist has since seen or read signs of the cotton plant in the ancient times of Egyptian civilization. How is it that a plant so striking, remarkable for its flowers and seed, was not described nor preserved habitually in the tombs if it were cultivated? How is it that Herodotus, Dioscorides, and Theophrastus made no mention of it when writing of Egypt? The cloths in which all the mummies are wrapt, and which were formerly supposed to be cotton, are always linen according to Thompson and many other observers who are familiar with the use of the microscope. Hence I conclude that if the seeds found by Rosellini were really ancient they were a rarity, an exception to the common custom, perhaps the product of a tree cultivated in a garden, or perhaps they came from Upper Egypt, a country where we know the tree-cotton to be wild. Pliny[3] does not say that cotton was cultivated in Lower Egypt; but here is a translation of his very remarkable passage, which is often quoted. "The upper part of Egypt, towards Arabia, produces a shrub which some

---

[1] Rosellini, *Monumenti dell' Egizia*, p. 2; *Mon. Civ.*, i. p. 60.
[2] Parlatore, *Specie dei Cotoni*, p. 16.
[3] Pliny, *Hist. Plant.*, lib. xix. cap. 1.

call *gossipion* and others *xylon,* whence the name *xylina* given to the threads obtained from it. It is low-growing, and bears a fruit like that of the bearded nut, and from the interior of this is taken a wool for weaving. None is comparable to this in softness and whiteness." Pliny adds, "The cloth made from it is used by preference for the dress of the Egyptian priests." Perhaps the cotton destined to this purpose was sent from Upper Egypt, or perhaps the author, who had not seen the fabrication, and did not possess a microscope, was mistaken in the nature of the sacerdotal raiment, as were our contemporaries who handled the grave-cloths of hundreds of mummies before suspecting that they were not cotton. Among the Jews, the priestly robes were commanded to be of linen, and it is not likely that their custom was different to that of the Egyptians.

Pollux,[1] born in Egypt a century later than Pliny, expresses himself clearly about the cotton plant, of which the thread was used by his countrymen; but he does not say whence the shrub came, and we cannot tell whether it was *Gossypium arboreum* or *G. herbaceum.* It does not even appear whether the plant was cultivated in Lower Egypt, or if the cotton came from the more southern region. In spite of these doubts, it may be suspected that a cotton plant, probably that of Upper Egypt, had recently been introduced into the Delta. The species which Prosper Alpin had seen cultivated in Egypt in the sixteenth century was the tree-cotton. The Arabs, and afterwards Europeans, preferred and transported into different countries the herbaceous cotton rather than the tree-cotton, which yields a poorer product and requires more heat.

Regarding the two cottons of the old world, I have made as little use as possible of arguments based upon Greek names, such as βυσσος, σινδον, ξυλον, Oθων, etc., or Sanskrit names, and their derivatives, as *carbasa,* *carpas,* or Hebrew names, *schesch, buz,* which are doubtfully attributed to the cotton tree. This has been a

---

[1] Pollux, *Onomasticon,* quoted by C. Ritter, *ubi supra,* p. 26.

fruitful subject of discussion,[1] but the clearer distinction
of species and the discovery of their origin greatly
diminishes the importance of these questions—to naturalists, at least, who prefer facts to words. Moreover,
Reynier, and after him C. Ritter, arrived in their researches at a conclusion which we must not forget: that
these same names were often applied by ancient peoples
to different plants and tissues—to linen and cotton, for
example. In this case as in others, modern botany
explains ancient words where words and the com-
mentaries of philologists may mislead.

Barbados Cotton—*Gossypium barbadense*, Linnæus.

At the time of the discovery of America, the Spaniards
found the cultivation and use of cotton established from
the West India Islands to Peru, and from Mexico to
Brazil. The fact is proved by all the historians of the
epoch. But it is still very difficult to tell what were the
species of these American cottons and in what countries
they were indigenous. The botanical distinction of the
American species or varieties is in the last degree confused. Authors, even those who have seen large collections of growing cotton plants, are not agreed as to the
characters. They are also embarrassed by the difficulty
of deciding which of the specific names of Linnæus should
be retained, for the original definitions are insufficient.
The introduction of American seed into African and
Asiatic plantations has given rise to further complica-
tions, as botanists in Java, Calcutta, Bourbon, etc., have
often described American forms as species under different
names. Todaro admits ten American species; Parlatore
reduced them to three, which answer, he says, to *Gossypium hirsutum*, *G. barbadense*, and *G. religiosum* of
Linnæus; lastly, Dr. Masters unites all the American
forms into a single species which he calls *G. barbadense,*
giving as the chief character that the seed bears only

[1] Reynier, *Économie des Arabes et des Juifs.*, p. 363 ; Bertoloni, *Nov.
Act. Acad. Bonon.*, ii. p. 213, and *Miscell. Bot.*, 6 ; Viviani, in *Bibl. Ital.*,
vol. lxxxi. p. 94 ; C. Ritter, *Géogr. Verbreitung der Baumwolle*, in 4to. ;
Targioni, *Cenni Storici*, p. 93 ; Brandis, *Der Baumwolle in Alterthum,*
in 8vo, 1880.

long hairs, whereas the species of the old world have a short down underneath the longer hairs.[1] The flower is yellow, with a red centre. The cotton is white or yellow. Parlatore strove to include fifty or sixty of the cultivated forms under one or other of the three heads he admits, from the study of plants in gardens or herbaria. Dr. Masters mentions but few synonyms, and it is possible that certain forms with which he is not acquainted do not come under the definition of his single species.

Where there is such confusion it would be the best course for botanists to seek with care the *Gossypia*, which are wild in America, to constitute the one or more species solely upon these, leaving to the cultivated species their strange and often absurd and misleading names. I state this opinion because with regard to no other genus of cultivated plants have I felt so strongly that natural history should be based upon natural facts, and not upon the artificial products of cultivation. If we start from this point of view, which has the merit of being a truly scientific method, we find unfortunately that our knowledge of the cottons indigenous in America is still in a very elementary state. At most we can name only one or two collectors who have found *Gossypia* really identical with or very similar to certain cultivated forms.

We can seldom trust early botanists and travellers on this head. The cotton plant grows sometimes in the neighbourhood of plantations, and becomes more or less naturalized, as the down on the seeds facilitates accidental transport. The usual expression of early writers—such a cotton plant *grows* in such a country—often means a cultivated plant. Linnæus himself in the eighteenth century often says of a cultivated species, " *habitat*," and he even says it sometimes without good ground.[2] Hernandez, one of the most accurate among sixteenth-century authors, is quoted as having described and figured a wild *Gossypium* in Mexico, but the text

---

[1] Masters, in Oliver, *Flora of Trop. Africa*, i. p. 322 ; and in Hooker, *Flora of Brit. India*, i. p. 347.

[2] He says, for instance, of *Gossypium herbaceum*, which is certainly of the old world, as facts known before his time show, " *habitat in America*."

suggests some doubts as to the wild condition of this plant,[1] which Parlatore believes to be *G. hirsutum,* Linnæus. Hemsley,[2] in his catalogue of Mexican plants, merely says of a *Gossypium* which he calls *barbadense,* " wild and cultivated." He gives no proof of the former condition. Macfadyen[3] mentions three forms wild and cultivated in Jamaica. He attributes specific names to them, and adds that they possibly all may be included in Linnæus' *G. hirsutum.* Grisebach[4] admits that one species, *G. barbadense,* is wild in the West Indies. As to the specific distinctions, he declares himself unable to establish them with certainty.

With regard to New Grenada, Triana[5] describes a *Gossypium* which he calls *G. barbadense,* Linnæus, and which he says is " cultivated and half wild along the Rio Seco, in the province of Bogota, and in the valley of the Cauca near Cali; " and he adds a variety, *hirsutum,* growing (he does not say whether spontaneously or no) along the Rio Seco. I cannot discover any similar assertion for Peru, Guiana, and Brazil;[6] but the flora of Chili, published by Cl. Gay,[7] mentions a *Gossypium,* "almost wild in the province of Copiapo," which the writer attributes to the variety *G. peruvianum,* Cavanilles. Now, this author does not say the plant is wild, and Parlatore classes it with *G. religiosum,* Linnæus.

An important variety of cultivation is that of the cotton with long silky down, called by Anglo-Americans *sea island,* or *long staple cotton,* which Parlatore ranks with *G. barbadense,* Linnæus. It is considered to be of American origin, but no one has seen it wild.

In conclusion, if historical records are positive in all that concerns the use of cotton in America from a time far earlier than the arrival of Europeans, the natural

---

[1] *Nascitur in calidis humidisque cultis præcipue locis* (Hernandez, *Novæ Hispaniæ Thesaurus,* p. 308).
[2] Hemsley, *Biologia Centrali-Americana,* i. p. 123.
[3] Macfadyen, *Flora of Jamaica,* p. 72.
[4] Grisebach, *Flora of Brit. W. India Is.,* p. 86.
[5] Triana and Planchon, *Prodr. Fl. Novo-Granatensis,* p. 170.
[6] The Malvaceæ have not yet appeared in the *Flora Brasiliensis.*
[7] Cl. Gay, *Flora Chilena,* i. p. 312.

wild habitation of the plant or plants which yield this
product is yet but little known. We become aware on
this occasion of the absence of floras of tropical America,
similar to those of the Dutch and English colonies of
Asia and Africa.

Mandubi, Pea-nut, Monkey-nut — *Arachis hypogœa*,
Linnæus.

Nothing is more curious than the manner in which
this leguminous plant matures its fruits. It is cultivated
in all hot countries, either for the seed, or for the oil
contained in the cotyledons.[1] Bentham has given, in
his *Flora of Brazil*, in folio, vol. xv. pl. 23, complete
details of the plant, in which may be seen how the
flower-stalk bends downwards and plunges the pod into
the earth to ripen.

The origin of the species was disputed for a century,
even by those botanists who employ the best means to
discover it. It is worth while to show how the truth
was arrived at, as it may serve as a guide in similar
cases. I will quote, therefore, what I wrote in 1855,[2]
giving in conclusion new proofs which allow no possibility of further doubt.

"Linnæus[3] said of the *Arachis*, 'it inhabits Surinam,
Brazil, and Peru.' As usual with him, he does not specify
whether the species was wild or cultivated in these
countries. In 1818, R. Brown[4] writes: 'It was probably introduced from China into the continent of India,
Ceylon, and into the Malay Archipelago, where, in spite
of its now general cultivation, it is thought not to be
indigenous, particularly from the names given to it. I
consider it not improbable that it was brought from
Africa into different parts of equatorial America, although,
however, it is mentioned in some of the earliest writings
on this continent, particularly on Peru and Brazil. According to Sprengel, it is mentioned by Theophrastus as

[1] The *Gardener's Chronicle* of Sept. 4, 1880, gives details about the
cultivation of this plant, the use of its seeds, and the extensive exportation of them from the west coast of Africa, Brazil, and India to Europe.
[2] A. de Candolle, *Géographie Botanique Raisonnée*, p. 962.
[3] Linnæus, *Species Plantarum*, p. 1040.
[4] R. Brown, *Botany of Congo*, p. 53.

cultivated in Egypt, but it is not at all evident that the
Arachis is the plant to which Theophrastus alludes in
the quoted passage. If it had been formerly cultivated
in Egypt it would probably still exist in that country,
whereas it does not occur in Forskal's catalogue nor in
Delile's more extended flora. There is nothing very
unlikely,' continues Brown, 'in the hypothesis that the
Arachis is indigenous both in Africa and America; but
if it is considered as existing originally in one of these
continents only, it is more probable that it was brought
from China through India to Africa, than that it took
the contrary direction.' My father in 1825, in the *Prodromus* (ii. p. 474), returned to Linnæus' opinion, and
admitted without hesitation the American origin. Let
us reconsider the question " (I said in 1855) "with the
aid of the discoveries of modern science.

"*Arachis hypogæa* was the only species of this singular
genus known. Six other species, all Brazilian, have
since been discovered.[1] Thus, applying the rule of probability of which Brown first made great use, we incline
à priori to the idea of an American origin. We must
remember that Marcgraf[2] and Piso[3] describe and figure
the plant as used in Brazil, under the name *mandubi*,
which seems to be indigenous. They quote Monardes, a
writer of the end of the sixteenth century, as having
indicated it in Peru under a different name, *anchic*.
Joseph Acosta[4] merely mentions an American name,
mani, and speaks of it with other species which are not
of foreign origin in America. The *Arachis* was not
ancient in Guiana, in the West Indies, and in Mexico.
Aublet[5] mentions it as a cultivated plant, not in Guiana,
but in the Isle of France. Hernandez does not speak of
it. Sloane[6] had seen it only in a garden, grown from
seeds brought from Guinea. He says that the slave-
dealers feed the negroes with it on their passage from

---

[1] Bentham, in *Trans. Linn. Soc.*, xviii. p. 159; Walpers, *Repertorium*,
i. p. 727.

[2] Marcgraf and Piso, *Brasil.*, p. 37, edit. 1648.

[3] *Ibid.*, edit. 1658, p. 256.

[4] Acosta, *Hist. Nat. Ind.*, French. trans., 1598, p. 165.

[5] Aublet, *Pl. Guyan*, p. 765.   [6] Sloane, *Jamaica*, p. 184.

Africa, which indicates a then very general cultivation in Africa. Pison, in his second edition (1658, p. 256), not in that of 1648, gives a figure of a similar fruit imported from Africa into Brazil under the name *mandobi*, very near to the name of the Arachis, *mundubi*. From the three leaflets of the plant it would seem to be the *Voandzeia*, so often cultivated; but the fruit seems to me to be longer than in this genus, and it has two or three seeds instead of one or two. However this may be, the distinction drawn by Piso between these two subterranean seeds, the one Brazilian, the other African, tends to show that the *Arachis* is Brazilian.

"The antiquity and the generality of its cultivation in Africa is, however, an argument of some force, which compensates to a certain degree its antiquity in Brazil, and the presence of six other *Arachis* in the same country. I would admit its great value if the *Arachis* had been known to the ancient Egyptians and to the Arabs; but the silence of Greek, Latin, and Arab authors, and the absence of the species in Egypt in Forskal's time, lead me to think that its cultivation in Guinea, Senegal,[1] and the east coast of Africa [2] is not of very ancient date. Neither has it the marks of a great antiquity in Asia. No Sanskrit name for it is known,[3] but only a Hindustani one. Rumphius [4] says that it was imported from Japan into several islands of the Indian Archipelago. It would in that case have borne only foreign names, like the Chinese name, for instance, which signifies only ' earth-bean.' At the end of the last century it was generally cultivated in China and Cochin-China. Yet, in spite of Rumphius's theory of an introduction into the islands from China or Japan, I see that Thunberg does not speak of it in his *Japanese Flora*. Now, Japan has had dealings with China for sixteen centuries, and cultivated plants, natives of one of the two countries, were commonly early introduced into the other. It is not mentioned by Forster among the plants employed in the

---

[1] Guillemin and Perrottet, *Fl. Senegal.*   [2] Loureiro, *Fl. Cochin.*
[3] Roxburgh, *Fl. Ind.*, iii. p. 280 ; Piddington, *Index.*
[4] Rumphius, *Herb. Amb.*, **v.** p. 426.

small islands of the Pacific. All these facts point to an American, I might even say a Brazilian, origin. None of the authors I have consulted mentions having seen the plant wild, either in the old or the new world. Those who indicate it in Africa or Asia are careful to say the plant is cultivated. Marcgraf does not say so, writing of Brazil, but Piso says the species is planted."

Seeds of *Arachis* have been found in the Peruvian tombs at Ancon,[1] which shows some antiquity of existence in America, and supports the opinion I expressed in 1855. Dr. Bretschneider's study of Chinese works [2] over-sets Brown's hypothesis. The *Arachis* is not mentioned in the ancient works of this country, nor even in the *Pent-sao*, published in the sixteenth century. He adds that he believes the plant was only introduced in the last century.

All the recent floras of Asia and Africa mention the species as a cultivated one, and most authors believe it to be of American origin. Bentham, after satisfying himself that it had not been found wild in America or elsewhere, adds that it is perhaps a form derived from one of the six other species wild in Brazil, but he does not say which. This is probable enough, for a plant provided with an efficacious and very peculiar manner of germinating does not seem of a nature to become extinct. It would have been found wild in Brazil in the same condition as the cultivated plant, if the latter were not a product of cultivation. Works on Guiana and other parts of America mention the species as a cultivated one; Grisebach [3] says, moreover, that in several of the West India islands it becomes naturalized from cultivation.

A genus of which all the well-known species are thus placed in a single region of America can scarcely have a species common to both hemispheres; it would be too

---

[1] Rochebrune, from the extract in the *Botanisches Centralblatt*, 1880, p. 1634.

[2] *Study and Value of Chinese Botanical Works*, p. 18.

[3] Grisebach, *Fl. Brit. W. Ind. Is.*, p. 189.

great an exception to the law of geographical botany. But then how did the species (or cultivated variety) pass from the American continent to the old world? This is hard to guess, but I am inclined to believe that the first slave-ships carried it from Brazil to Guinea, and the Portuguese from Brazil into the islands to the south of Asia, in the end of the fifteenth century.

Coffee—*Coffea arabica*, Linnæus.

This shrub, belonging to the family of the Rubiaceæ, is wild in Abyssinia,[1] in the Soudan,[2] and on the coasts of Guinea and Mozambique.[3] Perhaps in these latter localities, so far removed from the centre, it may be naturalized from cultivation. No one has yet found it in Arabia, but this may be explained by the difficulty of penetrating into the interior of the country. If it is discovered there it will be hard to prove it wild, for the seeds, which soon lose their faculty of germinating, often spring up round the plantations and naturalize the species. This has occurred in Brazil and the West India Islands,[4] where it is certain that the coffee plant was never indigenous.

The use of coffee seems to be very ancient in Abyssinia. Shehabeddin Ben, author of an Arab manuscript of the fifteenth century (No. 944 of the Paris Library), quoted in John Ellis's excellent work,[5] says that coffee had been used in Abyssinia from time immemorial. Its use, even as a drug, had not spread into the neighbouring countries, for the crusaders did not know it, and the celebrated physician Ebn Baithar, born at Malaga, who had travelled over the north of Africa and Syria at the beginning of the thirteenth century of the Christian era, does not mention coffee.[6] In 1596 Bellus sent to de l'Ecluse some seeds from which the Egyptians ex-

---

[1] Richard, *Tentamen Fl. Abyss.*, i. p. 349; Oliver, *Fl. Trop. Afr.*, iii. p. 180.

[2] Ritter, quoted in *Flora*, 1846, p. 704.

[3] Meyen, *Géogr. Bot.*, English trans., p. 384; Grisebach, *Fl. of Brit. W. Ind. Is.*, p. 338.

[4] H. Welter, *Essai sur l'Histoire du Café*, 1 vol. in 8vo, Paris, 1868.

[5] Ellis, *An Historical Account of Coffee*, 1774.

[6] Ebn Baithar, Sondtheimer's trans., 2 vols. 8vo, 1842.

tracted the drink *cavé*.[1] Nearly at the same time Prosper
Alpin became acquainted with coffee in Egypt itself. He
speaks of the plant as the "arbor *bon,* cum fructu suo
buna." The name *bon* recurs also in early authors under
the forms *bunnu, buncho, bunca.*[2] The names *cahue,
cahua, chaubé,*[3] *cavé,*[4] refer rather in Egypt and Syria to
the prepared drink, whence the French word *café.* The
name *bunnu,* or something similar, is certainly the primitive name of the plant which the Abyssinians still call
boun.[5]

If the use of coffee is more ancient in Abyssinia than
elsewhere, that is no proof that its cultivation is very
ancient. It is very possible that for centuries the berries
were sought in the forests, where they were doubtless very
common. According to the Arabian author quoted above,
it was a mufti of Aden, nearly his contemporary, who,
having seen coffee drunk in Persia, introduced the practice at Aden, whence it spread to Mocha, into Egypt, etc.
He says that the coffee plant grew in Arabia.[6] Other
fables or traditions exist, according to which it was
always an Arabian priest or a monk who invented the
drink,[7] but they all leave us in uncertainty as to the
date of the first cultivation of the plant. However this
may be, the use of coffee having been spread first in
the east, afterwards in the west, in spite of a number
of prohibitions and absurd conflicts,[8] its production
became important to the colonies. Boerhave tells us
that the Burgermeister of Amsterdam, Nicholas Witsen,
director of the East India Company, urged the Governor
of Batavia, Van Hoorn, to import coffee berries from Arabia
to Batavia. This was done, and in 1690 Van Hoorn sent
some living plants to Witsen. These were placed in the
Botanical Gardens of Amsterdam, founded by Witsen,
where they bore fruit. In 1714, the magistrates of the

---

[1] Bellus, *Epist. ad Clus.,* p. 309.   [2] Rauwolf, Clusius.
[3] Rauwolf; Bauhin, *Hist.,* i. p. 422.   [4] Bellus, *ubi supra.*
[5] Richard, *Tentamen Fl. Abyss.,* p. 350.
[6] An extract from the same author in Playfair, *Hist. of Arabia
Felix,* Bombay, 1859, does not mention this assertion.
[7] *Nouv. Dict. d'Hist. Nat.,* iv. p. 552.
[8] Ellis, *ubi supra; Nouv. Dict., ibid.*

town sent a flourishing plant covered with fruit to Louis XIV., who placed it in his garden at Marly. Coffee was also grown in the hothouses of the king's garden in Paris. One of the professors of this establishment, Antoine de Jussieu, had already published in 1713, in the *Mémoires de l'Académie des Sciences*, an interesting description of the plant from one which Pancras, director of the Botanical Garden at Amsterdam, had sent to him.

The first coffee plants grown in America were introduced into Surinam by the Dutch in 1718. The Governor of Cayenne, de la Motte-Aigron, having been at Surinam, obtained some plants in secret and multiplied them in 1725.[1] The coffee plant was introduced into Martinique by de Clieu,[2] a naval officer, in 1720, according to Deleuze;[3] in 1723, according to the *Notices Statistiques sur les Colonies Françaises*.[4] Thence it was introduced into the other French islands, into Guadaloupe, for instance, in 1730.[5] Sir Nicholas Lawes first grew it in Jamaica.[6] From 1718 the French East India Company had sent plants of Mocha coffee to Bourbon;[7] others say[8] that it was even in 1717 that a certain Dufougerais-Grenier had coffee plants brought from Mocha into this island. It is known how the cultivation of this shrub has been extended in Java, Ceylon, the West Indies, and Brazil. Nothing prevents it from spreading in nearly all tropical countries, especially as the coffee plant thrives

---

[1] This detail is borrowed from Ellis, *Diss. Caf.*, p. 16. In the *Notices Statistiques sur les Colonies Françaises* (ii. p. 46) I find : "About 1716 or 1721, fresh seeds of the coffee having been brought secretly from Surinam, in spite of the precautions of the Dutch, the cultivation of this colonial product became naturalized at Cayenne."

[2] The name of this sailor has been spelt in several ways—Declieux, Duclieux, Desclieux. From the information supplied me at the *ministère de la guerre*, I learn that de Clieu was a gentleman, and a connection of the Comte de Maurepas. He was born in Normandy, went into the navy in 1702, and retired in 1760, after a distinguished career. He died in 1775. The official reports have not neglected to mention the important fact that he introduced the coffee plant into the French colonies.

[3] Deleuze, *Hist. du Muséum*, i. p. 20.

[4] *Not. Stat. Col. Franç.*, i. p. 30.   *Ibid.*, i. p. 209.

[6] Martin, *Stat. Col. Brit. Emp.*   [7] *Nouv. Dict. Hist. Nat.*, iv. p. 135.

[8] *Not. Stat. Col. Franç.*, ii. p. 84.

on sloping ground and in poor soils where other crops
cannot flourish.   It corresponds in tropical agriculture to
the vine in Europe and tea in China.

Further details may be found in the volume published
by H. Welter [1] on the economical and commercial history
of coffee.   The author adds an interesting chapter on
the various fair or very bad substitutes used for a com-
modity which it is impossible to overrate in its natural
condition.

Liberian Coffee—*Coffea liberica*, Hiern.[2]

Plants of this species have for some years been sent
from the Botanical Gardens at Kew into the English
colonies.   It grows wild in Liberia, Angola, Golungo
Alto,[3] and probably in several other parts of western
tropical Africa.

It is of stronger growth than the common coffee, and
the berries, which are larger, yield an excellent product.
The official reports of Kew Gardens by the learned
director, Sir Joseph Hooker, show the progress of this
introduction, which is very favourably received, especially
in Dominica.

Madia—*Madia sativa*, Molina.

The inhabitants of Chili before the discovery of
America cultivated this annual species of the Composite
family, for the sake of the oil contained in the seed.
Since the olive has been extensively planted, the madia
is despised by the Chilians, who only complain of the
plant as a weed which chokes their gardens.[4]   The
Europeans began to cultivate it with indifferent success,
owing to its bad smell.

The madia is indigenous in Chili and also in California.[5]   There are other examples of this disjunction of
habitation between the two countries.[6]

[1] H. Welter, *Essai sur l'Histoire du Café*, 1 vol. 8vo, Paris, 1868.

[2] In Hiern, *Trans. Linn. Soc.*, 2nd series, vol. i. p. 171, pl. 24.   This
plate is reproduced in the Report of the Royal Botanical Gardens at
Kew for 1876.

[3] Oliver, *Fl. Trop. Afr.*, iii. p. 181.

[4] Cl. Gay, *Fl. Chilena*, iv. p. 268.

[5] Asa Gray, in Watson, *Bot. of California*, i. p. 359.

[6] A. de Candolle, *Géogr. Bot. Rais.*, p. 1047.

Nutmeg—*Myristica fragrans*, Houttuyn.

The nutmeg, a little tree of the order *Myristiceæ*, is wild in the Moluccas, principally in the Banda Islands.[1] It has long been cultivated there, to judge from the considerable number of its varieties. Europeans have received the nutmeg by the Asiatic trade since the Middle Ages, but the Dutch long possessed the monopoly of its cultivation. When the English owned the Moluccas at the end of the last century, they carried live nutmeg trees to Bencoolen and into Prince Edward's Islands.[2] It afterwards spread to Bourbon, Mauritius, Madagascar, and into some of the colonies of tropical America, but with indifferent success from a commercial point of view.

Sesame—*Sesamum indicum*, de Candolle ; *S. indicum* and *S. orientale*, Linnæus.

Sesame has long been cultivated in the hot regions of the old world for the sake of the oil extracted from the seeds.

The order *Pedalineæ* to which this annual belongs is composed of several genera distributed through the tropical parts of Asia, Africa, and America. Each genus has only a small number of species. Sesamum, in the widest sense of the name,[3] has ten, all African except perhaps the cultivated species whose origin we are about to seek. The latter forms alone the true genus Sesamum, which is a section in Bentham and Hooker's work. Botanical analogy points to an African origin, but the area of a considerable number of plants is known to extend from the south of Asia into Africa. Sesame has two *races*, the one with black, the other with white seed, and several varieties differing in the shape of the leaf. The difference in the colour of the seeds is very ancient, as in the case of the poppy.

The seeds of sesame often sow themselves outside plantations, and more or less naturalize the species. This has been observed in regions very remote one from the

---

[1] Rumphius, *Amboin.*, ii. p. 17 ; Blume, *Rumphia*, i. p. 180.
[2] Roxburgh, *Fl. Indica*, iii. p. 845.
[3] Bentham and Hooker, *Genera Pl.*, ii. p. 1059.

other; for instance, in India, the Sunda Isles, Egypt, and even in the West India Islands, where its cultivation is certainly of modern introduction.[1] This is perhaps the reason that no author asserts he has found it in a wild state except Blume,[2] a trustworthy observer, who mentions a variety with redder flowers than usual growing in the mountains of Java. This is doubtless an indication of origin, but we need others to establish a proof. I shall seek them in the history of its cultivation. The country where this began should be the ancient habitation of the species, or have had dealings with this ancient habitation.

That its cultivation dates in Asia from a very early epoch is clear from the diversity of names. Sesame is called in Sanskrit *tila*,[3] in Malay *widjin*, in Chinese *moa* (Rumphius) or *chi-ma* (Bretschneider), in Japanese *koba*.[4] The name *sesam* is common to Greek, Latin, and Arabic, with trifling variations of letter. Hence it might be inferred that its area was very extended, and that the cultivation of the plant was begun independently in several different countries. But we must not attribute too much importance to such an argument. Chinese works seem to show that sesame was not introduced into China before the Christian era. The first certain mention of it occurs in a book of the fifth or sixth century, entitled *Tsi-min-yao-chou*.[5] Before this there is confusion between the name of this plant and that of flax, of which the seed also yields an oil, and which is not very ancient in China.[6]

Theophrastus and Dioscorides say that the Egyptians cultivated a plant called sesame for the oil contained in its seed, and Pliny adds that it came from India.[7] He

---

[1] Pickering, *Chronol. History of Plants*, p. 223; Rumphius, *Herb. Amb.*, v. p. 204; Miquel, *Flora Indo-Batava*, ii. p. 760; Schweinfurth and Ascherson, *Aufzählung*, p. 273; Grisebach, *Fl. Brit. W. Ind. Is.*, p. 458.

[2] Blume, *Bijdragen*, p. 778.

[3] Roxburgh, *Fl. Ind.*, edit. 1832, vol. iii. p. 100; Piddington, *Index.*

[4] Thunberg, *Fl. Jap.*, p. 254.

[5] Bretschneider, letter of Aug. 23, 1801.

[6] *Ibid.*, *On Study*, etc., p. 16.

[7] Theophrastus, lib. viii. cap. 1, 5; Dioscorides, lib. ii. cap. 121; Pliny, *Hist.*, lib. xviii. cap. 10.

also speaks of a sesame wild in Egypt from which oil was extracted, but this was probably the castor-oil plant.[1] It is not proved that the ancient Egyptians before the time of Theophrastus cultivated sesame. No drawing or seeds have been found in the monuments. A drawing from the tomb of Rameses III. show the custom of mixing small seeds with flour in making pastry, and in modern times this is done with sesame seeds, but others are also used, and it is not possible to recognize in the drawing those of the sesame in particular.[2] If the Egyptians had known the species at the time of the Exodus, eleven hundred years before Theophrastus, there would probably have been some mention of it in the Hebrew books, because of the various uses of the seed and especially of the oil. Yet commentators have found no trace of it in the Old Testament. The name *semsem* or *simsim* is clearly Semitic, but only of the more recent epoch of the Talmud,[3] and of the agricultural treatise of Alawwam,[4] compiled after the Christian era began. It was perhaps a Semitic people who introduced the plant and the name *semsem* (whence the *sesam* of the Greeks) into Egypt after the epoch of the great monuments and of the Exodus. They may have received it with the name from Babylonia, where Herodotus says[5] that sesame was cultivated.

An ancient cultivation in the Euphrates valley agrees with the existence of a Sanskrit name, *tila*, the *tilu* of the Brahmans (Rheede, *Malabar*, i., ix., pp. 105–107), a word of which there are traces in several modern languages of India, particularly in Ceylon.[6] Thus we are carried back to India in accordance with the origin of which Pliny speaks, but it is possible that India itself may have received the species from the Sunda Isles before the arrival of the Aryan conquerors. Rumphius gives

---

[1] Pliny, *Hist.*, lib. xv. cap. 7.

[2] Wilkinson, *Manners and Customs of Ancient Egyptians*, vol. ii. ; Unger, *Pflanzen des Alten Ægyptens*, p. 45.

[3] Reynier, *Écon. Pub. des Arabes et des Juifs*, p. 431 ; Löw, *Aramäische Pflanzennamen*, p. 376.

[4] E. Meyer, *Geschichte der Botanik*, iii. p. 75.

[5] Herodotus, lib. i. cap. 193.   [6] Thwaites, *Enum.*, p. 209.

three names for the sesame in these islands, very different
one from the other, and from the Sanskrit word, which
supports the theory of a more ancient existence in the
archipelago than on the continent.

In conclusion, from the fact that the sesame is wild in
Java, and from historical and philological arguments,
the plant seems to have had its origin in the Sunda Isles.
It was introduced into India and the Euphrates valley
two or three thousand years ago, and into Egypt at a less
remote epoch, from 1000 to 500 B.C. It was transported
from the Guinea coast to Brazil by the Portuguese,[1] but
it is unknown how long it has been cultivated in the rest
of Africa.

Castor-oil Plant—*Ricinus communis*, Linnæus.

The most modern works and those in highest repute
consider the south of Asia to be the original home of this
Euphorbiacea; sometimes they indicate certain varieties
in Africa or America without distinguishing the wild
from the cultivated plant. I have reason to believe that
the true origin is to be found in tropical Africa, in
accordance with the opinion of Ball.[2]

The difficulties with which the question is attended
arise from the antiquity of cultivation in different
countries, from the facility with which the plant sows
itself and becomes naturalized on rubbish-heaps and in
waste ground, lastly from the diversity of its forms, which
have often been described as species. This latter point
need not detain us, for Dr. J. Müller's careful monograph[3]
proves the existence of sixteen varieties, scarcely hereditary, which pass one into the other by many transitions,
and constitute, therefore, but one species.

The number of varieties is the sign of a very ancient
cultivation. They differ more or less as to capsules,
seeds, inflorescence, etc. Moreover, they are small trees
in hot countries, but they do not endure frost, and
become annuals north of the Alps and in similar regions.
They are in such cases planted in gardens for ornament,

---

[1] Piso, *Brazil.*, edit. 1658, p. 211.
[2] Ball, *Floræ Maroccanæ Spicilegium*, p. 664.
[3] Müller, *Argov.*, in D.C., *Prodromus*, vol. xv. part 2, p. 1017.

while in the tropics, and even in Italy, they are grown
for the sake of the oil contained in the seed. This oil,
which is more or less purgative, is used for lamps in
Bengal and elsewhere.

In no country has the species been found wild with
such certainty as in Abyssinia, Sennaar, and the Kordofan.
The expressions of authors and collectors are distinct on
this head. The castor-oil plant is common in rocky
places in the valley of Chiré, near Goumalo, says Quartin
Dillon; it is wild in those parts of Upper Sennaar which
are flooded during the rains, says Hartmann.[1] I have
a specimen from Kotschy, No. 243, gathered on the
northern slope of Mount Kohn, in the Kordofan. The
indications of travellers in Mozambique and on the coast
of Guinea are not so clear, but it is possible that the
natural area of the species covers a great part of tropical
Africa. As it is a useful species, and one very conspicuous
and easily propagated, the negroes must have early
diffused it. However, as we draw near the Mediterranean,
it is no longer said to be indigenous. In Egypt, Schweinfurth and Ascherson[2] say the species is only cultivated
and naturalized. Probably in Algeria, Sardinia, and
Morocco, and even in the Canaries, where it is principally
found in the sand on the sea-shore, it has been naturalized
for centuries. I believe this to be the case with specimens brought from Djedda, in Arabia, by Schimper,
which were gathered near a cistern. Yet Forskal[3]
gathered the caster-oil plant in the mountains of Arabia
Felix, which may signify a wild station. Boissier[4]
indicates it in Beluchistan and the south of Persia,
but as "subspontaneous," as in Syria, Anatolia, and
Greece.

Rheede[5] speaks of the plant as cultivated in Malabar
and growing in the sand, but modern Anglo-Indian
authors do not allow that it is wild. Some make no

---

[1] Richard, *Tentamen Fl. Abyss.*, ii. p. 250; Schweinfurth, *Plantæ
Niloticæ a Hartmann*, etc., p. 13.
[2] Schweinfurth and Ascherson, *Aufzählung*, p. 262.
[3] Forskal, *Fl. Arabica*, p. 71.   [4] Boissier, *Fl. Orient.*, iv. p. 1143.
[5] Rheede, *Malabar*, ii. p. 57, t. 32.

mention of the species. A few speak of the facility with
which the species becomes naturalized from cultivation. Loureiro had seen it in Cochin-China and in
China "cultivated and uncultivated," which perhaps
means escaped from cultivation. Lastly, for the Sunda
Islands, Rumphius [1] is as usual one of the most
interesting authorities. The castor-oil plant, he says,
grows especially in Java, where it forms immense fields
and produces a great quantity of oil. At Amboyna, it is
planted here and there, near dwellings and in fields,
rather for medicinal purposes. The wild species grows
in deserted gardens (*in desertis hortis*); it is doubtless
sprung from the cultivated plant (*sine dubio degeneratio
domestica*). In Japan the castor-oil plant grows among
shrubs and on the slopes of Mount Wuntzen, but
Franchet and Savatier add,[2] "probably introduced."
Lastly, Dr. Bretschneider mentions the species in his
work of 1870, p. 20; but what he says here, and in
a letter of 1881, does not argue an ancient cultivation
in China.

The species is cultivated in tropical America. It
becomes easily naturalized in clearings, on rubbish-heaps,
etc.; but no botanist has found it in the conditions of
a really indigenous plant. Its introduction must have
taken place soon after the discovery of America, for a
common name, *lamourou*, exists in the West India
Islands; and Piso gives another in Brazil, *nhambu-
guacu, figuero inferno* in Portuguese. I have received
the largest number of specimens from Bahia; none are
accompanied by the assertion that it is really indigenous.

In Egypt and Western Asia the culture of the species
dates from so remote an epoch that it has given rise to
mistakes as to its origin. The ancient Egyptians practised
it extensively, according to Herodotus, Pliny, Diodorus,
etc. There can be no mistake as to the species, as its
seeds have been found in the tombs.[3] The Egyptian
name was *kiki*. Theophrastus and Dioscorides mention

---

[1] Rumphius, *Herb. Amb.*, vol. iv. p. 93.
[2] Franchet and Savatier, *Enum. Japon.*, i. p. 424.
[3] Unger, *Pflanzen des Alten Ægyptens*, p. 61.

it, and it is retained in modern Greek,[1] while the Arabs have a totally different name, *kerua, kerroa, charua*.[2]

Roxburgh and Piddington quote a Sanskrit name, *eranda, erunda,* which has left descendants in the modern languages of India. Botanists do not say from what epoch of Sanskrit this name dates ; as the species belongs to hot climates, the Aryans cannot have known it before their arrival in India, that is at a less ancient epoch than the Egyptian monuments.

The extreme rapidity of the growth of the castor-oil plant has suggested different names in Asiatic language, and that of *Wunderbaum* in German. The same circumstance, and the analogy with the Egyptian name *kiki*, have caused it to be supposed that the *kikajon* of the Old Testament,[3] the growth, it is said, of a single night, was this plant.

I pass a number of common names more or less absurd, as *palma Christi, girasole,* in some parts of Italy, etc., but it is worth while to note the origin of the name *castor oil,* as a proof of the English habit of accepting names without examination, and sometimes of distorting them. It appears that in the last century this plant was largely cultivated in Jamaica, where it was once called *agno casto* by the Portuguese and the Spaniards, being confounded with *Vitex agnus castus,* a totally different plant. From *casto* the English planters and London traders made *castor*.[4]

Walnut—*Juglans regia,* Linnæus.

Some years ago the walnut tree was known to be wild in Armenia, in the district to the south of the Caucasus and of the Caspian Sea, in the mountains of the north and north-east of India, and in Burmah.[5]

---

[1] Theophrastus, *Hist.*, lib. i. cap. 19; Dioscorides, lib. iv. cap. 171; Fraas, *Syn. Fl. Class.*, p. 92.

[2] Nemnich, *Polyglott. Lexicon;* Forskal, *Fl. Ægypt.*, p. 75.

[3] Jonah iv. 6. Pickering, *Chron. Hist. Plants*, p. 225, writes *kykwyn.*

[4] Flückiger and Hanbury, *Pharmacographia,* p. 511.

[5] A. de Candolle, *Prodr.*, xvi. part 2, p. 136; Tchihatcheff, *Asie Mineure,* i. p. 172; Ledebour, *Fl. Ross.*, i. p. 507; Roxburgh, *Fl. Ind.*, iii. p. 630; Boissier, *Fl. Orient.*, iv. p. 1160; Brandis, *Forest Flora of N.W. India,* p. 498; Kurz, *Forest Flora of Brit. Burmah,* p. 390.

C. Koch[1] denied that it was indigenous in Armenia and to the south of the Caucasus, but this has been proved by several travellers. It has since been discovered wild in Japan,[2] which renders it probable that the species exists also in the north of China, as Loureiro and Bunge said,[3] but without particularizing its wild character. Heldreich[4] has recently placed it beyond a doubt that the walnut is abundant in a wild state in the mountains of Greece, which agrees with passages in Theophrastus[5] which had been overlooked. Lastly, Heuffel saw it, also wild, in the mountains of Banat.[6] Its modern natural area extends, then, from eastern temperate Europe to Japan. It once existed in Europe further to the west, for leaves of the walnut have been found in the quaternary tufa in Provence.[7] Many species of Juglans existed in our hemisphere in the tertiary and quaternary epochs ; there are now ten, at most, distributed throughout North America and temperate Asia.

The use of the walnut and the planting of the tree may have begun in several of the countries where the species was found, and cultivation extended gradually and slightly its artificial area. The walnut is not one of those trees which sows itself and is easily naturalized. The nature of its fruit is perhaps against this ; and, moreover, it needs a climate where the frosts are not severe and the heat moderate. It scarcely passes the northern limit of the vine, and does not extend nearly so far south.

The Greeks, accustomed to olive oil, neglected the walnut until they received from Persia a better variety, called *karuon basilikon*,[8] or *Persikon*.[9] The Romans

---

[1] C. Koch, *Dendrologie*, i. p. 584.

[2] Franchet and Savatier, *Enum. Plant. Jap.*, i. 453.

[3] Loureiro, *Fl. Cochin.*, p. 702 ; Bunge, *Enum.*, p. 62.

[4] Heldreich, *Verhandl. Bot. Vereins Brandenb.*, 1879, p. 147.

[5] Theophrastus, *Hist. Plant.*, lib. iii. cap. 3, 6. These passages, and others of ancient writers, are quoted and interpreted by Heldreich better than by Hehn and other scholars.

[6] Heuffel, *Abhandl. Zool. Bot. Ges. in Wien*, 1853, p. 194.

[7] De Saporta, *33rd Sess. du Congres Scient. de France.*

[8] Dioscorides, lib. i. cap. 176.

[9] Pliny, *Hist. Plant.*, lib. xv. cap. 22.

cultivated the walnut from the time of their kings; they considered it of Persian origin.[1] They had an old custom of throwing nuts in the celebration of weddings.

Archæology confirms these details. The only nuts which have hitherto been found under the lake-dwellings of Switzerland, Savoy, or Italy are confined to a single locality near Parma, called Fontinellato, in a stratum of the iron age.[2] Now, this metal, very rare at the time of the Trojan war, cannot have come into general use among the agricultural population of Italy until the fifth or sixth century before Christ, an epoch at which even bronze was perhaps still unknown to the north of the Alps. In the station at Lagozza, walnuts have been found in a much higher stratum, and not ancient.[3] Evidently the walnuts of Italy, Switzerland, and France are not descended from the fossil plants of the quaternary tufa of which I spoke just now.

It is impossible to say at what period the walnut was first planted in India. It must have been early, for there is a Sanskrit name, akschôda, akhoda, or akhôta. Chinese authors say that the walnut was introduced among them from Thibet, under the Han dynasty, by Chang-kien, about the year 140–150 B.C.[4] This was per-haps a perfected variety. Moreover, it seems probable, from the actual records of botanists, that the wild walnut is rare in the north of China, and is perhaps wanting in the east. The date of its cultivation in Japan is un-known.

The walnut tree and walnuts had an infinite number of names among ancient peoples, which have exercised the science and imagination of philologists,[5] but the origin of the species is so clear that we need not stay to consider them.

Areca—*Areca Catechu*, Linnæus.

[1] Pliny, *Hist. Plant.*, lib. xv. cap. 22.
[2] Heer, *Pflanzen der Pfahlbauten*, p. 31.
[3] Sordelli, *Sulle piante della torbiera*, etc., p. 39.
[4] Bretschneider, *Study and Value*, etc., p. 16; and letter of Aug. 23, 1881.
[5] Ad. Pictet, *Origines Indo-Europ.*, edit. 2, vol. i. p. 289; Hehn, *Cul-turpflanzen und Hausthiere*, edit. 3, p. 341.

The areca palm is much cultivated in the countries where it is a custom to chew betel, that is to say throughout Southern Asia. The nut, or rather the almond which forms the principal part of the seed contained in the fruit, is valued for its aromatic taste; chopped, mixed with lime, and enveloped in a leaf of the pepper-betel, it forms an agreeable stimulant, which produces a flow of saliva and blackens the teeth to the satisfaction of the natives.

The author of the principal work on the order Palmaceæ, de Martius,[1] says of the origin of this species, "Its country is uncertain (*non constat*); probably the Sunda Isles." We may find it possible to affirm something positive by referring to more modern authors.

On the continent of India, in Ceylon and Cochin-China, the species is always indicated as cultivated.[2]   So in the Sunda Isles, the Moluccas, etc., to the south of Asia. Blume,[3] in his work entitled *Rumphia*, says that the "habitat" of the species is the Malay Peninsula, Siam, and the neighbouring islands. Yet he does not appear to have seen the indigenous plants of which he speaks. Dr. Bretschneider[4] believes that the species is a native of the Malay Archipelago, principally of Sumatra, for he says those islands and the Philippines are the only places where it is found wild. The first of these facts is not confirmed by Miquel, nor the second by Blanco,[5] who lived in the Philippines. Blume's opinion appears the most probable, but we must still say with Martius, "The country is not proved." The existence of a number of Malay names, *pinang, jambe*, etc., and of a Sanskrit name, *gouvaka*, as well as very numerous varieties, show the antiquity of cultivation. The Chinese received it, 111 B.C., from the south, with the Malay name, *pin-lang*.

---

[1] Martius, *Hist. Nat. Palmarum*, in folio, vol. iii. p. 170 (published without date, but before 1851).

[2] Roxburgh, *Fl. Ind.*, iii. p. 616; Brandis, *Forest Fl. of India*, p. 551; Kurz, *Forest Fl. of Brit. Burmah*, p. 537; Thwaites, *Enum. Zeylan.*, p. 327; Loureiro, *Fl. Cochin-Ch.*, p. 695.

[3] Blume, *Rumphia*, ii. p. 67; Miquel, *Fl. Indo-Batava.*, iii. p. 9; *uppl. de Sumatra*, p. 253.

[4] Bretschneider, *Study and Value*, etc., p. 28.

[5] Blanco, *Fl. di Filipinas*, edit. 2.

The Telinga name, *arek,* is the origin of the botanical name Areca.

Elæis—*Elæis guineensis,* Jacquin.

Travellers who visited the coast of Guinea in the first half of the sixteenth century [1] already noticed this palm, from which the negroes extracted oil by pressing the fleshy part of the fruit. The tree is indigenous on all that coast.[2] It is also planted, and the exportation of palm-oil is the object of an extensive trade. As it is also found wild in Brazil and perhaps in Guiana,[3] a doubt arose as to the true origin. It seems the more likely to be American that the only other species which with this one constitutes the genus *Elæis* belongs to New Granada.[4] Robert Brown, however, and the authors who have studied the family of palms, are unanimous in their belief that *Elæis guineensis* was introduced into America by the negroes and slave-traders in the traffic between the Guinea coast and the coast of America. Many facts confirm this opinion. The first botanists who visited Brazil, Piso and Marcgraf and others, do not mention the Elæis. It is only found on the littoral, from Rio di Janeiro to the mouth of the Amazon, never in the interior. It is often cultivated, or has the appearance of a species escaped from the plantations. Sloane,[5] who explored Jamaica in the seventeenth century, relates that this tree was introduced in his time into a plantation which he names, from the coast of Guinea. It has since become naturalized in some of the West India Islands.[6]

Cocoa-nut Palm—*Cocos nucifera,* Linnæus.

The cocoa-nut palm is perhaps, of all tropical trees, the one which yields the greatest variety of products. Its

---

[1] Da Mosto, in Ramusio, i. p. 104, quoted by R. Brown.

[2] Brown, *Bot. of Congo,* p. 55.

[3] Martius, *Hist. Nat. Palmarum,* ii. p. 62; Drude, in *Fl. Brasil.,* fasc. 85, p. 457. I find no author who asserts that this palm is wild in Guiana, as Martius affirms it to be in Brazil.

[4] *Elæis melanocarpa,* Gærtner. The fruit also contains oil, but it does not appear that the species is cultivated, as the number of oleaginous plants is considerable in all countries.

[5] Sloane, *Nat. Hist. of Jamaica,* ii. p. 113.

[6] Grisebach, *Flora of Brit. W. Ind. Is.,* p. 522.

wood and fibres are utilized in various ways. The sap
extracted from the inner part of the inflorescence yields a
much-prized alcoholic drink. The shell of the nut forms
a vessel, the milk of the half-ripe fruit is a pleasant drink,
and the nut itself contains a great deal of oil. It is not
surprising that so valuable a tree has been a good deal
planted and transported. Besides, its dispersion is aided
by natural causes. The woody shell and fibrous envelope
of the nut enable it to float in salt water without injury
to the germ. Hence the possibility of its transportation
to great distances by currents and its naturalization on
coasts where the temperature is favourable. Unfortunately, this tree requires a warm, damp climate, such as
exists only in the tropics, or in exceptional localities just
without them. Nor does it thrive at a distance from
the sea.

The cocoa-nut abounds on the littoral of the warm
regions of Asia, of the islands to the south of this continent, and in analogous regions of Africa and America,
but it may be asserted that it dates in Brazil, the West
Indies, and the west coast of Africa from an introduction
which took place about three centuries ago. Piso and
Marcgraf[1] seem to admit that the species is foreign to
Brazil without saying so positively. De Martius,[2] who
has published a very important work on the Palmaceæ,
and has travelled through the provinces of Bahia, Per-
nambuco, and others, where the cocoa-nut abounds, does
not say that it is wild. It was introduced into Guiana
by missionaries.[3] Sloane[4] says it is an exotic in the
West Indies. An old author of the sixteenth century,
Martyr, whom he quotes, speaks of its introduction. This
probably took place a few years after the discovery of
America, for Joseph Acosta[5] saw the cocoa-nut palm
at Porto Rico in the sixteenth century. De Martius
says that the Portuguese introduced it on the coast of
Guinea. Many travellers do not even mention it in this

---

[1] Piso, *Brasil.*, p. 65; Marcgraf, p. 138.
[2] Martius, *Hist. Nat. Palmarum*, 3 vols. in folio; see vol. ii. p. 125.
[3] Aublet, *Guyane*, suppl., p. 102.   [4] Sloane, *Jamaica*, ii. p. 9.
[5] J. Acosta, *Hist. Nat. des Indes*, French trans., 1598, p. 178.

region, where it is apparently of no great importance.
More common in Madagascar and on the east coast, it
is not, however, named in several works on the plants of
Zanzibar, the Seychelles, Mauritius, etc., perhaps because
it is considered as cultivated in these parts.

Evidently the species is not of African origin, nor of
the eastern part of tropical America. Eliminating these
countries, there remain western tropical America, the
islands of the Pacific, the Indian Archipelago, and the
south of Asia, where the tree abounds with every appearance of being more or less wild and long established.

The navigators Dampier and Vancouver[1] found it
at the beginning of the seventeenth century, forming
woods in the islands near Panama, not on the mainland,
and in the isle of Cocos, situated at three hundred miles
from the continent in the Pacific. At that time these
islands were uninhabited. Later the cocoa-nut palm was
found on the western coast from Mexico to Peru, but
usually authors do not say that it was wild, excepting
Seemann,[2] however, who saw this palm both wild and
cultivated on the Isthmus of Panama. According to
Hernandez,[3] in the sixteenth century the Mexicans called
it *coyolli*, a word which does not seem to be native.

Oviedo,[4] writing in 1526, in the first years of the conquest of Mexico, says that the cocoa-nut palm was abundant on the coast of the Pacific in the province of the
Cacique Chiman, and he clearly describes the species.
This does not prove the tree to be wild. In southern
Asia, especially in the islands, the cocoa-nut is both wild
and cultivated. The smaller the islands, and the lower
and the more subject to the influence of the sea air, the
more the cocoa-nut predominates and attracts the attention of travellers. Some take their name from the tree,
among others two islands close to the Andamans and one
near Sumatra.

[1] Vafer, *Voyage de Dampier*, edit. 1705, p. 186 ; Vancouver, French
edit., p. 325, quoted by de Martius, *Hist. Nat. Palmarum*, i. p. 188.

[2] Seemann, *Bot. of Herald.*, p. 204.

[3] Hernandez, *Thesaurus Mexic.*, p. 71.   He attributes the same name,
p. 75, to the cocoa-nut palm of the Philippine Islands.

[4] Oviedo, Ramusio's trans., iii. p. 53.

The cocoa-nut occurring with every appearance of an ancient wild condition at once in Asia and western America, the question of origin is obscure. Excellent authors have solved it differently. De Martius believes it to have been transported by currents from the islands situated to the west of Central America, into those of the Asiatic Archipelago. I formerly inclined to the same hypothesis,[1] since admitted without question by Grisebach;[2] but the botanists of the seventeenth century often regarded the species as Asiatic, and Seemann,[3] after a careful examination, says he cannot come to a decision. I will give the reasons for and against each hypothesis.

In favour of an American origin, it may be said—

1. The eleven other species of the genus Cocos are American, and all those which de Martius knew well are Brazilian.[4] Drude,[5] who has studied the Palmaceæ, has written a paper to show that each genus of this family is proper to the ancient or to the new world, excepting the genus Elæis, and even here he suspects a transport of the *E. guineensis* from America into Africa, which is not at all probable. (See above, p. 429.) The force of this argument is somewhat diminished by the circumstance that *Cocos nucifera* is a tree which grows on the littoral and in damp places, while the other species live under different conditions, frequently far from the sea and from rivers. Maritime plants, and those which grow in marshes or damp places, have commonly a more vast habitation than others of the same genus.

2. The trade winds of the Pacific, to the south and yet more to the north of the equator, drive floating bodies from America to Asia, a direction contrary to that of the general currents.[6] It is known, moreover, from the un-

---

[1] A. de Candolle, *Géogr. Bot. Raisonnée*, p. 976.

[2] Grisebach, *Vegetation der Erde*, pp. 11, 323.

[3] Seemann, *Flora Vitiensis*, p. 275.

[4] The cocoa-nut called Maldive belongs to the genus Lodoicea. *Coco mamillaris*, Blanco, of the Philippines is a variety of the cultivated *Cocos nucifera*.

[5] Drude, in *Bot. Zeitung*, 1876, p. 801; and *Flora Brasiliensis*, fasc. 85, p. 405.

[6] Stieler, *Hand Atlas*, edit. 1867 map 3

expected arrival of bottles containing papers on different coasts, that chance has much to do with these transports.

The arguments in favour of an Asiatic, or contrary to an American origin, are the following :—

1. A current between the third and fifth parallels, north latitude, flows from the islands of the Indian Archipelago to Panama.[1] To the north and south of this are currents which take the opposite direction, but they start from regions too cold for the cocoa-nut, and do not touch Central America, where it is supposed to have been long indigenous.

2. The inhabitants of the islands of Asia were far bolder navigators than the American Indians. It is very possible that canoes from the Asiatic Islands, containing a provision of cocoa-nuts, were thrown by tempests or false manœuvres on to the islands or the west coast of America. The converse is highly improbable.

3. The area for three centuries has been much vaster in Asia than in America, and the difference was yet more considerable before that epoch, for we know that the cocoa-nut has not long existed in the east of tropical America.

4. The inhabitants of the islands of Asia possess an immense number of varieties of this tree, which points to a very ancient cultivation. Blume, in his *Rumphia*, enumerates eighteen varieties in Java and the adjacent islands, and thirty-nine in the Philippines. Nothing similar has been observed in America.

5. The uses of the cocoa-nut are more varied and more habitual in Asia. The natives of America hardly utilize it except for the contents of the nut, from which they do not extract the oil.

6. The common names, very numerous and original in Asia, as we shall presently see, are rare, and often of European origin in America.

7. It is not probable that the ancient Mexicans and inhabitants of Central America would have neglected to spread the cocoa-nut in several directions, had it existed among them from a very remote epoch. The trifling

[1] Stieler, *ibid.*, map 9.

breadth of the Isthmus of Panama would have facilitated the transport from one coast to the other, and the species would soon have been established in the West Indies, at Guiana, etc., as it has become naturalized in Jamaica, Antigua,[1] and elsewhere, since the discovery of America. 8. If the cocoa-nut in America dated from a geological epoch more ancient than the pleiocene or even eocene deposits in Europe, it would probably have been found on both coasts, and the islands to the east and west equally. 9. We cannot find any ancient date of the existence of the cocoa-nut in America, but its presence in Asia three or four thousand years ago is proved by several Sanskrit names. Piddington in his index only quotes one, *narikela*. It is the most certain, since it recurs in modern Indian languages. Scholars count ten of these, which, according to their meaning, seem to apply to the species or its fruit.[2] *Narikela* has passed with modifications into Arabic and Persian.[3] It is even found at Otahiti in the form *ari* or *haari*,[4] together with a Malay name.

10. The Malays have a name widely diffused in the archipelago—*kalâpa, klâpa, klôpo.* At Sumatra and Nicobar we find the name *njîor, nieor ;* in the Philippines, *niog ;* at Bali, *niuh, njo ;* at Tahiti, *niuh ;* and in other islands, *nu, nidju, ni ;* even at Madagascar, *wua-niu.*[5] The Chinese have *ye,* or *ye-tsu* (the tree is *ye*). With the principal Sanskrit name this constitutes four different roots, which show an ancient existence in Asia. However, the uniformity of nomenclature in the archipelago as far as Tahiti and Madagascar indicates a transport by human agency since the existence of known languages.

The Chinese name means head of the king of Yuë, referring to an absurd legend of which Dr. Bretschneider speaks.[6] This savant tells us that the first mention of the cocoa-nut occurs in a poem of the second century before

---

[1] Grisebach, *Flora of Brit. W. Indies,* p. 552.

[2] Eugène Fournier has indicated to me, for instance, *drdapala* (with hard fruit), *palakecara* (with hairy fruit), *jalakajka.* (water-holder), etc.

[3] Blume, *Rumphia,* iii. p. 82.

[4] Forster, *De Plantis Esculentis,* p. 48 ; Nadeaud, *Enum. des Plantes de Taiti,* p. 41.

[5] Blume, *ubi supra.*   [6] Bretschneider, *Study and Value,* etc., p. 24.

Christ, but the most unmistakable descriptions are in works later than the ninth century of our era. It is true that the ancient writers scarcely knew the south of China, the only part of the empire where the cocoa-nut palm can live.

In spite of the Sanskrit names, the existence of the cocoa-nut in Ceylon, where it is well established on the coast, dates from an almost historical epoch. Near Point de Galle, Seemann tells us may be seen carved upon a rock the figure of a native prince, Kotah Raya, to whom is attributed the discovery of the uses of the cocoa-nut, unknown before him ; and the earliest chronicle of Ceylon, the *Marawansa*, does not mention this tree, although it carefully reports the fruits imported by different princes. It is also noteworthy that the ancient Greeks and Egyptians only knew the cocoa-nut at a late epoch as an Indian curiosity. Apollonius of Tyana saw this palm in Hindustan, at the beginning of the Christian era.[1]

From these facts the most ancient habitation in Asia would be in the archipelago, rather than on the continent or in Ceylon ; and in America in the islands west of Panama. What are we to think of this varied and contradictory evidence ? I formerly thought that the arguments in favour of Western America were the strongest. Now, with more information and greater experience in similar questions, I incline to the idea of an origin in the Indian Archipelago. The extension towards China, Ceylon, and India dates from not more than three thousand or four thousand years ago, but the transport by sea to the coasts of America and Africa took place perhaps in a more remote epoch, although posterior to those epochs when the geographical and physical conditions were different to those of our day.

[1] Seemann, *Fl. Vitiensis*, p. 276; Pickering, *Chronol. Arrangement*, p. 428.

PART III.

Summary and Conclusion.

CHAPTER I.

GENERAL TABLE OF SPECIES, WITH THEIR ORIGIN AND
THE EPOCH OF THEIR EARLIEST CULTIVATION.

THE following table includes a few species of which a
detailed account has not been given, because their origin
is well known, and they are of little importance.

Explanation of the signs used in the table: (1)
annual, (2) biennial, ⚷ perennial, 5 small shrub, 5 shrub
5 small tree, 5 tree. The letters indicate the certain
or probable date of earliest cultivation. For the species
of the old world: A, a species cultivated for more than
four thousand years (according to ancient historians, the
monuments of ancient Egypt, Chinese works, and botanical
and philological indications); B, cultivated for more than
two thousand years (indicated in Theophrastus, found
among lacustrine remains, or presenting various signs, such
as possessing Hebrew or Sanskrit names); C, cultivated for
less than two thousand years (mentioned by Dioscorides
and not by Theophrastus, seen in the frescoes at Pompeii,
introduced at a known date, etc.). For American species:
D, cultivation very ancient in America (from its wide
area and number of varieties); E, species cultivated
before the discovery of America, without showing signs
of a great antiquity of culture; F, species only cultivated
since the discovery of America.

SPECIES NATIVE TO THE OLD WORLD.

CULTIVATED FOR THE SUBTERRANEAN PARTS.

Name and duration.	Date.	Origin.
Radish—Raphanus sativus (1).	B.	Temperate Asia.[1]
Horse-Radish—Cochlearia Armoracia, ♃.	C.	Eastern temperate Europe.
Turnip—Brassica Rapa (2).	A.	Europe, western Siberia (?).
Rape—Brassica Napus (2).	A.	Europe, western Siberia (?).
Carrot—Daucus Carota (2).	B.	Europe, western temperate Asia (?).
Parsnip—Pastinaca sativa (2).	C.	Central and southern Europe.
Tuberous Chervil — Chærophyllum bulbosum (2).	C.	Central Europe, Caucasus.
Skirret—Sium Sisarum, ♃.	C.	Altaic Siberia, northern Persia.
Madder—Rubia tinctorum, ♃	B.	Western temperate Asia, south-east of Europe.
Salsify—Tragopogon porrifolium (2).	C. (?)	South-east of Europe, Algeria.
Scorzonera—Scorzonera hispanica.	C.	South-west of Europe, south of the Caucasus.
Rampion — Campanula Rapunculus (2).	C.	Temperate and southern Europe.
Beet—Beta vulg. (2), ♃. ⎰ Vegetable.	B.	Canaries, Mediterranean basin, western temperate Asia.
⎱ Root.	B.	A result of cultivation.
Garlic—Allium sativum, ♃.	B.	Desert of the Kirghis, in western temperate Asia.
Onion—Allium Cepa (2).	A.	Persia, Afghanistan, Beluchistan, Palestine (?).
Welsh Onion—Allium fistulosum, ♃.	C.	Siberia (from the land of the Kirghis to Baikal).
Shallot—Allium ascalonicum, ♃.	C.	Modification of *A. cepa* (?), unknown wild.
Rocambole—Allium Scorodoprasum ♃.	C.	Temperate Europe.
Chives—Allium Schœnoprasum, ♃.	C. (?)	Temperate and northern Europe, Siberia, Khamschatka, North America (Lake Huron).
Taro—Colocasia antiquorum, ♃.	B.	India, Malay Archipelago, Polynesia.

[1] Dr. Bretschneider writes to me from Pekin, Dec. 22, 1882, that the species is mentioned in the *Ryd*, a work of the year 1100 B.C. I do not know if we must suppose the original habitat to be China or western Asia.

Name and duration.	Date.	Origin.
Apé—Alocasia macrorrhiza, ♃.	(?)	Ceylon, Malay Archipelago, Polynesia.
Konjak—Amorphophallus Konjak, ♃.	(?)	Japan (?).
Dioscorea sativa, ♃.	B. (?)	Southern Asia [especially Malabar (?), Ceylon (?),. (Java (?)].
Yams—Dioscorea Batatas, ♃.	B. (?)	China (?).
Dioscorea japonica, ♃.	(?)	Japan (?).
Dioscorea alata, ♃.	(?)	East of the Asiatic Archipelago.

CULTIVATED FOR THE STEMS OR LEAVES.

1. *Vegetables.*

Cabbage — Brassica oleracea (1), (2), 5.	A.	Europe.
Chinese Cabbage—Brassica chinensis (2).	(?)	China (?), Japan (?).
Water-Cress—Nasturtium officinale, ♃.	(?)	Europe, northern Asia.
Garden-Cress—Lepidium sativum (1).	B.	Persia (?).
Sea Kale—Crambe maritima, ♃.	C.	Western temperate Europe.
Purslane—Portulaca oleracea (1).	A.	From the western Himalayas to southern Russia and Greece.
New Zealand Spinach — Tetragonia expansa (1).	C.	New Zealand and New Holland.
Garden Celery — Apium graveolens (2).	B.	Temperate and southern Europe, northern Africa, western Asia.
Chervil—Anthriscus cerefolium (1).	C.	South-east of Russia, western temperate Asia.
Parsley—Petroselinum sativum (2).	C.	Southern Europe, Algeria, Lebanon.
Alexanders—Smyrnium Olus-atrum (2).	C.	Southern Europe, Algeria, western temperate Asia.
Corn Salad—Valerianella olitoria (1).	C.	Sardinia, Sicily.
Artichoke—CynaraCardunculus (2), ♃. { Cardoon.	C.	Southern Europe, northern Africa, Canaries, Madeira.
{ Artichoke.	C.	Derived from the cardoon.
Lettuce—Latuca Scariola (1), (2).	B.	Southern Europe, northern Africa, western Asia.
Wild Chicory—Cichorium Intybus, ♃.	C.	Europe, northern Africa, western temperate Asia.
Endive —Cichorium Endivia (1).	C.	Mediterranean basin, Caucasus, Turkestan.
Spinach—Spinacia oleracea (1).	C.	Persia (?).
Orach—Atriplex hortensis (1).	C.	Northern Europe and Siberia

Name and duration.	Date.	Origin.
Amaranth—Amarantus gangeticus (1).	(?)	Tropical Africa, India (?).
Sorrel—Rumex acetosa, ♃ (1).	(?)	Europe, northern Asia, mountains of India.
Patience Dock—Rumex patientia, ♃.	(?)	Turkey in Europe, Persia.
Asparagus—Asparagus officinalis, ♃.	B.	Europe, western temperate Asia.
Leek—Allium ampeloprasum, ♃.	B.	Mediterranean basin.

2. Fodder.

Lucern—Medicago sativa, ♃.	B.	Western temperate Asia.
Sainfoin—Onobrychis sativa, ♃.	C.	Temperate Europe, south of the Caucasus.
French Honeysuckle — Hedysarum coronarium, ♃.	C.	Centre and west of the Mediterranean basin.
Purple Clover—Trifolium pratense, ♃.	C.	Europe, Algeria, western temperate Asia.
Alsike Clover—Trifolium hybridum (1).	C.	Temperate Europe.
Italian Clover—Trifolium incarnatum (1).	C.	Southern Europe.
Egyptian Clover — Trifolium alexandrinum (1).	C.	Syria, Anatolia.
Ervilla—Ervum Ervilia (1).	B.	Mediterranean basin.
Vetch—Vicia sativa (1).	B.	Europe, Algeria, south of the Caucasus.
Flat-podded Pea—Lathyrus Cicera (1).	B.	From Spain and Algeria to Greece.
Chickling Vetch—Lathyrus sativus (1).	B.	South of the Caucasus.
Ochrus—Lathyrus ochrus (1).	B.	Italy, Spain.
Fenugreek — Trigonella fœnumgræcum (1).	B.	North-east of India and western temperate Asia.
Bird's-Foot—Ornithopus sativus (1).	B. (?)	Portugal, south of Spain, Algeria.
Nonsuch—Medicago lupulina (1), (2).	C.	Europe, north of Africa (?), temperate Asia.
Corn Spurry—Spergula arvensis (1).	B. (?)	Europe.
Guinea Grass—Panicum maximum, ♃.	C. (?)	Tropical Africa.

3. Various Uses.

Tea—Thea sinensis, ♄.	A.	Assam, China, Mantschuria.
Flax anciently cultivated—Linum angustifolium, ♃ (2), (1).	A.	Mediterranean basin.
Flax now cultivated—Linum usitatissimum (1).	A. (?)	Western Asia (?), derived from the preceding (?).
Jute—Corchorus capsularis (1).	C. (?)	Java, Ceylon.

Name and duration.	Date.	Origin.
Jute—Corchorus olitorius (1).	C. (?)	North-west of India, Ceylon.
Sumach—Rhus coriaria, 5.	C.	Mediterranean basin, western temperate Asia.
Khât—Celastrus edulis, ♄.	(?)	Abyssinia, Arabia (?).
Indigo—Indigofera tinctoria, ♄.	B.	India (?).
Silver Indigo—Indigofera argentea,♄.	(?)	Abyssinia, Nubia, Kordofan, Senaar, India (?).
Henna—Lawsonia alba, ♄.	A.	Western tropical Asia, Nubia (?).
Blue Gum—Eucalyptus globulus, 5.	C.	New Holland.
Cinnamon — Cinnamonum zeylanicum, 5.	C.	Ceylon, India.
China Grass —Bœhmeria nivea, ♃, ♄.	(?)	China, Japan.
Hemp—Cannabis sativa (1).	A.	Dahuria, Siberia.
White Mulberry—Morus alba, 5.	A. (?)	India, Mongolia.
Black Mulberry—Morus nigra, 5.	B. (?)	Armenia, northern Persia.
Sugar-Cane — Saccharum officinarum, ♃.	B.	Cochin-China (?), south-west of China.

CULTIVATED FOR THE FLOWERS OR THEIR ENVELOPES.

Clove—Carophyllus aromaticus, 5.	(?)	Moluccas.
Hop—Humulus lupulus, ♃.	C.	Europe, western temperate Asia, Siberia.
Carthamine—Carthamus tinctorius (1).	A.	Arabia (?).
Saffron—Crocus sativus, ♃.	A.	Southern Italy, Greece, Asia Minor.

CULTIVATED FOR THE FRUITS.

Shaddock—Citrus decumana, 5.	B.	Pacific Islands, to the east of Java.
Citron, Lemon—Citrus medica, 5.	B.	India.
Bitter Orange — Citrus Aurantium Bigaradia, 5.	B.	East of India.
Sweet Orange — Citrus Aurantium sinense, 5.	C.	China and Cochin-China.
Mandarin—Citrus nobilis, 5.	(?)	China and Cochin-China.
Mangosteen — Garcinia mangostana, 5.	(?)	Sunda Islands, Malay Peninsula.
Ochro—Hibiscus esculentus (1).	C.	Tropical Africa.
Vine—Vitis vinifera, ♄.	A.	Western temperate Asia, Mediterranean basin.
Common Jujube— Ziziphus vulgaris, 5.	B.	China.
Lotus Jujube—Ziziphus lotus, 5.	(?)	Egypt to Marocco.

Name and duration.	Date.	Origin.
Indian Jujube—Zizyphus Jujuba, 5.	A. (?)	Burmah, India.
Mango—Mangifera indica, 5.	A. (?)	India.
Tahiti Apple—Spondias dulcis, 5.	(?)	Society, Friendly, and Fiji Isles.
Raspberry—Rubus idæus, ♃.	C.	Temperate Europe and Asia.
Strawberry—Fragaria vesca, ♃.	C.	Temperate Europe and western Asia, east of North America.
Bird-Cherry—Prunus avium, 5.	B.	Western temperate Asia, temperate Europe.
Common Cherry—Prunus cerasus, 5.	B.	From the Caspian to western Anatolia.
Plum—Prunus domestica, 5.	B.	Anatolia, south of the Caucasus, north of Persia.
Plum—Prunus insititia, ♃.	(?)	Southern Europe, Armenia, south of the Caucasus, Talysch.
Apricot—Prunus Armeniaca, 5.	A.	China.
Almond—Amygdalus communis, 5.	A.	Mediterranean basin, western temperate Asia.
Peach—Amygdalus Persica, 5.	A.	China.
Common Pear—Pyrus communis, 5.	A.	Temperate Europe and Asia.
Chinese Pear—Pyrus sinensis, 5.	(?)	Mongolia, Mantschuria.
Apple—Pyrus Malus, 5.	A.	Europe, Anatolia, south of the Caucasus.
Quince—Cydonia vulgaris, 5.	A.	North of Persia, south of the Caucasus, Anatolia.
Loquat—Eriobotrya japonica, 5.	(?)	Japan.
Pomegranate—Punica granatum, 5.	A.	Persia, Afghanistan, Beluchistan.
Rose Apple—Jambosa vulgaris, 5.	B.	Malay Archipelago, Cochin-China, Burmah, north-east of India.
Malay Apple—Jambosa malaccensis, 5.	B.	Malay Archipelago, Malacca.
Bottle Gourd—Cucurbita lagenaria (1).	C.	India, Moluccas, Abyssinia.
Spanish Gourd—C. maxima (1).	C. (?)	Guinea.
Melon—Cucumis Melo (1).	C.	India, Beluchistan, Guinea.
Water-Melon—Citrullus vulgaris (1).	A.	Tropical Africa.
Cucumber—Cucumis sativus (1).	A.	India.
West Indian Gherkin—Cucumis Anguria (1).	C. (?)	Tropical Africa (?).
White Gourd-Melon—Benincasa hispida (1).	(?)	Japan, Java.
Towel Gourd—Luffa cylindrica (1).	C.	India.
Angular Luffa—Luffa acutangula (1).	C.	India, Malay Archipelago.
Snake Gourd—Trichosanthes anguina (1).	C.	India (?).

Name and duration.	Date.	Origin.
Gooseberry—Ribes grossularia, ♄.	C.	Temperate Europe, north of Africa, Caucasus, western Himalayas.
Red Currant—Ribes rubrum, ♄.	C.	Northern and temperate Europe, Siberia, Caucasus, Himalayas, north-east of the United States.
Black Currant—Ribes nigrum, ♄.	C.	Northern and central Europe, Armenia, Siberia, Mantschuria, western Himalayas.
Kaki—Diospyros Kaki, 5.	(?)	Japan, northern China.
Date Plum—Diospyros lotos, 5.	(?)	China, India, Afghanistan, Persia, Armenia, Anatolia.
Olive—Olea europea, 5.	A.	Syria, southern Anatolia and neighbouring islands.
Aubergine—Solanum melongena (1).	A.	India.
Fig—Ficus Carica, 5.	A.	Centre and south of the Mediterranean basin, from Syria to the Canaries.
Bread-Fruit—Artocarpus incisa, ♄.	(?)	Sunda Isles.
Jack-Fruit—Artocarpus integrifolia, ♄.	B. (?)	India.
Date-Palm—Phœnix dactylifera, ♄.	A.	Western Asia and Africa, from the Euphrates to the Canaries.
Banana—Musa sapientum, 5.	A.	Southern Asia.
Oil Palm—Elæis guineensis, ♄.	(?)	Guinea.

CULTIVATED FOR THE SEEDS.

1. *Nutritive.*

Litchi—Nephelium Litchi, 5.	(?)	Southern China, Cochin-China.
Longan—Nephelium longana, 5.	(?)	India, Pegu.
Rambutan—Nephelium lappaceum,5.	(?)	India, Pegu.
Pistachio—Pistacia vera, ♄.	C.	Syria.
Bean—Faba vulgaris (1).	A.	South of the Caspian (?).
Lentil—Ervum lens (1).	A.	Western temperate Asia, Greece, Italy.
Chick-Pea—Cicer arietinum (1).	A.	South of the Caucasus and of the Caspian.
Lupin—Lupinus albus (1).	B.	Sicily, Macedonia, south of the Caucasus.
Egyptian Lupin — Lupinus termis (1).	A.	From Corsica to Syria.
Field-Pea—Pisum arvense (1).	C. (?)	Italy.

Name and duration.	Date.	Origin.
Garden-Pea—Pisum sativum (1).	B.	From the south of the Caucasus to Persia (?) northern India (?).
Soy—Dolichos soja (1).⸍	A.	Cochin-China, Japan, Java.
Pigeon-Pea—Cajanus indicus, ♃.	C.	Equatorial Africa.
Carob—Ceratonia siliqua, ♃.	A. (?)	Southern coast of Anatolia, Syria, Cyrenaica (?).
Moth—Phaseolus aconitifolius (1).	C.	India.
Three-lobed Kidney Bean—Phaseolus trilobus, ♃ (1).	B.	India, tropical Africa.
Green Gram—Phaseolus Mungo (1).	B. (?)	India.
Wall—Phaseolus Lablab, ♃ (1).	B.	India.
Lubia—Phaseolus Lubia (1).	C.	Western Asia (?).
Bambarra Ground Nut—Voandzeia subterranea (1).	(?)	Intertropical Africa.
Buckwheat — Fagopyrum esculentum (1).	C.	Mantschuria, central Siberia.
Tartary Buckwheat — Fagopyrum tartaricum (1).	C.	Tartary, Siberia to Dahuria.
Notch-seeded Buckwheat—Fagopyrum emarginatum (1).	(?)	Western China, eastern Himalayas.
Kiery—Amarantus frumentaceus (1).	(?)	India.
Chestnut—Castanea vulgaris, ♃.	(?)	From Portugal to the Caspian Sea, eastern Algeria. Varieties: Japan, North America.
Wheat — Triticum vulgare and varieties (?), (1).	A.	Region of the Euphrates.
Spelt—Triticum spelta (1).	A.	Derived from the preceding (?).
One-grained Wheat—Triticum monococcum (1).	(?)	Servia, Greece, Anatolia (if the identity with the Triticum bœoticum be admitted).
Two-rowed Barley — Hordeum distichon (1).	A.	Western temperate Asia.
Common Barley—Hordeum vulgare (1).	(?)	Derived from the preceding (?).
Six-rowed Barley—Hordeum hexastichon (1).	A.	Derived from the preceding (?).
Rye—Secale cereale (1).	B.	Eastern temperate Europe(?).
Common Oats—Avena sativa (1).	B.	Eastern temperate Europe(?).
Eastern Oats—Avena orientalis (1).	C. (?)	Western Asia (?).
Common Millet—Panicum miliaceum (1)	A.	Egypt, Arabia.
Italian Millet—Panicum italicum (1).	A.	China, Japan, Indian Archipelago (?)
Sorghum—Holcus sorghum (1).	A.	Tropical Africa (?).

Name and duration.	Date.	Origin.
Sweet Sorghum—Holcus saccharatus (1).	(?)	Tropical Africa (?).
Coracan—Eleusine coracana (1).	B.	India.
Rice—Oryza sativa (1).	A.	India, southern China (?).

2. *Various Uses.*

Poppy—Papaver somniferum (1).	B.	Derived from *P. setiferum* of the Mediterranean basin.
White Mustard—Sinapis alba (1).	B.	Temperate and southern
Black Mustard—Sinapis nigra (1).	B.	Europe, north of Africa, western temperate Asia.
Gold of Pleasure—Camelina sativa (1).	B. (?)	Temperate Europe, Caucasus, Siberia.
Herbaceous Cotton—Gossypium herbaceum, ♄ (1).	B.	India.
Tree Cotton—Gossypium arboreum, ♄.	B. (?)	Upper Egypt.
Arabian Coffee—Coffea arabica, ♄.	C.	Tropical Africa, Mozambique, Abyssinia, Guinea.
Liberian Coffee—Coffea liberica, ♄.	C.	Guinea Angola.
Sesame—Sesamum indicum (1).	A.	Sunda Isles.
Nutmeg—Myristica fragrans, ♄.	B.	Moluccas.
Castor-Oil Plant — Ricinus communis, ♄.	A.	Abyssinia, Sennaar, Kordofan.
Walnut—Juglans regia, ♄.	(?)	Eastern temperate Europe, temperate Asia.
Black Pepper—Piper nigrum, ♄.	B.	India.
Long Pepper—Piper longum, ♄.	B.	India.
Medicinal Pepper — Piper officinalis, ♄.	B.	Malay Archipelago.
Betel Pepper—Piper Betle, ♄.	B.	Malay Archipelago.
Areca Nut—Areca Catechu, ♄.	B.	Malay Archipelago.
Cocoa Nut—Cocos nucifera, ♄.	(?)	Malay Archipelago (?), Polynesia (?).

SPECIES OF AMERICAN ORIGIN.

Cultivated for the Underground Parts.

Arracacha—Arracacha esculenta, ♃ (1).	E.	New Granada (?).
Jerusalem Artichoke — Helianthus tuberosus, ♃.	E. (?)	North America (Indiana).
Potato—Solanum tuberosum, ♃.	E.	Chili, Peru (?).
Sweet Potato—Convolvulus batatas, ♃.	D.	Tropical America (where ?).
Manioc—Manihot utilissima, ♄.	E.	East of tropical Brazil.
Arrowroot—Maranta arundinacea, ♃.	(?)	Tropical (continental ?) America.

CULTIVATED FOR THE STEMS OR LEAVES.

Name and duration.	Date.	Origin.
Maté—Ilex paraguariensis, 5.	D.	Paraguay and western Brazil.
Coca—Erythroxylon Coca, 5.	D.	East of Peru and Bolivia.
Quinine—Cinchona Calisaya, 5.	F.	Bolivia, southern Peru.
Crown Bark—Cinchona officinalis, 5.	F.	Ecuador (province of Loxa).
Red Cinchona Bark—Cinchona succirubra, 5.	F.	Ecuador (province of Cuenca).
Tobacco—{ Nicotiana Tabacum (1).	D.	Ecuador and neighbouring countries.
Nicotiana rustica (1).	E.	Mexico (?), Texas (?), California (?).
American Aloe—Agave americana, 5.	E.	Mexico.

CULTIVATED FOR THE FRUITS.

Sweet Sop—Anona squamosa, 5.	(?)	West India Isles.
Sour Sop—Anona muricata, 5.	(?)	West India Isles.
Custard Apple—Anona reticulata, 5.	(?)	West India Isles, New Granada.
Chirimoya—Anona Cherimolia, 5.	E.	Ecuador, Peru (?).
Mammee Apple — Mammea americana, 5.	(?)	West India Isles.
Cashew Nut—Anacardium occidentale, 5.	(?)	Tropical America.
Virginian Strawberry—Fragaria virginiana, ♃.	F.	Temperate North America.
Chili Strawberry—Fragaria chiloensis, ♃.	F.	Chili.
Guava—Psidium guayava, 5.	E.	Continental tropical America.
Pumpkin and Squash — Cucurbita Pepo and Melopepo (1).	E.	Temperate North America.
Prickly Pear — Opuntia ficus indica, 5.	E.	Mexico.
Chocho—Sechium edule (1).	E.	Mexico (?), Central America.
Star-Apple—Chrysophyllum Caïnito, 5.	E.	West India Isles, Panama.
Caimito—Lucuma Caimito, 5.	E.	Peru.
Marmalade Plum — Lucuma mammosa, 5.	E.	Valley of the Orinoco.
Sapodilla—Sapota achras, 5̃.	E.	Campeachy, Isthmus of Panama, Venezuela.
Persimmon — Diospyros virginiana, 5.	F.	Eastern States of America.
Annual Capsicum—Capsicum annuum (1).	E.	Brazil (?).
Shrubby Capsicum—Capsicum frutescens, 5̃.	E.	From the east of Peru to Bahia.

Name and duration.	Date.	Origin.
Tomato—Lycopersicum esculentum (1).	E.	Peru.
Avocado Pear—Persea gratissima, 5.	E.	Mexico.
Papaw—Papaya vulgaris, 5.	E.	West Indies,Central America.
Pine-Apple—Ananassa sativa, ♃.	E.	Mexico, Central America, Panama, New Granada, Guiana (?), Bahia (?).

CULTIVATED FOR THE SEEDS.

1. Nutritious.

Cacao—Theobroma Cacao, 5.	D.	Amazon and Orinoco Valley, Panama (?), Yucatan (?).
Sugar Bean—Phaseolus lunatus, ♃.	E.	Brazil.
Quinoa—Chenopodium quinoa (1).	E.	New Granada, Peru (?), Chili (?).
Maize—Zea mays (1).	D.	New Granada (?).

2. Various Uses.

Arnotto—Bixa orellana.	D.	Tropical America.
Barbados Cotton—Gossypium barbadense, 5.	(?)	New Granada (?), Mexico (?), West Indies.
Earth Nuts—Arachis hypogæa (1).	E.	Brazil (?).
Madia—Madia sativa (1).	E.	Chili, California.

CRYPTOGAM CULTIVATED FOR THE WHOLE PLANT.

Mushroom—Agaricus campestris, ♃. | C. | Northern hemisphere.

SPECIES OF UNKNOWN OR ENTIRELY UNCERTAIN ORIGIN.

Common Haricot—Phaseolus vulgaris (1).
Musk Gourd—Cucurbita moschata (1).
Fig-leaved Gourd—Cucurbita ficifolia, ♃.

CHAPTER II.

GENERAL OBSERVATIONS AND CONCLUSIONS.

Article I.—Regions where Cultivated Plants originated.

IN the beginning of the nineteenth century, the origin of most of our cultivated species was unknown. Linnæus made no efforts to discover it, and subsequent authors merely copied the vague or erroneous expressions by which he indicated their habitations. Alexander von Humboldt expressed the true state of the science in 1807, when he said, " The origin, the first home of the plants most useful to man, and which have accompanied him from the remotest epochs, is a secret as impenetrable as the dwelling of all our domestic animals. . . . We do not know what region produced spontaneously wheat, barley, oats, and rye. The plants which constitute the natural riches of all the inhabitants of the tropics, the banana, the papaw, the manioc, and maize, have never been found in a wild state. The potato presents the same phenomenon." [1]

At the present day, if a few cultivated species have not yet been seen in a wild state, this is not the case with the immense majority. We know at least, most frequently, from what country they first came. This was already the result of my work of 1855, which modern more extensive research has confirmed in almost all points. This research has been applied to 247 species,[2]

---

[1] *Essai sur la Géographie des Plantes*, p. 28.
[2] Counting two or three forms which are perhaps rather very distinct races.

cultivated on a large scale by agriculturists, or in
kitchen gardens and orchards. I might have added a
few rarely cultivated or but little known, or of which
the cultivation has been abandoned; but the statistical
results would be essentially the same.

Out of the 247 species which I have studied, the old
world has furnished 199, America 45, and three are still
uncertain.

No species was common to the tropical and austral
regions of the two hemispheres before cultivation.
Allium schœnoprasum, the hop (*Humulus lupulus*),
the strawberry (*Fragaria visca*), the currant (*Ribes
rubrum*), the chestnut (*Castanea vulgaris*), and the
mushroom (*Agaricus campestris*), were common to the
northern regions of the old and new worlds. I have
reckoned them among the species of the old world, since
their principal habitation is there, and there they were
first cultivated.

A great number of species originated at once in
Europe and Western Asia, in Europe and Siberia, in the
Mediterranean basin and Western Asia, in India and
the Asiatic archipelago, in the West Indies and Mexico,
in these two regions and Columbia, in Peru and Brazil,
or in Peru and Columbia, etc., etc. They may be counted
in the table. This is a proof of the impossibility of sub-
dividing the continents and of classing the islands in
well-defined natural regions. Whatever be the method
of division, there will always be species common to two,
three, four, or more regions, and others confined to a
small portion of a single country. The same facts may
be observed in the case of uncultivated species.

A noteworthy fact is the absence in some countries
of indigenous cultivated plants. For instance, we have
none from the arctic or antarctic regions, where, it is
true, the floras consist of but few species. The United
States, in spite of their vast territory, which will soon
support hundreds of millions of inhabitants, only yields,
as nutritious plants worth cultivating, the Jerusalem
artichoke and the gourds. *Zizana œquatica*, which
the natives gathered wild, is a grass too inferior to

our cereals and to rice to make it worth the trouble of planting it. They had a few bulbs and edible berries, but they have not tried to cultivate them, having early received the maize, which was worth far more.

Patagonia and the Cape have not furnished a single species. Australia and New Zealand have furnished one tree, *Eucalyptus globulus*, and a vegetable, not very nutritious, the *Tetragonia*. Their floras were entirely wanting in graminæ similar to the cereals, in leguminous plants with edible seeds, in Cruciferæ with fleshy roots.[1] In the moist tropical region of Australia, rice and *Alocasia macrorhiza* have been found wild, or perhaps naturalized, but the greater part of the country suffers too much from drought to allow these species to become widely diffused.

In general, the austral regions had very few annuals, and among their restricted number none offered evident advantages. Now annual species are the easiest to cultivate. They have played a great part in the ancient agriculture of other countries.

In short, the original distribution of cultivated species was very unequal. It had no proportion with the needs of man or the extent of territory.

Article II.—Number and Nature of Cultivated Species at Different Epochs.

The species marked A in the table on pp. 437–446 must be regarded as of very ancient cultivation. They are forty-four in number. Some of the species marked B are probably as ancient, though it is impossible to prove it. The five American species marked D are probably cultivated as early as those in the category C, or the most ancient in the category B.

As might be supposed, the species A are especially plants provided with roots, seeds, and fruits proper for the food of man. Afterwards come a few species having

---

[1] See the list of the useful plants of Australia by Sir J. Hooker, *Flora Tasmania*, p. cx. ; and Bentham, *Flora Australiensis*, vii. p. 156.

fruits agreeable to the taste, or textile, tinctorial, oil-producing plants, or yielding stimulating drinks by infusion or fermentation. There are among these only two green vegetables, and no fodder. The orders which predominate are the Cruciferæ, Leguminosæ, and Graminaceæ.

The number of annuals is twenty-two out of the forty-four, or fifty per cent. Out of five American species marked D, two are annuals. In the category A, there are two biennials, and D has none. Among all the Phanerogams the annuals are not more than fifty per cent., and the biennials one or at most two per cent. It is clear that at the beginning of civilization plants which yield an immediate return are most prized. They offer, moreover, this advantage, that their cultivation is easily diffused or increased, either because of the abundance of seed, or the same species may be grown in summer in the north, and in winter or all the year round in the tropics.

Herbaceous perennial plants are rare in categories A and D. They are only from two to four per cent., unless we include *Brassica oleracea,* and the variety of flax which is usually perennial (*L. angustifolium*), cultivated by the Swiss lake-dwellers. In nature herbaceous perennials constitute about forty per cent. of the Phanerogams.[1]

A and D include twenty ligneous species out of forty-nine, that is about forty-one per cent. They are in the proportion of forty-three per cent. of the Phanerogams.

Thus the earliest husbandmen employed chiefly annuals or biennials, rather fewer woody species, and far fewer herbaceous perennials. These differences are due to the relative facility of cultivation, and the proportion of the evidently useful species in each division.

The species of the old world marked B have been in cultivation for more than two thousand years, but perhaps some of them belong to category A. The American

---

[1] The proportions which I give for the Phanerogams collectively are based upon an approximative calculation, made with the aid of the first two hundred pages of Steudel's *Nomenclator.* They are justified by the comparison with several floras.

species marked E were cultivated before the discoveries of Columbus, perhaps for more than two thousand years. Many other species marked (?) in the table date probably from an ancient epoch, but as they chiefly exist in countries without a literature and without archæological records we do not know their history. It is useless to insist upon such doubtful categories; on the other hand, the plants which we know to have been first cultivated in the old world less than two thousand years ago, and in America since its discovery, may be compared with plants of ancient cultivation.

These species of modern cultivation number sixty-one in the old world, marked C, and six in America, marked F; sixty-seven in all.

Classed according to their duration, they number thirty-seven per cent. annuals, seven to eight per cent. biennials, thirty-three per cent. herbaceous perennials, and twenty-two to twenty-three per cent. woody species.

The proportion of annuals or biennials is also here larger than in the whole number of plants, but it is not so large as among species of very ancient cultivation. The proportions of perennials and woody species are less than in the whole vegetable kingdom, but they are higher than among the species A, of very ancient cultivation.

The plants cultivated for less than two thousand years are chiefly artificial fodders, which the ancients scarcely knew; then bulbs, vegetables, medicinal plants (Cinchonas); plants with edible fruits, or nutritious seeds (buckwheats) or aromatic seeds (coffee).

Men have not discovered and cultivated within the last two thousand years a single species which can rival maize, rice, the sweet potato, the potato, the bread-fruit, the date cereals, millets, sorghums, the banana, soy. These date from three, four, or five thousand years, perhaps even in some cases six thousand years. The species first cultivated during the Græco-Roman civilization and later nearly all answer to more varied or more refined needs. A great dispersion of the ancient species from one country to another took place, and at the same time a selection of the best varieties developed in each species. The introduc-

tions within the last two thousand years took place in a
very irregular and intermittent manner.   I cannot quote
a single species cultivated for the first time after that date
by the Chinese, the great cultivators of ancient times.
The peoples of Southern and Western Asia innovated in
a certain degree by cultivating the buckwheats, several
cucurbitaceæ, a few alliums, etc.   In Europe, the Romans
and several peoples in the Middle Ages introduced the
cultivation of a few vegetables and fruits, and that of
several fodders.   In Africa a few species were then first
cultivated separately.   After the voyages of Vasco di
Gama and of Columbus a rapid diffusion took place of
the species already cultivated in either hemisphere.
These transports continued during three centuries without any introduction of new species into cultivation.
In the two or three hundred years which preceded the
discovery of America, and the two hundred which followed, the number of cultivated species remained almost
stationary.   The American strawberries, *Diospyros virginiana,* sea-kale, and *Tetragonia expansa* introduced in
the eighteenth century, have but little importance.   We
must come to the middle of the present century to find
new cultures of any value from the utilitarian point of
view, such as *Eucalyptus globulus* of Australia and the
Cinchonas of South America.

The mode of introduction of the latter species shows
the great change which has taken place in the means of
transport.   Previously the cultivation of a plant began
in the country where it existed, whereas the Australian
Eucalyptus was first planted and sown in Algeria, and
the Cinchonas of America in the south of Asia.   Up to
our own day botanical or private gardens had only
diffused species already cultivated somewhere; now they
introduce absolutely new cultures.   The royal garden at
Kew is distinguished in this respect, and other botanical
gardens and acclimatization societies in England and elsewhere are making similar attempts.   It is probable that
tropical countries will greatly profit by this in the course
of a century.   Others will also find their advantage from
the growing facility in the transport of commodities.

When a species has been once cultivated, it is rarely, perhaps never completely, abandoned. It continues to be here and there cultivated in backward countries, or those whose climate is especially favourable. I have passed over some of these species which are nearly abandoned, such as dyer's woad (*Isatis tinctoria*), mallow (*Malva sylvestris*), a vegetable used by the Romans, and certain medicinal plants formerly much used, such as fennel, cummin, etc., but it is certain that they are still grown in some places.

The competition of species causes the cultivation of some to diminish, of others to increase; besides, vegetable dyes and medicinal plants are rivalled by the discoveries of chemists. Woad, madder, indigo, mint, and several simples must give way before the invasion of chemical products. It is possible that men may succeed in making oil, sugar, and flour, as honey, butter, and jellies are already made, without employing organic substances. Nothing, for instance, would more completely change agricultural conditions than the manufacture of flour from its known inorganic elements. In the actual state of science, there are still products which will be more and more required of the vegetable kingdom; these are textile substances, tan, indiarubber, gutta-percha, and certain spices. As the forests where these are found are gradually destroyed, and these substances are at the same time more in demand, there will be the greater inducement to cultivate certain species.

These usually belong to tropical countries. It is in these regions also, particularly in South America, that fruit trees will be more cultivated—those of the order Anonaceæ for instance, of which the natives and botanists already recognize the value. Probably the number of plants suitable for fodder, and of forest trees which can live in hot dry countries, will be increased. The additions will not be numerous in temperate climates, nor especially in cold regions.

From these data and reflections it is probable that at the end of the nineteenth century men will cultivate on a large scale and for use about three hundred species.

This is a small proportion of the one hundred and twenty
or one hundred and forty thousand in the vegetable
kingdom ; but in the animal world the proportion of
creatures subject to the will of man is far smaller.
There are not perhaps more than two hundred species of
domestic animals—that is, reared for our use,—and the
animal kingdom reckons millions of species.   In the
great class of molluscs the oyster alone is cultivated, and
in that of the Articulata, which counts ten times more
species than the vegetable kingdom, we can only name
the bee and two or three silk-producing insects.   Doubtless the number of species of animals and vegetables
which may be reared or cultivated for pleasure or
curiosity is very large : witness menageries and zoological and botanical gardens, but I am only speaking here
of useful plants and animals, in general and customary
employment.

Article III.—Cultivated Plants known or not known in a Wild State.

Science has succeeded in discovering the geographical
origin of nearly all cultivated species; but there is less
progress in the knowledge of species in a natural state—
that is wild, far from cultivation and dwellings.   There
are species which have not been discovered in this
condition, and others whose specific identity and truly
wild condition are doubtful.

In the following enumeration I have classed the
species according to the degree of certainty as to the
wild character, and the nature of the doubts where such
exist.[1]

1. Spontaneous species, that is wild, seen by several
botanists far from dwellings and cultivation, with every
appearance of indigenous plants, and under a form identical
with one of the cultivated varieties.   These are the

----

[1] The species in italics are of very ancient cultivation (A or D),
those marked with an asterisk have been less than two thousand years
in cultivation (C or F).

species which are not enumerated below; they are 169 in number.

Among these 169 species, 31 belong to the categories A and D, of very ancient cultivation, 56 have been in cultivation less than two thousand years, C, and the others are of modern or unknown date.

2. Seen and gathered in the same conditions, but by a single botanist in a single locality. Three species.

Cucurbita maxima, *Faba vulgaris, Nicotiana Tabacum.*

3. Seen and mentioned but not gathered in the same conditions by one or two authors and botanists, more or less ancient, who may have been mistaken. Two species.

Carthamus tinctorius, Triticum vulgare.

4. Gathered wild by botanists in several localities under a form slightly different to those which are cultivated, but which most authors have no hesitation in classing with the species. Four species.

Olea europœa, Oryza sativa, Solanum tuberosum, *Vitis vinifera.*

5. Wild, gathered by botanists in several localities under forms considered by some botanists as constituting different species, while others treat them as varieties. Fifteen species.

Allium ampeloprasum porrum, Cichorium Endivia, var., *Crocus sativus, var.,* *Cucumis melo, Cucurbita Pepo, Helianthus tuberosus, Latuca scariola sativa, *Linum usitatissimum annuum,* Lycopersicum esculentium, Papaver somniferum, Pyrus nivalis var., *Ribes grossularia, Solanum Melongena, *Spinacia oleracea var., Triticum monococcum.

6. Subspontaneous, that is half-wild, similiar to one or other of the cultivated forms, but possibly plants escaped from cultivation, judging from the locality. Twenty-four species.

Agava americana, Amarantus gangeticus, *Amygdalus persica,* Areca catechu, *Avena orientalis, Avena sativa, *Cajanus indicus, *Cicer arietinum,* Citrus decumana, Cucurbita moschata, Dioscorea japonica, Ervum Ervilia, *Ervum lens,* Fagopyrum emarginatum, Gossypium barbadense, Holcus saccharatus, *Holcus sorghum,* Indigofera

tinctoria, Lepidum sativum, Maranta arundinacea, Nicotiana rustica, *Panicum miliaceum,* Raphanus sativus,
Spergula arvensis.

7. Subspontaneous like the preceding, but different
enough from the cultivated varieties to lead the majority
of authors to regard them as distinct species. Three
species.

*Allium ascalonicum (variety of *A. cepa ?*), Allium
scorodoprasum (variety of A. sativum ?), Secale cereale
(variety of one of the perennial species of Secale ?).

8. Not discovered in a wild state nor even half-wild,
derived perhaps from cultivated species at the beginning
of agriculture, but too different not to be commonly
regarded as distinct species. Three species.

Hordeum hexastichon (derived from *H. distichon ?*),
Hordeum vulgare (derived from *H. distichon ?*), *Triticum
spelta* (derived from *T. vulgare ?*)

9. Not discovered in a wild state nor even half-wild,
but originating in countries which are not completely
explored, and belonging perhaps to little-known wild
species of these countries. Six species.

Arachis hypogea, Carophyllus aromaticus, *Convolvulus
batatas,* *Dolichos lubia, Manihot utilissima, Phaseolus
vulgaris.

10. Not found in a wild state, nor even half-wild,
but originating in countries which are not sufficiently
explored, or in similar countries which cannot be defined,
more different than the latter from known wild species.
Eighteen species.

Amorphophallus konjak, Arracacha esculenta, Brassica chinensis, Capsicum annuum, Chenopodium quinoa,[1]
Citrus nobilis, Cucurbita ficifolia, Dioscorea alata, Dioscorea Batatas, Dioscorea sativa, Eleusine coracana, Lucuma
mammosa, Nephelium Litchi, *Pisum sativum, Saccharum
officinarum, Sechium edule, *Tricosanthes anguina, *Zea
mays.*

Total 247 species.

[1] Since this list was printed, I have been informed that the quinoa
is wild in Chili. Some of the figures need modification in consequence
of this error.

These figures show that there are 193 species known to be wild, 27 doubtful, as half-wild, and 27 not found wild.

I believe that these last will be found some time or other, if not under one of the cultivated forms, at least in an allied form called species or variety according to the author. To attain this result tropical countries will have to be more thoroughly explored, collectors must be more attentive to localities, and more floras must be published of countries now little known, and good monographs of certain genera based upon the characters which vary least in cultivation.

A few species having their origin in countries fairly well explored, and which it is impossible to confound with others because each is unique in its genus, have not been found wild, or only once, which leads us to suppose that they are extinct in nature, or rapidly becoming so. I allude to maize and the bean (see pp. 387 and 316). I mention also in Article IV. other plants which appear to be becoming extinct in the last few thousand years. These last belong to genera which contain many species, which renders the hypothesis less probable ;[1] but, on the other hand, they are rarely seen at a distance from cultivated ground, and they hardly ever become naturalized, that is wild, which shows a certain feebleness or a tendency to become the prey of animals and parasites.

The 67 species cultivated for less than two thousand years (C, F) are all found wild, except the species marked with an asterisk, which have not been found or which are subject to doubts. This is a proportion of eighty-three per cent.

What is more remarkable is that the great majority of species cultivated for more than four thousand years (A), or in America for three thousand or four thousand years (D), still exist wild in a form identical with some one of the cultivated varieties. Their number is thirty-one out of forty-nine, or sixty-three per cent. In categories 9 and 10 there are only two of these species of

---

[1] For reasons which I cannot here express, monotypical genera are for the most part in process of extinction.

very ancient cultivation, or four per cent., and these are
two species which probably exist no longer as wild plants.

I believed, *à priori,* that a great number of the
species cultivated for more than four thousand years
would have altered from their original condition to such
a degree that they could no longer be recognized among
wild plants. It appears, on the contrary, that the forms
anterior to cultivation have commonly remained side by
side with those which cultivators employed and propagated from century to century. This may be explained
in two ways: 1. The period of four thousand years
is short compared to the duration of most of the specific
forms in phanerogamous plants. 2. The cultivated
species receive, outside of cultivated ground, continual
reinforcements from the seeds which man, birds, and
different natural agents disperse and transport in a
thousand ways. Naturalizations produced in this manner
often confound the wild plants with the cultivated ones,
and the more easily that they fertilize each other since
they belong to the same species. This fact is clearly
demonstrated in the case of a plant of the old world
cultivated in America, in gardens, and which, later,
becomes naturalized on a large scale in the open country
or the woods, like the cardoon at Buenos Ayres, and the
oranges in several American countries. Cultivation
widens areas, and supplements the deficits which the
natural reproduction of the species may present. There
are, however, a few exceptions, which are worth mentioning in a separate article.

Article. IV.—Cultivated Plants which are Extinct, or becoming Extinct in a Wild State.

These species to which I allude present three remarkable characters :—

1. They have not been found wild, or only once or
twice, and often doubtfully, although the regions whence
they come have been visited by several botanists.

2. They have not the faculty of sowing themselves,
and propagating indefinitely outside cultivated ground.

In other terms, in such cases they do not pass out of the condition of adventitious plants.

3. It cannot be supposed that they are derived within historic times from certain allied species.

These three characters are found united in the following species:—Bean (*Faba vulgaris*), chick-pea (*Cicer arietinum*), ervilla (*Ervum Ervilia*), lentil (*Ervum lens*), tobacco (*Nicotiana tabacum*), wheat (*Triticum vulgare*), maize (*Zea mays*). The sweet potato (*Convolvulus batatas*) should be added if the kindred species were better known to be distinct, and the carthamine (*Carthamus tinctorius*) if the interior of Arabia had been explored, and we had not found a mention of the plant in an Arabian author.

All these species, and probably others of little-known countries or genera, appear to be extinct or on their way to become so. Supposing they ceased to be cultivated, they would disappear, whereas the majority of cultivated plants have become somewhere naturalized, and would persist in a wild state.

The seven species mentioned just now, excepting tobacco, have seeds full of fecula, which are the food of birds, rodents, and different insects, and have not the power of passing entire through their alimentary canal. This is probably the sole or principal cause of their inferiority in the struggle for existence.

Thus my researches into cultivated plants show that certain species are extinct or becoming extinct since the historical epoch, and that not in small islands but on vast continents without any great modifications of climate. This is an important result for the history of all organic beings in all epochs.

Article V.—Concluding Remarks.

1. Cultivated plants do not belong to any particular category, for they belong to fifty-one different families. They are, however, all phanerogamous except the mushroom (*Agaricus campestris*).

2. The characters which have most varied in cultivation are, beginning with the most variable : *a*. The size, form, and colour of the fleshy parts, whatever organ they belong to (root, bulb, tubercle, fruit, or seed), and the abundance of fecula, sugar, and other substances which are contained in these parts ; *b*. The number of seeds, which is often in inverse ratio to the development of the fleshy parts of the plant ; *c*. The form, size, or pubescence of the floral organs which persist round the fruits or seeds ; *d*. The rapidity of the phenomena of vegetation—whence often results the quality of ligneous or herbaceous plants, and of perennial, biennial, or annual.

The stems, leaves, and flowers vary little in plants cultivated for those organs. The last formations of each yearly or biennial growth vary most; in other terms, the results of vegetation vary more than the organs which cause vegetation.

3. I have not observed the slightest indication of an adaptation to cold. When the cultivation of a species advances towards the north (maize, flax, tobacco, etc.), it is explained by the production of early varieties, which can ripen before the cold season, or by the custom of cultivating in the north, in summer, the species which in the south are sown in winter. The study of the northern limits of wild species had formerly led me to the same conclusion, for they have not changed within historic times although the seeds are carried frequently and continually to the north of each limit. Periods of more than four or five thousand years, or changements of form and duration, are needed apparently to produce a modification in a plant which will allow it to support a greater degree of cold.

4. The classification of varieties made by agriculturists and gardeners are generally based on those characters which vary most (form, size, colour, taste of the fleshy parts, beard in the ears of corn, etc.). Botanists are mistaken when they follow this example; they should consult those more fixed characters of the organs for the sake of which the species are not cultivated.

5. A non-cultivated species being a group of more or

less similar forms, among which subordinate groups may often be distinguished (races, varieties, sub-varieties), it may have happened that two or more of these slightly differing forms may have been introduced into cultivation. This must have been the case especially when the habitation of a species is extensive, and yet more when it is disjunctive. The first case is probably that of the cabbage (*Brassica*), of flax, bird-cherry (*Prunus avium*), the common pear, etc. The second is probably that of the gourd, the melon, and trefoil haricot, which existed previous to cultivation both in India and Africa.

6. No distinctive character is known between a naturalized plant which arose several generations back from a cultivated plant, and a wild plant sprung from plants which have always been wild. In any case, in the transition from cultivated plant to wild plant, the particular features which are propagated by grafting are not preserved by seedlings. For instance, the olive tree which has became wild is the *oleaster*, the pear bears smaller fruits, the Spanish chestnut yields a common fruit. For the rest, the forms naturalized from cultivated species have not yet been sufficiently observed from generation to generation. M. Sagot has done this for the vine. It would be interesting to compare in the same manner with their cultivated forms Citrus, Persica, and the cardoon, naturalized in America, far from their original home, as also the Agave and the prickly pear, wild in America, with their naturalized varieties in the old world. We should know exactly what persists after a temporary state of cultivation.

7. A species may have had, previous to cultivation, a restricted habitation, and subsequently occupy an immense area as a cultivated and sometimes a naturalized plant.

8. In the history of cultivated plants, I have noticed no trace of communication between the peoples of the old and new worlds before the discovery of America by Columbus. The Scandinavians, who had pushed their excursions as far as the north of the United States, and the Basques of the Middle Ages, who followed whales

perhaps as far as America, do not seem to have transported a single cultivated species. Neither has the Gulf Stream produced any effect. Between America and Asia two transports of useful plants perhaps took place, the one by man (the Batata, or sweet potato) the other by the agency of man or of the sea (the cocoa-nut palm).

INDEX

A

Abi, 285
Agava americana, 153
Alexanders, 91
Alexandrine clover, 107
Alligator pear, 292
Allium Ampeloprasum, 101
—— Ascalonicum, 68
—— Cepa, 66
—— fistulosum, 68
—— sativum, 63
—— Schœnoprasum, 72
—— Scorodoprasum, 71
Almond, 218
Alocasia macrorhiza, 75
Aloe, American, 153
Amarantus frumentaceus, 352
—— gangeticus, 100
American Aloe, 153
—— indigoes, 137
Amorphophallus Konjak, 76
—— Rivieri, 76
Amygdalus communis, 218
—— Persica, 221
Anacardium occidentale, 198
Ananassa sativa, 311
Andropogon saccharatus, 382
—— Sorghum, 380
Angular Luffa, 371
Angurian cucumber, 267
Annual capsicum, 289
Anona Cherimolia, 174
—— muricata, 168, 173
—— reticulata, 174
—— squamosa, 168
Anthriscus Cerefio lum, 90

Apé, 75
Apium graveolens, 90
Apple, 233
——, custard, 168, 174
——, Malay, 241
——, mammee, 189
——, pine, 311
——, star, 285
——, sugar, 168
——, Tahiti, 202
Apricot, 215
Arab tea, 134
Arachis hypogæa, 411
Areca catechu, 427
Armeniaca vulgaris, 215
Arnotto, 401
Arracacha esculenta, 40
Arrowroot, 81
Artichoke, 92
——, Jerusalem, 42
Artocarpus incisa, 298
—— integrifolia, 299
Arum esculentum, 73
—— macrorhizon, 75
Aubergine, 287
Avena orientalis, 373
—— sativa, 373
—— strigosa, 375
Avocado pear, 292

B

Bambarra ground-nut, 347
Banana, 304
Barbados cotton, 408
Barleys, 367

Batatas edulis, 53
Batata mammosa, 57
Bean, broad, 316
——, kidney, 338
Beetroot, 58
Benincasa, 268
Beta vulgaris, 58
Bird-cherry, 205
Bird's foot, 113
Bitter orange, 183
Bixa Orellana, 401
Black currant, 278
Brassica campestris, 36
—— Napus, 36
—— oleracea, 36, 83
—— Rapa, 36
Bread-fruit, 298
Broad bean, 316
Bromelia Ananas, 311
Buckwheat, common, 348
——, notch-seeded, 351
——, Tartary, 353
Bullace, 214
Bullock's heart, 174

C

Cabbage, 83
Cacao, 313
Caïmito, 285
Calabash, 245
Cannabis sativa, 148
Capsicum annuum, 289
—— frutescens, 290
Cardoon, 92
Carica Papaya, 273
Carob, 334
Carthamine, 164
Caryophyllus aromaticus, 161
Cashew, 198
Cassis, 278
Castanea vulgaris, 353
Castor-oil plant, 422
Catha edulis, 134
Celery, 89
Cerasus vulgaris, 207
Ceratonia Siliqua, 334
Chayote, 273
Chenopodium Quinoa, 351
Cherry, bird, 205
——, sour, 207

Chervil, 90
Chestnut, 353
Chickling vetch, 110
Chick-pea, 323
Chicorium Endivia, 97
—— Intybus, 96
Chicory, 96
China grass, 146
Chinese pear, 233
Chirimoya, 174
Chives, 72
Chocho, 273
Chrysophyllum Caïmito, 285
Cinnamon, 146
Cinnamonum zeylanicum, 146
Citron, 178
Citrullus vulgaris, 262
Citrus Aurantium, 188
—— decumana, 177
—— medica, 178
—— nobilis, 188
Clove, 161
Clover, crimson, 106
——, Egyptian, 107
——, purple, 105
Coca, 135
Cochlearia Armoracia, 33
Cocoa-nut palm, 429
Cocos nucifera, 429
Coffee, 415
Coffea arabica, 418
—— liberica, 418
Colocasia, 73
Convolvolus Batatas, 53
—— mammosa, 57
Corchorus capsularis, 130
—— olitorius, 130
Corn salad, 91
Corn spurry, 114
Cotton, Barbados, 408
——, herbaceous, 452
——, tree, 408
Cress, garden, 166
Crocus sativum, 86
Cucumber, 264
Cucumis Anguria, 267
—— Melo, 258
—— sativas, 264
Cucurbita citrullus, 262
—— ficifolia, 257
—— Lagenaria, 245
—— maxima, 249

Cucurbita Melopepo, pepo, 253
—— moschata, 257
Currant, black, 278
——, red, 277
Custard apple, 168
Cydonia vulgaris, 236
Cynara Cardunculus, 92
—— Cytisus Cajan, 332
—— Scolymus, 92

D

Date-palm, 301
Dioscorea, 76
Dolichos Lablab, 346
—— Lubia, 347
—— Soja, 330
Dyer's indigo, 136

E

Egyptian clover, 107
—— lupin, 327
—— wheat, 259
Elæis guineensis, 429
Eleusine Coracana, 384
Endive, 97
Ervilla, 107
Ervum Ervilia, 107
—— lens, 321
Erythroxylon Coca, 135
Eugenia Jambos, 240
—— malaccensis, 241

F

Faba vulgaris, 316
Fagopyrum emarginatum, 351
—— esculentum, 348
—— tataricum, 350
Fenugreek, 112
Ficus Carica, 295
Field-pea, 327
Fig, 295
Fig-leaved pumpkin, 257
Fig, Indian, 274
Flat-podded pea, 109
Flax, 119
Fragaria chiloensis, 205
—— vesca, 203
—— virginiana, 205
French honeysuckle, 104

G

Garcinia Mangostana, 118
Garden cress, 86
—— pea, 328
Garlic, 63
Glycine soya, 330
—— subterranea, 347
Gombo, 189
Gooseberry, 276
Gossypium arboreum, 408
—— barbadense, 408
—— herbaceum, 402
Gourd, 245, 249
——, snake, 273
——, towel, 269
Grass, China, 146
Grass, guinea, 115
Green gram, 346
Guava, 241

H

Haricot bean, 338
Hedysarium coronarium, 104
Helianthus tuberosus, 42
Hemp, 148
Henna, 138
Hibiscus esculentus, 189
Holcus saccharatus, 382
—— Sorghum, 380
Hop, 162
Hordeum distichon, 367
—— hexastichon, 369
—— vulgare, 368
Horse-radish, 33
Humulus Lupulus, 162

I

Ilex paraguariensis, 135
Indian fig, 274
Indigo, American, 137
——, dyer's, 136
——, silver, 137
Indigofera argentea, 137
—— cerulea, 137
—— tinctoria, 136
Ipomea mammosa, 57

J

Jack-fruit, 299
Jambosa Malaccensis, 241
—— vulgaris, 240
Jatropha manihot, 59
Jerusalem artichoke, 42
Juglans regia, 425
Jujube, common, 194
——, Indian, 197
——, Lotus, 196
Jute, 130

K

Kidney bean, 338
——, moth, 344
——, three-lobed, 345
Kiery, 352
Khât, 134
Konjak, 76

L

Lablab, 347
Lagenaria vulgaris, 245
Lamb's lettuce, 91
Lathyrus Cicera, 109
—— Ochrus, 110
—— sativus, 111
Lattuca scariola, 95
Lawsonia alba, 138
Leek, 101
Lemon, 178
Lens esculenta, 221
Lentil, 321
Lepidum sativum, 86
Lettuce, 95
——, lamb's, 91
Linum usitatissimum, 119
Litchi, 314
Longan, 315
Lotos jujube, 196
Lubia, 347
Lucern, 102
Lucuma Caïmito, 285
—— mammosa, 286
Lupin, 325
Lupinus albus, 325
—— termis, 327
Lycopersicum esculentum, 290

M

Madder, 41
Madia sativa, 418
Maize, 387
Malay apple, 241
Mammee, 199
—— americana, 189
—— Sapota, 286
Mandarin, 188
Mandubi, 411
Mangifera indica, 200
Mango, 200
Mangosteen, 188
Manioc, 59
Manihot utilissima, 59
Maranta arundinacea, 81
Marmalade plum, 286
Maté, 135
Medicago sativa, 102
Melon, 258
——, pumpkin, 256
——, water, 262
——, white gourd, 268
Millet, common, 276
——, Italian, 278
Momordica cylindrica, 269
Monkey-nut, 411
Morus alba, 149
—— nigra, 152
Mulberry, 149
Mung, 346
Musk pumpkin, 356
Myristica fragrans, 419

N

Nephelium lappaceum, 315
—— litchi, 314
—— longana, 315
New Zealand spinach, 89
Nicotiana tabacum, 139
Nutmeg, 419

O

Oats, 372
Ochro, 189
Ochrus, 111
Oil-palm, 429
Olea europea, 279

Olive, 279
Onion, 66
——, spring or Welsh, 66
Onobrychis sativa, 104
Opuntia ficus Indica, 274
Orange, 181
——, bitter, 185
——, sweet, 183
Ornithopus sativus, 113
Oryza sativa, 385

P

Palm, cocoa-nut, 429
—— oil, 429
Panicum italicum, 378
—— maximum, 115
—— miliaceum, 376
Papava somniferum, 397
Papaw, 293
Papaya vulgaris, 293
Parsley, 90
Pea, 327
——, field, 327
——, garden, 328
—— nut, 411
——, pigeon, 382
Peach, 221
Pear, 229
——, avocado, 272
——, Chinese, 233
——, prickly, 274
——, sand, 233
——, snowy, 232
Pepper, red, 288
Persea gratissima, 292
Persica vulgaris, 221
Petroselinum sativum, 90
Phaseolus aconitifolius, 345
—— lunatus, 344
—— Mungo, 346
—— vulgaris, 338
Phœnix dactylifera, 301
Pigeon-pea, 332
Pine-apple, 311
Pistachio nut, 316
Pistacia vera, 316
Pisum arvense, 327
—— Ochrus, 111
—— sativum, 328
Plum, 211

Polygonum emarginatum, 351
—— fagopyrum, 348
—— tataricum, 353
Pomegranate, 327
Poppy, 397
Portulaca oleracea, 87
Potato, 45
——, sweet, 83
Prickly pear, 274
Prunus Amygdalus, 218
—— Armeniaca, 215
—— avium, 205
—— Cerasus, 207
—— domestica, 212
—— insititia, 214
—— Persica, 221
Psidium guayava, 241
Pumpkin, fig-leaved, 257
——, musk or melon, 256
Punica Granatum, 237
Purslane, 87
Pyrus communis, 229
—— malus, 233
—— nivalis, 233
—— sinensis, 233

Q

Quince, 236
Quinoa, 351

R

Radish, 29
——, horse, 33
Rambutan, 315
Raphanus sativus, 29
Rhus Coriaria, 133
Ribes Grossularia, 276
—— nigrum, 278
—— rubrum, 277
—— Uva-crispa, 276
Rice, 385
Ricinus communis, 422
Rocambole, 72
Rose-apple, 240
Rubia tinctorum, 41
Rye, 370

S

Saccharatum officinale, 154
Saffron, 166
Sainfoin, 104
——, Spanish, 104
Salsify, 44
Sapodilla, 286
Sapota achras, 286
Scandix cerefolium, 90
Scorzonera hispanica, 44
Secale cereale, 370
Sechium edule, 272
Sesame, 419
Sesamum indicum, 419
Setaria Italica, 380
Shaddock, 177
Shallot, 68
Sium Sisarum, 39
Skirret, 39
Smyrnium Olus-atrum, 91
Snake gourd, 272
Solanum Commersonii, 46
—— immite, 49
—— maglia, 49
—— tuberosum, 45
—— verrucosus, 49
Sorghum saccharatus, 382
—— vulgaris, 380
Sour sop, 173
Soy, 330
Spanish sainfoin, 104
Spelt, 362
Spergula arvensis, 114
Spinach, 98
——, New Zealand, 87
Spinacia oleracea, 98
Spondias dulcis, 202
Spurry, corn, 114
Strawberry, 203
——, Chili, 205
——, Virginian, 205
Sugar apple, 168
—— cane, 154
Sumach, 133
Sweet potato, 83
—— sop, 168

T

Tahiti apple, 202
Tare, 108
Tea, 117
Tetragonia expansa, 89
Thea sinensis, 117
Theobroma Cacao, 313
Tobacco, 139
Towel gourd, 269
Trigonella Fœnum-græcum, 112
Trifolium Alexandrinum, 107
—— incarnatum, 146
—— pratense, 105
Triticum æstivum, 354
—— compositum, 359
—— dicoccum, 365
—— durum, 360
—— hybernum, 354
—— moncoccum, 365
—— polonicum, 361
—— spelta, 262
—— vulgare, 354
Turnip, 36

V

Valerianella olitoria, 89
Vetch, chickling, 110
——, common, 108
Vicia ervilla, 107
—— sativa, 108
Vine, 191
Vitis vinifera, 191
Voandzeia subterranea, 347

W

Walnut, 245
Wheats, 354

Y

Yams, 76

Z

Zea Mays, 387
Zizyphus jujube, 197
—— Lotus, 196
—— vulgaris, 194